IN THE AGGREGATES

FORTY-TWO YEARS IN THE SANDBOX

By Andrew F. Haumesser

PUBLISHED BY THE
SOCIETY FOR MINING, METALLURGY & EXPLORATION

Society for Mining, Metallurgy & Exploration (SME)
12999 E. Adam Aircraft Circle
Englewood, Colorado, USA 80112
(303) 948-4200 / (800) 763-3132
www.smenet.org

SME advances the worldwide mining and minerals community through information exchange and professional development. SME is the world's largest association of mining and minerals professionals.

ISBN 978-0-87335-503-2
Ebook 978-0-87335-504-9

Library of Congress Cataloging-in-Publication Data has been applied for.

Contents

Preface

The book title *In the Aggregates* is a play on the phrase "in the aggregate," which loosely means to consider something in its whole or entirety. I distinctly remember a day more than 30 years ago when I was working with a colleague on a mine plan while perusing several aerial photographs. He suddenly began to laugh out loud, and when I asked what struck him as humorous, he said he had just that moment realized why he had worked in the aggregates industry for the past 30 years. He proceeded to explain he had realized that we were just little boys still playing in the sandbox, only now our sandbox was 600 acres (243 hectares) and our trucks now held 50-ton loads instead of 15 ounces (425 grams). I thought of that conversation many times over my career. The memory still makes me chuckle and, in retrospect, I agree with his analysis.

I wrote this book in the hope that it might inspire and motivate some young person seeking a career to find their way into a rewarding profession within the industry that I grew to love. Someone asked my wife about the book I was writing and she described it as a technical memoir, which may be an entirely new genre of writing. The book contains a history of aggregates and discusses the importance of the industry to the economy as well as modern society. Within it I have attempted to describe exploration techniques, real estate acquisition, permitting, mining and production methodology, management philosophy, and sustainability. Throughout the book I have attempted to add a little humor and many personal experiences to enhance the readability beyond just the technical issues, graphs, and formulas. But I hope it will also serve as a technical reference to help employees new to the industry become successful in their own careers.

I would like to acknowledge the assistance of four colleagues who were kind enough to act as peer reviewers for this book: Dick Brannigan, John Harris, Cory Kiser, and Roy Wagner. I very much appreciate the time and effort they expended in providing me this courtesy. I also want to acknowledge my wife Debbie for always holding down the fort during the long, long hours and the many nights I needed to be out of town on business. She was always understanding and I couldn't have done it without her.

About the Author

Photo by F. Erdmann

Andrew F. Haumesser graduated from college with a B.S. in geology in the early 1970s. Following an adventure-filled summer traveling across Canada and up the Alaska Highway, he began working as a field geologist for a large international environmental consulting firm. His consulting projects ranged from appraisal of the groundwater resources of buried glacial valleys to analysis of potential sites for the underground storage of natural gas. By the mid-1970s, he was employed in the aggregates industry by a division of a large engineering/construction company as staff geologist, mapping a portion of three states for potential aggregates reserves, evaluating underground limestone formations for potential sources of limestone for flue gas desulfurization, managing acquisitions, and designing reclamation plans.

Upon moving to another large regional aggregates producer as a staff geologist, he was heavily involved in exploration, mine planning, permitting, and ultimately underground mine design. Becoming bored with staff positions, Haumesser requested a transfer to operations management, where he performed the duties of plant superintendent, general superintendent, and operations manager. The company was purchased and resold four times in three years and was ultimately acquired by a large international producer of construction materials, which opened opportunities to travel throughout the United States and to evaluate numerous other operations as a member of a benchmarking team. After a three-year stint as regional manager of a consulting firm specializing in advising aggregates producers, he returned to operations management for a large midwestern heavy highway contractor and asphalt/aggregates producer. He was instrumental in growing their small aggregates division into a regional powerhouse with eighteen aggregates mining operations, a mineral filler plant, and several rail unloading facilities in a three-state area.

Writing a book about the aggregates industry is a longtime dream of the author's that came to fruition only after retirement and with the advent of the COVID-19 pandemic. Haumesser has been married for forty-seven years and has two adult sons and one teenaged grandson. He and his wife retired to their beloved Rocky Mountains in 2018 along with their 104-pound German shepherd dog.

CHAPTER 1

Introduction

Mark Twain was the author of numerous well-known classic books but he was also quite a homespun philosopher. One of my favorite quotes attributed to Twain is: "Find a job that you love and you'll never work another day in your life." I was one of the lucky few people I know to whom that quote applied. There were very few days in my career in the aggregates industry that I didn't look forward to the following day's challenges. Like any job, there were frustration, mental pressures, and some anxiety but never enough to discourage my enthusiasm for the industry. My purpose in writing this book is to inspire another generation to find a job that they can love enough to never work another day in their lives. The aggregates industry provided me with the sort of challenging environment that was variable enough on a daily basis to keep me intellectually stimulated and active enough to keep me physically stimulated, all while working with some of the greatest people in the world to keep me socially stimulated.

As I write this book, I have been actively retired for 4 years but continue to consult on a limited, part-time basis, which allows me to pursue various hobbies and other interests while still contributing to an industry I love. While I've thought about writing this book for the past several years, it was not until the COVID-19 pandemic hit the United States in early 2020 that I sat down in front of my computer to seriously attempt a first pass at putting my thoughts on paper. So, perhaps, even a pandemic has an upside!

When I entered this business in 1973 there was nowhere to turn for advice on technical issues except the experience of others in the industry. There were no reference books where you could look up how to plan a gravel pit, the best type of crusher to use in a granite quarry, the best methods for reclaiming a quarry and selling the real estate at a profit, or the environmental regulations that apply to water discharges from a quarry sump. These were the times that predated the internet and you couldn't Google your way to many answers like you can today. Some state geological surveys touched on certain subjects that were peripheral to the industry. The National Stone, Sand and Gravel Association (NSSGA), which at that time was called the National Stone Association, now publishes a comprehensive manual entitled *The Aggregates Handbook* designed to be a technical reference on industry issues and a guide to regulatory agencies (NSSGA 2013). Many state aggregates associations now publish environmental manuals explaining the myriad environmental rules and regulations pertaining to their individual states. NSSGA as well as some state aggregates associations are now offering seminars concerning operational matters to assist newcomers to the business, but none of this existed in my early days in the industry.

This book is not intended to be the end-all technical reference for all things aggregates, but a more general text discussing the various facets of a fascinating industry that might pique the interest of a young geologist, engineer, information technology specialist, or marketing graduate facing the same dilemma I faced 45 years ago when looking for a meaningful career. The remaining portion of this introduction consists of some of the more interesting projects that kept me engrossed, motivated, and stimulated over a long, rewarding career in the industry. It has also been an interesting trip down memory lane for me to put them in writing.

As a young college graduate with a bachelor of science in geology, I was faced with some difficult choices. Do I stay in college for 2–5 more years and pursue a graduate degree in geology, or do I attempt some other graduate major? Undergraduate degrees in a field of science generally don't portend a bright financial future. I went to college in the late 1960s and early 1970s with the specific intention of avoiding the Vietnam conflict. You needed to be enrolled in a serious college curriculum to get an educational deferment from the draft board. Like most 18-year-olds, I had no idea what I wanted to do with the rest of my life or any inkling of what career choices to make. I entered college as a political science major with the thought that I might someday pursue a career in law. To this day I'm glad that career path did not come to pass. As a freshman, I was required to take at least one science class and I chose geology because that subject had not been comprehensively covered in my previous education and I especially liked spending time outdoors. Unexpectedly, I fell in love with the subject after only a few months and was hooked.

As a child I had always maintained a keen interest in history, and what is more historical than the history of the formation of the planet on which we live? Geology 101 was an introduction to geology but also required a field class that met every Saturday morning at 8:00 AM. We would gather with our classmates and split up into numerous individuals' automobiles and proceed to various locations around a three-state area to assess the local outcrops, stream valleys, fossil beds, and even cemeteries where we were tasked with identifying the numerous rock types of the headstones. At other times, we'd tour the downtown metro area to identify the building stones on the facades of skyscrapers. In our classroom we would study theories of sedimentation and glaciation and then take a field trip to a gravel pit to see the results in the field or to identify various landforms resulting from glacial action. The cross-bedding and physical sorting of the sand and gravel sizes that had been explained in class were now apparent in our natural, outdoor laboratory. Maybe this was a throwback to my childhood when my friends and I would sneak into the local gravel pit and build rafts to float on and fish in the lakes. I can still see the overweight gravel pit foreman that used to curse at us and chase us away into the woods.

Classes in geomorphology, mineralogy, stratigraphy and sedimentation, historical geology, paleontology, and sedimentary, metamorphic, and igneous petrology were intermingled with the standard curriculum of science, language, literature, philosophy, and mathematics. Those of us wishing to receive a bachelor of science versus a bachelor of arts were required to complete two extra steps: a summer field course between our third and fourth years, plus we were required to write and defend a senior thesis. The thesis entailed choosing from a list several research projects made available by our professors and advisors designed to test our field, laboratory, and research skills. My senior thesis project consisted of mapping an area in Central Kentucky where deposits of barite, sphalerite, and galena had been exploited on an intermittent basis in the mid-1800s. Soil samples were taken throughout the area and analyzed in the lab to determine the concentrations of lead and zinc in the soil and predict trends in mineralization in the limestone host rock. My project also identified some previously unmapped faulting in the area.

A 6-week summer field course was also required that taught surveying and field mapping techniques, toured huge open pit mines, and required us to make detailed geological maps of several areas along the Rocky Mountain Front. This is probably the reason I chose to retire in Idaho near the Rocky Mountains I first fell in love with in 1970. After graduating with my B.S. degree, I decided I had had enough of education for a while and did not want to pursue a graduate degree at that time. Following graduation, a summer trip to Alaska via Canada and the Alaska Highway was spent camping, exploring, and panning for gold. Reality set in after the money ran out and it was time to get a job and face the real world.

My earliest professional employment was with a major international consulting firm specializing in geological and environmental engineering. As a young field geologist, I was required to spend the majority of my time sitting behind a drill rig (in all weather conditions including –40°F in Minnesota one February) recording what the drill was bringing to the surface, whether it was rock chips, rock core, or samples of sediments. There were projects where I was on call 24 hours per day to insert electric logging probes into the drill holes to record and log gamma ray/neutron, spontaneous potential, resistivity, and so on. Young field geologists were (and still are) the data gatherers of the industry, while the job of analysis and report writing was left to more senior personnel with master's and Ph.D. degrees. Projects included location of potential underground storage reservoirs for natural gas, evaluating glacial buried valleys for potential groundwater resources, and foundation drilling for power plants, just to name a few. Throughout it all there were many, many isolated nights spent in lonely motel rooms in unfamiliar towns around the country. Deciding that this itinerant life was not the life for me, I re-evaluated my future employment prospects.

Moving forward, I applied for work at a local sand and gravel company that was a subsidiary of a large, highly diversified engineering and construction consortium. I was initially told they had no jobs available but persisted and informed them I had a geology degree and wanted to pursue a career in an associated field. This piqued the interest of the general manager, who wanted to expand the reserve base of the company and enter some new markets. For the next several years I worked on mapping the aggregates resources of a portion of three states. During this period I was indoctrinated into many aspects of the aggregates business. I worked closely with and advised the quality control department and the land management/survey department, and became heavily involved with mine planning and reclamation.

Being in a staff position, I was exposed to many facets of the business. Very early on in my tenure, I was "elected" to become a second shift foreman when demand exceeded one shift of production capacity at one of our processing plants. My new young wife was not thrilled with my being gone overnight five nights per week but in her usual stoic way she supported me and my career. Production was an entirely different game that I had to learn the hard way by making mistakes. I was given a crew of green college kids on summer break with even less experience than myself and one "seasoned old hand" for advice. Much to my amazement before the summer was out, my "college boys" outproduced the first shift of experienced miners.

This was the mid-1970s when environmental regulations were in their infancy. Upon enactment of mining and reclamation legislation at the state level, I found myself part of an ad hoc committee formed by the state aggregates association to assist the state department of natural resources in negotiating and writing rules for this neophyte legislation. Being part of a larger engineering and construction company opened up other opportunities as well. Our parent company was taking advantage of new environmental regulations and developing

FIGURE 1.1 Dragline feeding a dump hopper loading rail cars

underground limestone mines for the production of high-calcium lime for use in flue gas desulfurization at regional coal-fired power plants along the Ohio River. This gave me the opportunity to assist in the exploration and reserve analysis on two underground limestone mining projects and introduced me to the fascinating new world of underground mining.

After several years, I moved on to work for a regional aggregates producer with 45 mining operations in a three-state area. I was hired as a staff geologist with geological, permitting, and mine planning responsibilities for two of the company's five divisions. My firsthand mining experience at this point was restricted to open pit sand and gravel mining. Little did I know where this new job was going to take me. My first assignment was to design a mine plan for a sand and gravel operation utilizing a 12-cubic-yard (9.2-cubic-meter) walking dragline feeding a large hopper that loaded rail cars for a 3-mile (4.8-kilometer) trip to the processing plant (Figure 1.1). This was a virgin deposit that lay across the river from the existing pit. This assignment entailed designing a constantly shifting railroad layout that included a bridge crossing a fairly large midwestern river. Upon receiving this assignment, my first thought was "What the hell do I know about designing railroad lines? I'm a geologist!" But, always up for a new challenge, I managed to educate myself on railroad curves and grades and dragline specifications, and developed a functional mine plan and ultimately moved on to another project.

One day I received a call from one of the district managers asking me to attend a meeting the following week with him and a "few of the neighbors" who were concerned about the water levels in their wells declining. They blamed our limestone quarry for dewatering their wells. The district manager wanted me to attend the meeting in a consulting capacity in case any technical questions concerning groundwater were raised. This was billed to me as a small meeting with two or three adjoining property owners. Upon arrival at the meeting venue, it was evident the meeting was not "as advertised;" we had been blindsided by the neighbors. Approximately 100 people awaited us along with a couple of township trustees, a county commissioner, a state senator, and two TV news crews.

I was tempted to remain in my car and immediately head for home; instead, I gritted my teeth, pulled up my breeches, and headed into the open jaws of a lynch mob. And, as can be expected, the meeting was chaotic, overly emotional, and entirely irrational with numerous people playing to the TV cameras. Spontaneous meetings with no one in control (i.e., a zoning board, a commissioners meeting) never come to any rational conclusion. After 3 hours of answering questions and attempting to maintain control of the audience, the district manager ended the meeting with no concrete conclusion and many hostile feelings. After several years of contention, controversy, depositions, and litigation, which began with this meeting, the groundwater laws of a major midwestern state were rewritten by the state legislature.

Another big challenge came in an adjoining state with an extremely large dewatering project. A 1,300-acre (526-hectare) property bordered on one side by almost 2 miles (3.2 kilometers) of a medium-sized river had been mined over the previous 30 years for its sand and gravel reserves. The glacial outwash sand and gravel deposit was 115 feet (35 meters) thick and contained as many as seven layers of interbedded glacial till (silt and clay) varying from 2 to 13 feet (0.6 to 3.96 meters) in thickness. Underlying all of these sediments was 30 to 50 feet (9 to 15 meters) of shale and 250 feet (76 meters) of minable limestone. The company had begun the process of recovering the underlying limestone about 2 years prior to my involvement. Two large walking draglines operating on different levels were removing the overlying sand and gravel deposits. The upper 70 feet (21 meters) was excavated by a 13-cubic-yard (9.9-cubic-meter) dragline while a smaller (10 cubic yards [7.6 cubic meters]) machine on the lower bench was utilized to sort out the clay from the sand and gravel and feed it to the larger machine on the upper level. A layer averaging about 25 feet (7.6 meters) of organic shale lay between the overlying sediments and the 250 feet (76 meters) of minable limestone. This shale layer was drilled, blasted, and hauled with trucks and wheel loaders to a waste pile that was 150 feet (46 meters) higher in elevation and approximately ¾ of a mile (1.2 kilometers) away (Figure 1.2).

The shale layer acted as an aquitard and kept the limestone relatively water free, but the porous glacial outwash deposits above the shale, constantly recharged by 2 miles (3.2 kilometers) of riverbed, produced enormous volumes of water. The first time I saw the limestone quarry it was approximately 600 feet (183 meters) wide, 1,250 feet (381 meters) long, and

FIGURE 1.2 Removal of the sand and gravel and intervening shale layer

FIGURE 1.3 Settling pond with three 16-inch (406-millimeter) vertical turbine pumps in the shale layer

40 feet (12.2 meters) deep with the slopes of the overlying sediments toeing out very close to the top edge of the quarry walls. Needless to say, water was pouring almost uncontrollably over the quarry walls and into the bottom of the newly excavated quarry. Wheel loaders and haul trucks worked in water above their axles as the quarry pumps strained to lift the water 200 feet (61 meters) vertically out of the hole.

This was an ongoing project for the next 5–6 years that consisted of establishing new sumps at higher elevations on top of the shale layer, utilizing large, more efficient, vertical turbine pumps, and installing settling ponds. The ultimate challenge was not to dewater the wells of a neighboring residential area and deprive the families of their drinking water. This required that we recirculate a large portion of the pumped water into a holding lake located along the property line between our quarry and the neighbors at a substantial cost in electrical power. A major part of the project was monitoring the cost of the power utilized for pumping purposes versus the power for processing. On a typical day, 40 million gallons (18,184,360 liters) of groundwater was pumped and large storm events could increase that 10-fold. And as the quarry increased in size, so did the pumping volume, the number of pumps, and the electrical consumption (Figure 1.3).

I was intimately involved with the move of our company into underground limestone mining when the overburden removal costs at one quarry became unsustainable. Going from surface mining to underground mining was a huge psychological hurdle for our corporate management to overcome. It took numerous meetings over about a 3-year time frame to instill management with enough confidence to proceed underground. I'm not sure if it was fear of the unknown or fear of the dark, but convincing upper management of such a radical culture change is very difficult, at least the first time. The learning curve was very large: new types of equipment to evaluate and purchase, new planning procedures, new blasting techniques, employee training, negotiations with union officials, and new scrutiny and regulation from the Mine Safety and Health Administration (MSHA).

FIGURE 1.4 Results of the Saturday roof fall

A consulting mining engineer affiliated with a university school of mining was hired to analyze and approve the initial room-and-pillar mine plan I had completed in-house. We were somewhat concerned about a dominant jointing pattern that had previously been identified in the surface quarry. The consultant advised orienting the rooms at 45° to the predominant jointing direction, utilizing 40 × 40 feet (12 × 12 meter) square pillars, 24-feet-high × 40-feet-wide (7 × 12 meters) headings, and ultimately developing a 40-foot-high (12 meter) bench. Eventually we broke ground on a mine portal and horizontal tunnel (adit) into an existing quarry highwall in close proximity to the primary crusher. To comply with the MSHA regulations for two means of egress, the tunnel was shaped like a horseshoe and a second tunnel daylighted into the quarry a few hundred feet away. The mining began in earnest, orienting the rooms as recommended by the consulting mining engineer.

Not too many years later, I requested a transfer from a corporate staff position to production management. I had come to the point in my career when I decided that a staff position was nowhere near as exciting as making day-to-day production decisions. My first management role was supervising the sales and production of two fairly small, rural limestone quarries that posed an interesting set of unique problems. I was assisted in this role by two foremen and a salesman. As time progressed, I was promoted into larger operations and later into management of multiple operations.

Approximately 2 years after starting the underground mine mentioned earlier, I took over management of that particular operation beginning on a Monday morning. The previous Saturday, a substantial section of the mine roof collapsed (Figure 1.4). Fortunately, no employees were in the mine at the time and no injuries occurred, but it made for a very exciting first week on a new job. My first meeting was, of course, with an MSHA underground mining inspector. After hiring a different mining consultant, the conclusion of our internal investigation into the roof fall was that a "swarm" of seven joints had coalesced at a crosscut intersection. This created a substantial zone of weakness down the middle of the drift. The joint swarm was at just the right angle that it was not supported by any pillars for a length of almost 200 feet (61 meters). After extensive joint mapping, we determined that to maintain a stable roof beam we needed to reorient the entire mine layout and utilize 35 × 60 foot

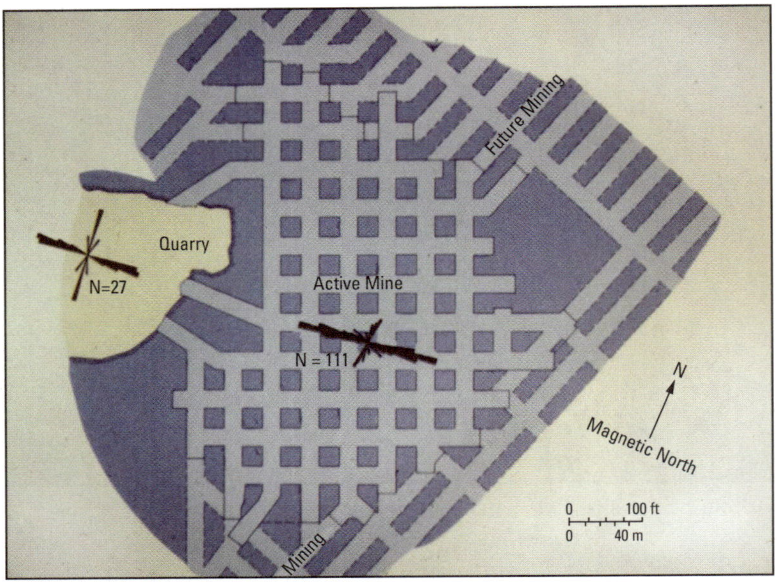

Courtesy of American Aggregates Corporation

FIGURE 1.5 Reorientation of the mine relative to joint patterns

(10 × 18 meter) rectangular pillars instead of 40 × 40 foot (12 × 12 meter) square pillars. The new angle of orientation was designed so that no predominant joint would be unsupported by a pillar for more than a span of 50 feet (15 meters). This ultimately eliminated the roof problems but the reorientation took over a year of production to accomplish (Figure 1.5). I later became involved in a second underground mine that was developed at another quarry for the same purpose of avoiding the cost of removing excessive overburden. The second mine was a two-level mine with a 60-foot-thick (18-meter) sill pillar between mining levels.

Labor relations can be a challenging issue especially when working in a union environment. In the early 1980s I was assigned to manage a workforce at a three-shift production operation staffed by 52 members of a large international union. I was assisted by five foremen with varying degrees of education, experience, and longevity. Almost one-third of the labor force happened to be members of a fairly violent, local motorcycle gang, many of whom were related (brothers, nephews, brothers-in-law, etc.). This was at a time when drug testing had not yet become mainstream in the aggregates industry. On more than one occasion I was awakened in the wee hours of the morning by a call from the night foreman reporting incidents of fights, employees having to be taken to the hospital after being hit in the head with a shovel, and so on.

Early in my tenure at this operation my boss called me and said that a university in an adjoining state was offering a 3-day seminar on how to deal with difficult employees and suggested I attend. I did as he recommended and during the opening seminar session, the attendees were asked to share some of their employee issues. I shared stories of various assaults, knife fights, and such when my turn came. I'll never forget the first day at lunch when I was approached by a meek young woman who sat next to me and said: "After hearing about your employees, I decided that I don't belong at this seminar." I asked her why not and she replied: "I'm here because several of the women in the typing pool that I supervise are constantly bickering; no one ever pulls knives on one another." I ended up playing the role of the belligerent union steward during the session on handling grievances and no one ever managed to defuse the arguments I had learned from experience.

The union steward (employee representative of the union) at this operation was the ringleader of the motorcycle gang, had several relatives employed at the operation, and liked nothing better than to stir up trouble within the workforce. He would, on a regular basis, approach an employee with a grievance form he had filled out and tell them to sign it in order to file the grievance (whether the employee felt aggrieved or not). He always had his "enforcer" with him during these grievance signing incidents. His enforcer was a fellow gang member with sinewy muscles and knife scars on both forearms. While the union steward coerced the innocent employee to sign the form, his enforcer would stand nearby cleaning his fingernails with an extremely large pocketknife. The employees easily got the message and grudgingly signed off on the grievances. Our labor agreement with the union had a three-step grievance procedure that required employee meetings, answering grievances in writing, meetings with the union bargaining agent, and ultimately arbitration if the issues were not resolved with the first three steps.

After my first year dealing with 223 grievances, I barely had enough time to manage the production operations because of continuing labor relation issues brought on by one union steward wanting to cause chaos. On one occasion I had a grievance filed against me personally over a ham sandwich. Following numerous consultations with our corporate labor relations manager, we decided that the only recourse was to take every grievance filed by the union to arbitration. We hired a young, very tenacious, attorney who specialized in labor law and spent the next several months in and out of arbitrators' offices, ultimately winning every arbitration. Our strategy finally succeeded when the local union officials ultimately decided they had paid enough legal fees. We hurt them in their pocketbooks. Our rogue union steward was finally "corralled" by the officials of the local union. As a footnote, the union steward's enforcer was ultimately murdered in a drug deal gone wrong.

I blame my predecessors at this operation for the labor problems. Historically, they had taken the easy route by paying off the "aggrieved employees" with additional overtime in response to the unfounded grievances. The rogue union steward's ploy worked well and cost the company dearly in additional labor costs over the years, not to mention the burden of hours of management time, effort, and frustration dealing with the labor problems. I am happy to say that the labor incidents related above are not typical of the industry in general. The majority of the people (both union and nonunion) I have worked with over the years in this industry are the salt of the earth: good, hardworking, capable, dependable Americans trying to earn a living to support their families. I have managed numerous other operations, both union and nonunion, without ever again encountering the type of labor problems related above.

I personally prefer dealing with individual employees firsthand, without the need for a union representative present. I believe in treating employees better than the union environment by providing them with better benefits and higher wages, and treating them with the respect they deserve. Many plants I've managed have set up their own employee associations and negotiated contracts directly with management without the auspices of union oversight. Signing contracts with employee associations, as long as they register as such with the U.S. Department of Labor, eliminates the possibility of one of the large unions attempting to organize the employees and allows for a simple contract (employee agreement) with reasonable rules and standards. This also saves the employees the dues or fees normally paid to the union for their representation.

In the course of my 42-year career, I worked for several aggregates producers, 7 years for a division of a large national construction engineering firm, 18 years for a regional producer with 45 operations in a three-state area, 4 years in a consulting capacity, and finally for a privately owned highway general contractor, asphalt producer, and aggregates producer.

The regional producer was fairly closely held, traded over-the-counter, and was extremely good to its employees; providing management with lucrative profit sharing and stock options. After about 15 years with this company, a hostile takeover attempt by a rival company outside the aggregates industry made it necessary for upper management to hire an investment banking firm to list the company for sale. In the mid-1980s, U.S. companies were undervalued compared to corporate entities in other countries and became prime targets for foreign takeovers. Part of this trend can also be attributed to the lure of a foreign company owning large tracts of U.S. real estate. The company was subsequently sold and resold three times in less than 4 years; first to a Dutch conglomerate, next to an international company incorporated in the U.K., and finally spun off to an Australian consortium with operations all over the world. This provided some of us with opportunities to expand our horizons into other areas of the country and the world. However, with these opportunities also came a lot of travel and time away from home.

After 18 years with this organization, I decided to try my hand at consulting for a brief stint and opened a new office for an established consulting company that specialized in advising aggregates producers on issues concerning environmental issues, geology, mine planning, processing, and marketing. My personality type enjoys challenges, action, decision-making, even a certain amount of stress, and the consulting business was a bit too passive for my tastes. I found myself understimulated and bored a good portion of the time. This, coupled with the fact that we had a few clients that asked for our input on projects, ultimately did not follow our recommendations, and returned to us for advice on the same project a year later, frustrated me mightily. When I expressed my frustration to one of my contemporaries in the company about repeating the same projects for the same client multiple times, he quipped: "Those are the best type of clients; they pay us multiple times to complete the same task over and over again: a cash cow."

I stayed with the consulting business for about 3 years and ultimately applied for permanent work with a local client I had developed. A family-held company in the highway construction, asphalt, and aggregates business hired me as the Manager of Business Development and Plant Systems for their aggregates operations, which at the time consisted of four active production facilities and several inactive reserve sights. When I was hired, the owner of the company promised that I would never be bored working for him. I have to say that he kept this promise in spades! His other comment when he hired me was that he wanted to get bigger in the aggregates business. Over the next 17 years we went from a small player in aggregates to a regional powerhouse with 18 production operations, one industrial mineral filler facility (ground calcium carbonate), two portable plants recycling asphalt and concrete, and three rail unloading yards, operating in three states with over 100 people employed in aggregates production. I ultimately ended up as operations manager responsible for all of the production facilities. This expansion was accomplished through the cooperation of numerous dedicated people both through acquisitions of existing producers as well as the development of new greenfield sites.

The challenges of this rapid expansion were almost constant while at the same time I was dealing with the day-to-day stimulation of numerous production issues, labor relations, budgets, capital projects, environmental and safety regulations, contractor meetings, and personnel. The expansion brought on many nights spent at public hearings on zoning, meetings with regulators, environmental permit negotiations, and due diligence evaluations. Upon acquisition of an existing operation, new employees had to be evaluated, retrained, and indoctrinated into our methods, standards, systems, and culture.

Transforming a highway-based quarry into a rail-focused operation was another interesting challenge. We gradually moved in this direction by developing our first rail-focused sales yard in the area of a major city that was fairly remote from any existing aggregates facilities. After negotiating with a short-line railroad to move rail cars overnight from our largest quarry (approximately 60 miles [96.5 kilometers]), we leased a piece of land and had the property rezoned to permit a resale aggregates yard. Our presentation to the city zoning board focused on the reduction of environmental impact on air pollution and increased highway safety based on a substantial reduction in over-the-road mileage of multiple dump trucks versus one overnight train 3 days per week.

The next step was to build the rail infrastructure at the resale yard as well as the quarry. New rail sidings and loading capabilities at the quarry and rail sidings, dump hoppers, feeders, and stacking conveyors at the sales yard were contracted for and constructed. As our sales force developed new customers for the resale rail yard, the sales volume increased and rail shipments followed suit. Ultimately, we developed a second unloading facility at one of our sister companies' asphalt plants in an adjoining state and a third site in a different rural market. At the time of my retirement a fourth resale yard was in development with plans for additional rail siding expansion at the quarry site.

There were many challenges over the years, frustrating at the time but enjoyable in retrospect. At the close of one season, we purchased an operating quarry about 2.5 hours from my residence. This rural quarry was first developed in the late 1940s to provide aggregates for a large federally funded project nearby. It had several owners over the years, and we purchased it when the existing owner wanted to retire. Very limited capital had been spent on the quarry over the last several years and the state of the equipment reflected that fact. Upon consummation of the purchase, we were officially supposed to take over ownership and management on January 2.

I got a call in the early hours of New Year's Day that the quarry was rapidly flooding and we needed to respond immediately. It so happened that after several inches of snow late in the year, the temperature rose and the rains began. The location was inundated with, not a 100-year flood event, but a 500-year flood event taking place over a 36-hour period. Add that to the snow already on the ground and we had a disaster in progress. What a holiday weekend that turned out to be; spent rousting employees from their holiday celebrations, waking up venders to rent any portable pumps we could locate, working 12-hour shifts in heavy rain, up to our knees in freezing water. After a 2-day battle we were able save our "new quarry" from being entirely inundated. The employees that came along with the purchase, none of whom we had met or interviewed, showed up without being called and offered their services on New Year's Day. After 12 hours of sloshing in the cold rain, one of our new employees asked me if

FIGURE 1.6 Moving a dredge using airbags

they'd be paid for today. When I assured him he would be paid, he said: "Well, when our old boss called us in to work on the weekends, he never paid us!" This is a good example of the caliber of most of the people in this industry. These men showed up and volunteered to help on a holiday weekend without knowing their new employers, and not sure if they were going to be paid.

On one occasion we needed to move a suction dredge from one lake to another over a half-mile (0.8-kilometer) distance because of a pipeline easement. The dredge weighed in excess of 3.5 million pounds (1,587,573 kilograms). The estimated cost to disassemble and reassemble the machine was more than $100,000. I had seen a presentation at an NSSGA seminar where another producer had moved a dredge across a road using airbags, and decided if they could do it so could we. I contracted with a manufacturer of airlift bags to make sausage-shaped airbags that were approximately 3 feet (0.9 meters) in diameter by 10 feet (3 meters) long with a double-walled construction. My boss was rather dubious when I asked to spend $30,000 on airbags but he eventually acquiesced. We utilized the airbags much like the ancient Egyptians are depicted in the movies, using logs to move the giant pyramid stones (Figure 1.6). Once the front of the dredge was lifted from the water by two large cranes, the airbags were inserted under the pontoons and a wheel loader was used to pull the machine. We kept four men busy rotating airbags from the rear to the front of the dredge. Two days of heavy labor during the hot summer months and the dredge was launched into the other lake.

The largest capital project in the history of the company began about 3 years prior to my retirement. A quarry the company had purchased in 1989 contained a deposit of high-calcium limestone that was too soft to make a quality aggregates product. This bench had historically been sold to the steel industry for flux stone and was also ground for use as mineral fillers and agricultural limestone. A new aggregates processing plant had been built close to the mining face on the quarry floor a few years before where two other benches were crushed into aggregates products. There was also an intervening shale layer between two of the limestone ledges that was routinely stripped and removed to waste piles.

The old pulverizing mill was almost 2 miles (3.2 kilometers) from the quarry face and had originally been built in the 1930s (the concrete silos were constructed by the Civilian Conservation Corps). Feed for the mill had to be hauled from the quarry with mine trucks on internal haul roads. The milling equipment had been upgraded once in the 1950s, again in the late 1980s, and once again in the early 2000s, with some minor improvements annually since that time. The old mill contained about a dozen bucket elevators, 30 screw conveyors, and numerous air slides just to move material within the two mill buildings. The primary ball mill produced one product directly but also a huge volume of oversize material that had be stored in silos and recirculated through the ball mill or through two roller mills on another shift. The infeed material had to be dried to less than 0.5% moisture content through an oversized, inefficient rotary dryer and the fuel costs were excessive. The electrical system consisted of two different current systems that were medium voltage and the power company was having difficulty locating appropriate transformers. The mills were operated by huge 4,160-volt synchronous motors that were the size of a Volkswagen Beetle.

The capital budget for the following year included over $1.2 million for silo and roofing repairs at the mill. The primary ball mill was in dire need of new drive gear that was 12 feet (3.6 meters) in diameter and cost twice what the entire mill cost when purchased in the 1930s. The gear had to be custom made and had a 46-week delivery timeline. Faced with these capital and mechanical challenges it was decided it was finally time for a new state-of-the-art pulverizing mill. This project had been discussed several times over the previous 10 years but the capital costs were very large and other capital projects always seemed to take priority and thus the capital. Now was the time! Over the years we had researched what type of milling system we wanted and proceeded to take bids from various manufacturers.

Throughout this process we acted as our own general contractor, purchasing components from numerous different manufacturers. Working with a steel fabricating contractor (with engineering capabilities) and the mill manufacturer we were able to design a pulverizing plant plan. We purchased two roller mill systems that were "closed systems," each consisting of a feeder, a mill, flash dryer, system fan, ducting, and cyclone. These two mill systems were interconnected via dilute-phase pneumatic conveying systems to five outdoor storage silos, each with its own truck scale to load pneumatic tanker trucks for transport to customers. No more recirculating load from the ball mill to recrush; the new system eliminated the recirculating loads with everything being produced into a salable product in one pass through the mills. No more 2-mile (3.2-kilometer) haul to get the mill feed to the plant; the new mill was located on the quarry floor near the aggregates processing plant. The plant was designed with modern 480-volt electrical system and controls, total plant automation, and efficient flash dryers, all providing substantial savings in utilities costs. What once required production on a 24 hours per day, 7 days per week schedule could now be produced on one shift in 6 days. The labor savings were also substantial.

A fair portion of our filler volume was sold as bagged products either in 50-pound (23-kilogram) paper bags or super sacks holding up to 3,000 pounds (1,360 kilograms). The project included a new bagging operation and a separate warehouse facility for the storage of bagged products. Throughout the project (both mill and bagging), we utilized the latest automation technology provided by another, separate contractor. The bagging system included a five-spout automated bagger, bagging conveyors, a robotic palletizer, and shrink wrapper (not to be confused with a rapper like L.L. Cool J). The fact that we produced products in the size range of 3–12 micrometers for median particle diameter meant that environmental dust control systems (bag houses) were an absolute necessity. (The state environmental protection agency officials were thrilled with the upgrade from our 1930s technology.) Functioning as our own general contractor we obtained separate contracts for the milling systems, structure steel and fabrication, silos, two different buildings (one over 90 feet [27 meters] tall), erection, electrical controls and wiring, automation, a 6,000-foot (1,829-meter) gas line installation, high-voltage electrical supply, and fiber-optic communication.

The entrepreneur that hired me for my last career position kept both of his promises to me. Together with him and some others, we substantially enlarged his stake in the aggregates industry and there was never a day while I worked for him that I was the least bit bored. At the party the company held to honor my retirement I made it a point to remind and thank him for keeping his promise and also for allowing me to spend so much of his money over the years. It was definitely an interesting, challenging, and enjoyable ride for 17.5 years. As I attempt to write this book, I've been retired for almost 4 years. My wife tells everyone that I retired on a Friday and I didn't get bored until the following Tuesday. I have to say that is the truth! The first 6 months of my retirement were a very difficult transition. And that's the downside to finding a job you love; it's also very hard to leave it behind.

CHAPTER 2

Importance of the Aggregates Industry

The *Ten Books of Architecture* written by Marcus Vitruvius Pollio in 25 BCE discusses the various types of aggregates utilized in Roman concrete. The Roman Empire is famous for its use of concrete in construction from marine underwater construction in port facilities to concrete domes like the Pantheon that has, over the centuries, survived thousands of earthquakes since its completion in Rome in 126 CE. There is even evidence in some much earlier Greek ruins of the use of concrete construction. Rome built its empire throughout Europe and North Africa by constructing a vast network of roads paved with aggregates for use by the Roman legions and as networks for commerce. The Romans provided their citizens with elaborate baths, plumbing, and fresh water supplied through miles of aqueducts, many of which were made from concrete. No one knows exactly when ancient humans first utilized aggregates by throwing some broken rocks into a mud puddle to provide dry footing.

Today, by volume, aggregates is the largest mining industry in the world. A 2018 article in *Pit & Quarry* forecast the worldwide sales of aggregates in that year to be 53.2 billion metric tons and predicted that the worldwide market will reach 66.2 billion metric tons by the year 2022 (PQ Staff 2018). Aggregates is defined in the *Aggregates Handbook* as "any combination of crushed stone, sand or gravel in their natural or processed state" (NSSGA 2013). Preliminary estimates of 2019 U.S. aggregates production by the U.S. Geological Survey (USGS) are 1.53 billion metric tons of crushed stone valued at $18.7 billion and 970 million metric tons of sand and gravel with a value of $9.0 billion. This equates to a total of 2.53 billion metric tons valued at $27.7 billion to the U.S. economy in 2019 (USGS, n.d.-b).

The above referenced statistics concluded that the production came from 10,270 pits and quarries throughout the United States operated by approximately 5,300 different companies. Crushed stone produced in 2019 consisted of 69% limestone and dolomite, 15% granite, 6% traprock, 5% miscellaneous, 3% sandstone and quartzite, and the remainder marble, volcanic cinder and scoria, marl, slate, and shell. The sand and gravel produced in 2019 was sold for the following uses: 46% as concrete products, 21% as road base, 13% as fill materials, 12% in asphalt products, and the remailing 8% a combination of filtration, roofing granules, plaster sand, and snow and ice control. Per the USGS (n.d.-b), in 2016 the top 10 states for crushed stone production in descending order were Texas, Pennsylvania, Florida, Missouri, Ohio, North Carolina, Virginia, Georgia, Illinois, and Kentucky. For the same year the top 10 states in sand and gravel production were California, Texas, Minnesota, Michigan, Arizona, Washington, Colorado, Ohio, Utah, and New York. The U.S. aggregates industry experienced a record-setting year in 2006. According to USGS (n.d.-b), during 2006 the industry directly employed 121,000 people, producing 3.04 billion metric tons of aggregates at an estimated value of $22.4 billion. The recession of 2007–2008 caused a substantial reduction in production and the volume produced in 2006 has yet to be surpassed. Figure 2.1 shows U.S. aggregates production for the 10-year period of 2009–2018 and Figure 2.2 gives the value of U.S. aggregates production for the same period.

The aggregates industry directly employs more than 100,000 people annually as indicated in Table 2.1.

A publication dated March 2017 entitled *The Economic Impact of the Natural Aggregates Industry: A National, State, and County Analysis* written by the Phoenix Center for Advanced Legal & Economic Public Policy Studies contains statistics concerning the economic impact of mineral aggregates on the U.S. economy. According to this study, "each job in the aggregates industry supports an additional 4.87 jobs throughout the economy; each dollar of earnings (wages) creates another $4.19 of earnings in other sectors; and each dollar in aggregate sales produces another $3.47 of sales in other industries" (Phoenix Center for Advanced Legal &

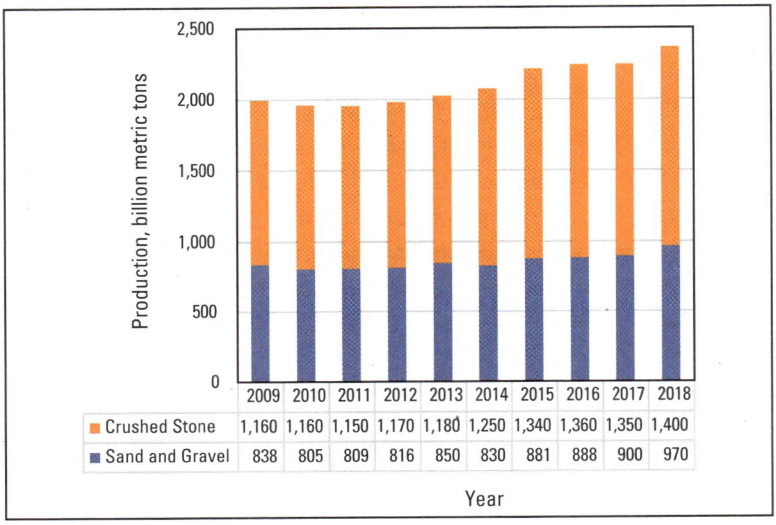

	2009	2010	2011	2012	2013	2014	2015	2016	2017	2018
Crushed Stone	1,160	1,160	1,150	1,170	1,180	1,250	1,340	1,360	1,350	1,400
Sand and Gravel	838	805	809	816	850	830	881	888	900	970

Data from USGS

FIGURE 2.1 U.S. aggregates production, 2009–2018

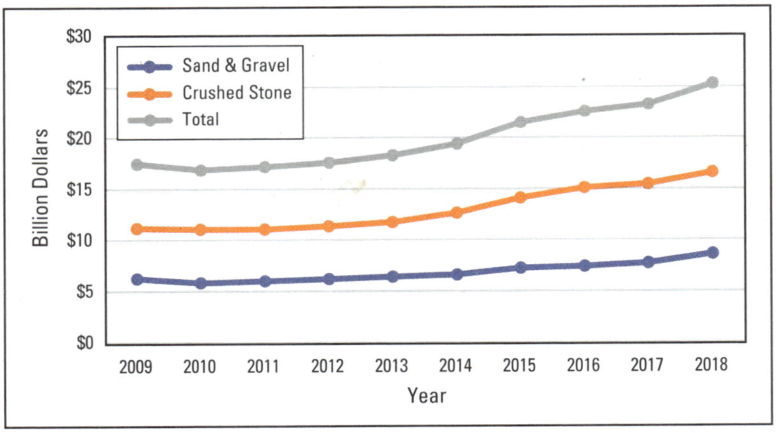

Data from USGS

FIGURE 2.2 Value of U.S. aggregates, 2009–2018

TABLE 2.1 Direct employment in the U.S. aggregates industry

Year	Sand and Gravel	Crushed Stone	Total
2009	30,800	70,300	101,100
2010	29,500	67,600	97,100
2011	29,800	67,000	96,800
2012	30,600	66,200	96,800
2013	30,000	65,900	95,900
2014	34,600	65,600	100,200
2015	34,800	67,100	101,900
2016	35,300	68,100	103,400
2017	36,500	68,600	105,100
2018	36,200	67,200	103,400

Data from USGS

TABLE 2.2 Impact of the aggregates industry on the U.S. economy

	Aggregates Industry	Impact on Other Industries	Total Impact
Jobs (Quarry)	61,024	297,275	358,317
Jobs (Total)	102,000	496,740	598,740
Earnings (Wages)	$6.1 billion	$25.6 billion	$31.7 billion
Sales	$27.2 billion	$94.4 billion	$121.6 billion

Source: Phoenix Center for Advanced Legal & Economic Public Policy Studies 2017

Economic Public Policy Studies 2017). The additional jobs created include delivery truck drivers, parts suppliers, heavy equipment manufacturers, toolmakers, steel manufacturers, equipment mechanics, explosives suppliers, and government regulators, to name just a few. The aggregates industry generates more than $570 million in taxes on an annual basis. The individual wages of aggregates workers average over $75,000 per year. According to the Phoenix Center for Advanced Legal & Economic Public Policy Studies (2017), these factors have the cumulative impact on the U.S. economy shown in Table 2.2.

The U.S. Geological Survey estimates that every American born after the year 2000 will require 3.5 million pounds (1,587,573 kilograms) of minerals, metals, and fuels in their lifetime. Each person will require 1.62 million pounds (725,747 kilograms) of stone, sand, and gravel; 80,454 gallons (365,751 liters) of petroleum; 64,750 pounds (29,307 kilograms) of cement; 5.7 million cubic feet (161,406 cubic meters) of natural gas; 32,990 pounds (14,964 kilograms) of iron ore; 923 pounds (419 kilograms) of lead; 4,999 pounds (2,267 kilograms) of bauxite (aluminum); 561,447 pounds (254,6687 kilograms) of coal; 22,224 pounds (10,081 kilograms) of phosphate rock; 1.458 troy ounces (45.3 grams) of gold; 1,615 pounds (749 kilograms) of copper; 21,448 pounds (9,279 kilograms) of clay; 29,530 pounds (13,394 kilograms) of salt; 846 pounds (384 kilograms) of zinc; and 49,601 pounds (22,498 kilograms) of other minerals and metals (USGS, n.d.-b). The construction of an average new American home requires more than 120 metric tons of aggregates. Approximately 85,000 metric tons of aggregates are needed to build 1 mile (1.6 kilometers) of four-lane interstate highway. The construction of the Denver International Airport in 1995 required in excess of 5 million metric tons of crushed stone and sand/gravel. According to the USGS, 94% of the material used to construct an interstate highway consists of natural aggregates, with cement, asphalt, and steel making up the remainder (USGS, n.d.-b).

More than half of all the aggregates sold in the United States are utilized in public works infrastructure projects financed directly or indirectly through tax dollars. Aggregates are bulk commodities, the cost of which can easily double as transportation costs are added. For each additional 10-mile truck haul, a community of 100,000 residents would pay an additional $1.3–$1.5 million annually for this basic resource. Limiting the distance that aggregates have to be transported is beneficial not only to the immediate consumer but also to the average taxpayer. Local sources of readily available, quality construction aggregates are a major financial benefit to local communities and their economy.

Whether it's crushed stone aggregates for asphalt highways, sand and gravel produced for concrete foundations, mineral fillers for use in PVC pipe, gypsum manufactured into wallboard, clay and shale formed and baked into bricks, or salt for the dinner table and deicing roadways, America's appetite for industrial minerals is insatiable. The United States' infrastructure requirements coupled with a burgeoning population will only increase the need for industrial minerals production in the 21st century and beyond. Industrial minerals will continue to play an integral role in maintaining a vital economy in the United States and the world.

The American Society of Civil Engineers (ASCE) releases a *Report Card for America's Infrastructure* assessing the status of 12 infrastructure categories: roads, bridges, mass transit, aviation, schools, drinking water, wastewater, dams, solid waste, hazardous waste, navigable waterways, and energy. According to the ASCE's *Report Card for America's Infrastructure,* the overall rating over the past 10 years has been a very disappointing D+ with an estimate of $2.0 billion annually in capital improvements needed to bring the infrastructure up to minimum acceptable levels (ASCE 2021). According to ASCE, the country's decaying transportation infrastructure costs consumers $147 billion annually: $109 billion in increased vehicle operating costs, $36 billion in time delays, $1.4 billion in safety costs, and $0.7 billion in environmental costs. This poor infrastructure rating indicates an even greater need to increase the production of aggregates in the United States over the next decade.

The federal government is required by the U.S. Constitution to fund highways. Since 1993, the Highway Trust Fund has been financed through a tax on gasoline and diesel fuel of $0.09/gallon with no inflation rider. This level of federal funding has not ever been increased. The meager funding has mostly gone to provide maintenance on existing highways with little money available for new highway projects. More than 20 states have taken it upon themselves to add more taxes on fuel to help with local highway funding. In the partisan political climate of the 21st century, the U.S. Congress has been unable to agree on any substantive funding increases. New highway funding needs to be tied to fuel costs but not on a specific price per gallon basis. Any new funding levels should be based on a percentage of the cost of fuel so that as the price of fuel fluctuates so does the tax level. Using the percentage calculation provides a built-in inflation factor and those using the roadways pay for that usage in the cost of their fuel. More thought will need to be given on how to fund roadways in the future. As more electric vehicles begin using the highways, a tax on diesel fuel and gasoline will not suffice as a usage tax and other taxing methods will need to be developed.

The industry has come a long way since the early days of picks and shovels and horse-drawn wagons hauling material that was sized by the power of human muscle wielding sledgehammers. The innovation of steam power revolutionized the industry in the 1830s–1860s. A large market for crushed stone railroad ballast developed as the transcontinental railroad system expanded to cities and towns throughout the country. Steam-powered locomotives (both on tracks and wheels) soon moved the run-of-mine rock to steam-powered crushing plants for further size reduction. In the early 1900s the industry began the transition from wood- and coal-generated steam to petroleum and electrical power. Heavy-duty (by 1920 standards) gasoline- and diesel-powered haul trucks were being built, and primitive attachments on tractors were developed to facilitate the loading of trucks (Figure 2.3). In 1905 the USGS first recognized the importance of aggregates and began publishing statistics on sales of sand and gravel and crushed stone in its publication *Mineral Resources of the United States* (USGS, n.d.-c).

Aggregates have many uses in the construction industries. Crushed stone is utilized as a structural material under roadways, parking areas, buildings, and sidewalks. It serves a similar function under railroad lines as ballast and sub-ballast. Structurally, the crushed particles in the base layer underneath a highway or railroad grade are compacted into a uniform layer thick enough to support the pavement and decrease the stress with depth caused by the traffic loading. A layer of subbase lies below the base layer and generally consists of a lower-quality granular fill material mixed to some degree with the underlying natural materials, leveled, graded, and compacted prior to constructing the base layer. Base material generally consists of a graded mixture of nonplastic, crushed stone or gravel with a maximum size of 2.0 inches (50.8 millimeters) and no more than 8% fines (minus 200 mesh). Where drainage is a problem, the fines can be reduced below the 8% threshold to allow for additional percolation. Local soil conditions (bearing capacity, plasticity, porosity, moisture content, etc.) generally dictate the amount and type of subbase required. Depending on soil conditions, stabilizing additives may be required to be mixed with the underlying soil prior to placing the subbase. In areas with "fat" clay soils, stabilizers such as quick or hydrated lime may be required to reduce moisture issues and prevent soil pumping.

Sub-ballast serves the same function in railroad construction as subbase under highways. Railroad ballast is the base material in the railroad construction industry but serves several other functions as well. The ballast supports the structural load under the railroad ties but also absorbs vibrations and shock from moving trains. The angular, crushed ballast tends to lock together and anchor the ties to prevent lateral and longitudinal movement of the track. Limestone and dolomite were once extensively utilized in railroad ballast but over the years railroad engineers found that the dust generated as the rock particles ground together had a tendency to recement over time, negatively affecting the drainage under the tracks. Most railroad ballast specifications today require the use of granite or traprock because these materials are harder and more abrasion resistant and do not contain calcium or magnesium carbonates, which have a tendency to recement the fines. Railroad ballast is shipped very long distances into areas devoid of the appropriate deposits of these harder, noncarbonate rocks.

Sixty-five to eighty percent of concrete is made up of aggregates, with the remaining portions being cement, water, additives, and air. There are several types of hydraulic cement available, but the most common used today is portland cement (PC), which is a manufactured form of calcium aluminum silicate. PC is manufactured from a combination of limestone and other additives such as clay, shale, sand, bauxite, fly ash, slag, and sometimes iron ore. Some impure, argillaceous limestones meet all the chemical requirements without any additives, but this is fairly rare. The combination of raw materials (generally 80%+ limestone) is initially heated in a kiln to approximately 1,200°F–1,700°F (649°C–927°C) to calcine the material and the temperature is then increased to 2,650°F (1,454°C) to sinter (fuse) the material into a "clinker," which is later ground into the final product, cement powder. Generally about 5% calcium sulfate (gypsum or anhydrate) is added with the clinker prior to grinding. Other additives may be utilized to make specialized varieties of cement. The clinker is ground to a size range of minus 5 micrometers and shipped in either pneumatic tankers or bagged and palletized. There are presently eight different types (formulations) of PC depending on the end use (e.g., early strength, low heat of hydration, air entrained). Table 2.3 summarizes U.S. cement (not concrete) production and sales from 2012 to 2016.

Courtesy of Greenville Manufacturing Company

FIGURE 2.3 A 1920s era one-man-operated road truck (OMORT)

TABLE 2.3 U.S. cement production and sales

	2012	2013	2014	2015	2016
Cement (million metric tons)	74.15	76.80	82.53	84.40	84.69
Clinker (million metric tons)	67.13	69.42	74.37	76.04	75.63
Sales (million metric tons)	78.30	81.70	88.90	92.00	94.30
Sales (billion dollars)	7.02	7.76	8.94	9.80	10.50

Data from USGS

The terms *cement* and *concrete*, although not synonymous, are commonly misapplied. Cement is a component of concrete, the binder in the mix; concrete is the entire mixture of cement, aggregates, water, and air. It is not uncommon to see terms like *cement sidewalks*, *cement slabs*, *cement blocks*, or *cement walls* misused in lieu of the proper term *concrete*.

Concrete is one of the most widely utilized construction materials in the world. In the process of making concrete, cement and water combine to form a paste that coats the aggregate particles. During cement manufacturing, the calcination process removes the carbon dioxide (CO_2) from the calcium carbonate ($CaCO_3$) to form calcium oxide (CaO). When water is added to the cement paste, an exothermic chemical reaction called hydration takes place, causing the paste to harden around the aggregates. Caution must be taken when pouring and finishing concrete because the heat generated by hydration can cause serious burns to the skin. Hydration can continue indefinitely for years, as long as there is a source of water to react with the cement. Concrete mixtures need to be cured for a period of time until the desired properties, such as strength and density, are achieved. Most concrete generally reaches its maximum practical strength after about 28 days. As the paste hardens, air voids (small bubbles) occur within the concrete. These air voids are beneficial in cold climates, providing a small amount of internal expansion during freeze–thaw cycles. Certain chemicals can be added to the process that provide additional entrained air if and when required.

There are many different concrete designs utilized in modern construction depending on the final results required. Varying the ratio of water to cement in a concrete mixture can have a dramatic effect on the final concrete product. The water/cement ratio is defined as

$$\frac{\text{pounds of cement}}{\text{pounds of water}} = \frac{\text{water}}{\text{cement ratio}}$$

The water/cement ratios in most concrete range from 0.35 to 0.75. More cement and less water in the mix provides a higher density of cement particles, which equates to a lower-permeability and higher-strength concrete. Cement, which sells at more than $100 per metric ton (10 times the cost of aggregates), is by far the highest cost component in a concrete mixture. The water/cement ratio directly affects the cost of the final concrete mix. If additional strength is not needed, a water/cement ratio of 0.4–0.5 makes for a much less costly mix. There are a number of admixtures utilized for specific purposes that affect the characteristics of concrete. There are additives that reduce water usage, retard setting time, accelerate setting time, increase early strength, entrain additional air, color the mixture, inhibit corrosion, modify viscosity, reduce shrinkage, and inhibit alkali-silica reactivity, among other properties.

Concrete is generally produced in a batch plant where the individual ingredients (cement, fine aggregates, coarse aggregates, water, and additives) are weighed to the mix design proportions and loaded into a ready-mix truck (Figure 2.4). The truck acts as the mixer, utilizing a rotating drum to mix and blend the ingredients as the concrete batch is being shipped to the jobsite. When the truck arrives on site, the speed of the drum rotation is increased and the plastic concrete mixture is discharged via chutes to the appropriate location for finishing or pumping. On very large concrete jobs (large factories, warehouses, airport runways, etc.), the concrete can be mixed in a central mixing plant on site and trucked in more conventional trucks the relatively shorter distances within the jobsite.

Aggregates are the largest component of concrete mixtures (65%–80%) and form the internal structure of the final product. It is therefore of upmost importance that quality aggregates are utilized in order to obtain a durable, strong, long-lasting final product. The aggregates

Courtesy of 4D Concrete

FIGURE 2.4 Example of a ready-mix concrete plant

need to be sound, nonreactive, resistant to abrasion, and properly graded to maximize the bond with the cement paste and produce a durable product. Adding more cement to the water/cement ratio can indeed give additional strength to concrete, but if the internal strength of the mix is comprised by utilizing inferior aggregates, nothing has been gained. Aggregates are much more than just a filler in the concrete mix.

ASTM International is one agency that defines specifications for concrete and concrete aggregate. ASTM C33 defines the various aggregate gradations for coarse and fine aggregates. A number of aggregate properties directly affect the performance of concrete. Among these properties are individual gradation of both coarse and fine aggregates, combined aggregates gradation, maximum particle size, particle shape, bulk density, specific gravity, water absorption, surface texture, freeze–thaw durability, pore structure, thermal expansion, deleterious materials, and organic impurities. These qualities are further discussed in Chapter 7 on quality control. Aggregates have a direct effect on the workability, density, stability, strength, and permeability of the concrete.

ASTM C33 also defines the following materials to be deleterious in concrete aggregates: clay lumps, friable particles, chert (amorphous silica), coal, lignite, and other lightweight materials. Certain aggregates can react chemically with the alkali (potassium and sodium oxide) components of cement. Certain naturally occurring siliceous materials contained in some aggregates such as chalcedony, strained quartz, opal, certain types of chert, and glassy materials adversely react with the alkalis in the cement in what is known as alkali-silica reactivity. Alkali-carbonate reactivity can also occur with certain dolomites and dolomitic limestones containing clay or silt or with a specific texture featuring fine, individual, dolomite crystals that react with alkali hydroxide in the cement. Both alkali-silica and alkali-carbonate reactions are similar in occurrence. The chemical reaction forms a gel within the concrete that expands with absorbed moisture. With a sufficient amount of reactive particles and adequate absorbed water, the gel can exert enough internal pressure over time to crack the concrete or deteriorate the surface with "pop-outs." There are seven different test methods to determine whether certain aggregates are reactive in concrete, including ASTM C227, ASTM C289, ASTM C295, ASTM C1260, AASHTO T 303, ASTM C1293, and ASTM C1567.

Other forms of concrete utilizing aggregates are manufactured in precast concrete facilities. Concrete blocks, concrete pipe, concrete burial vaults, precast culvert structures, and septic tanks are examples of precast concrete products. Prestressed concrete beams for bridges are

TABLE 2.4 Standard sieve sizes

Sieve Size	Opening (millimeters)	Opening (inches)	Sieve Size	Opening	Opening (inches)
3 inches	75	3.0	10 mesh	2.00 millimeters	0.0787
2 ½ inches	63	2.5	12 mesh	1.70 millimeters	0.0661
2 inches	50	2.0	16 mesh	1.18 millimeters	0.0469
1 ½ inches	37.5	1.5	20 mesh	850 micrometers	0.0331
1 inch	25	1.0	30 mesh	600 micrometers	0.0234
¾ inch	19	0.75	40 mesh	425 micrometers	0.0165
⅝ inch	16	0.625	50 mesh	300 micrometers	0.0117
½ inch	12.5	0.500	60 mesh	250 micrometers	0.0098
⅜ inch	9.5	0.375	70 mesh	212 micrometers	0.0083
¼ inch	6.3	0.250	100 mesh	150 micrometers	0.0059
4 mesh	4.75	0.187	120 mesh	125 micrometers	0.0049
6 mesh	3.35	0.132	140 mesh	106 micrometers	0.0010
8 mesh	2.36	0.0937	200 mesh	75 micrometers	0.0029

manufactured by stretching (tensioning) steel cables and pouring concrete around these pre-tensioned cables, forming high-strength structural members.

Hot mix asphalt or asphaltic concrete is another industry that utilizes large volumes of aggregates. An even higher percentage of aggregates (up to 90%+) makes up most asphaltic concrete mixes. Aggregates gradations are a fundamental part of asphalt mix design and performance. Aggregates gradation is essentially the distribution of particle sizes in an aggregates mixture and is defined as the percent of the mixture retained on an individual sieve (screen) size; for example, percent retained on the 1-inch (25.4-millimeter) screen, percent retained on the ⅜-inch (8.25-millimeter) screen, or percent retained on the 4-mesh screen. This is most often expressed as the reciprocal of percent retained being the percent "passing" the individual sieve sizes. Coarse aggregates are generally considered any size that is greater than (plus) the 4-mesh sieve and fine aggregates as any size that is smaller than (minus) the 4-mesh sieve. Table 2.4 shows the standard sieve sizes used in laboratory testing facilities for aggregates materials.

When discussing gradations, some other terms should be defined. *Well graded* refers to the fact that the aggregates particles are well distributed with similar quantities of coarse-, intermediate-, and fine-sized particles. *Uniformly graded aggregates* are a blend of aggregates that are mostly the same size. When an aggregates blend has the majority of particles that are either coarse or fine with very little intermediate-sized materials present, it is known as gap graded.

There are three principal types of hot mix asphalt: dense graded, open graded, and stone matrix asphalt. Dense-graded asphalt is a dense hot mix asphalt that utilizes a well-graded aggregates mixture. Open-graded asphalt mixtures are used to provide a porous base layer to encourage drainage or are designed with high-friction-resistant aggregates to act as an open-graded friction surface to increase skid resistance. Stone matrix asphalt is designed with a gap-graded aggregates mixture to provide stone-to-stone contact for high durability and resistance to rutting. Asphalt percentages generally range from 4.5% to 10% depending on the desired properties and the type of aggregates used.

Most of the same attributes that good-quality aggregates provide to PC concrete are applicable to asphaltic concrete mixes as well. Most asphalt pavement is placed and compacted in multiple layers with a base layer (4–8 inches [101–203 millimeters]), a thinner intermediate layer (2–3 inches [50.8–76.2 millimeters]), and an even thinner surface layer (0.75–1 inch [19–25.4 millimeters]). The base layer generally contains the coarsest aggregates, with each successive layer containing smaller-sized aggregates material. Base layers are generally open-graded, permeable mixtures and connected to a subsurface drainage system to provide rapid surface water discharge. The surface layer is generally a stone matrix design that provides stability, strength, and resistance to rutting, and, when utilizing skid-resistant aggregates, can help reduce skidding and hydroplaning.

The design of stable, high-performance asphalt pavements is a complicated engineering process. There are many different asphalt mix designs for all types of pavements from local supermarket parking lots and residential driveways to much more sophisticated applications such as interstate highways or airport runways. The properties of aggregates used in asphalt mixtures that affect the performance of the pavements are durability (soundness and expansion), particle shape, specific gravity, absorption, unit weight, gradation, deleterious materials, and resistance to abrasion. Aggregates particle shape is critical in the design of stable, durable asphalt pavements. Crushed particles that are roughly cubical in shape provide the maximum internal friction in the mix; they interlock to provide much higher stability. Rounded aggregates (e.g., uncrushed gravel) tend to act like ball bearings in asphalt mixes and encourage rutting of the pavement under heavy loads on truck tires. Flat and elongated particles (≥3:1) tend to increase voids in the mix, which can inhibit compaction and can easily fracture into smaller particles during compaction or traffic load. Good compaction in the field is critical to achieving a high density, which increases strength and stability of the pavement. Aggregate absorption can affect the amount of asphaltic cement (AC) required in the mix. Aggregates in a hot mix asphalt mix are coated by the AC and absorb a very small percentage of the binder. Too much oil absorption can have a deleterious effect on AC content and voids in mineral aggregate, which are both critical to stability.

Asphalt is a naturally occurring substance that can be found in certain pool deposits or in porous types of rock, but natural asphalts like those found in the famous La Brea Tar Pits in southern California are extremely rare. Most asphalts in commercial use today are a product or by-product of petroleum refining.

Many types of crude oil exist and have varying percentages of asphaltic material, some containing little or no heavy asphalt. Petroleum refining is a complex chemical distillation process with different types of crude oil producing different products, from petrochemicals to fuels. As the crude oil is being distilled, the lighter constituents of the petroleum (petrochemicals, gasoline, kerosene) are dissipated first, followed by the heavier, less pure portions (diesel fuel, jet fuel), with the heaviest constituents remaining being the asphalt products.

AC is known as thermoplastic because it is a semisolid at room temperature. The AC material needs to be heated to about 350°F (177°C) to be a usable liquid. In many if not most commercial applications the AC is mixed with various types of polymers to modify the viscosity and homogenize the blend. Numerous polymer-modified asphalts exist in today's commercial asphalt industry. Some asphalt mixtures are diluted with kerosene or naphtha and are known as cutback asphalts and other mixtures are emulsified in water and called asphalt emulsions. It is not my intention to delve into the myriad asphalt designs available to modern hot mix asphalt producers; that subject has been covered in several other books on asphalt production.

The process of making hot mix asphalt entails heating the AC mixture in a hot mix plant to a liquid phase at approximately 350°F (177°C) and at the same time heating the blended

Courtesy of Astec Industries

FIGURE 2.5 Typical hot mix asphalt plant

aggregates to the same temperature (Figure 2.5). The liquefied asphalt is then sprayed onto and mixed with the aggregates and mineral fillers until entirely coated. It is essential to maintain the temperature of the mix while transporting the asphalt to the jobsite. An ideal asphalt mix will be laid at a temperature of about 300°F (149°C). During the paving process the mixture is delivered to a paver via dump trucks, which distribute the asphalt in a layer of even thickness onto the surface to be paved. The mixture is then rolled and compacted by vibratory rollers to form an even layer while it begins the cooling process. As the mixture cools to ambient temperature, the thermoplastic nature of the asphalt once again becomes semisolid, forming the bonding agent to hold the pavement together. Because of the semisolid nature of the asphalt in the mix, it is known as a flexible pavement as opposed to PC concrete, which is considered to be rigid pavement.

Other uses of asphaltic material in paving are as surface treatments of existing pavements. In surface treating, a layer of asphaltic material is sprayed onto the existing roadway and aggregates are evenly spread over the liquid asphalt. The aggregates are then rolled into the liquid and as the liquid asphalt dries it bonds the aggregates to the roadway while providing a waterproofing membrane to the top of the pavement and filling in any minor cracks. Both chip sealing and slurry sealing are examples of surface treatments. These applications are most often done as maintenance and generally on less traveled sections of roads such as rural or residential roadways.

Eighty-five to ninety percent of aggregates are sold to the construction industry (USGS, n.d.-b). The remainder is utilized for other purposes in pollution control, water filtration, agriculture use, animal feed, pharmaceuticals, industrial fillers, and steel manufacturing. Sixty-nine percent of crushed stone production is crushed from carbonate rock (limestone and dolomite). The chemical composition of limestone is $CaCO_3$, while dolomite is chemically calcium magnesium carbonate ($CaMg(CO_3)_2$). Like most things found in nature, the chemical purity of the stone varies in a compositional continuum from high-calcium limestone (0%–2% $MgCO_3$) to magnesium limestone (2.1%–4.5% $MgCO_3$) to dolomitic limestone (4.5%–22.5% $MgCO_3$) to calcitic dolomite (22.5%–40.7% $MgCO_3$) to pure dolomite (40.7%–45.5% $MgCO_3$). The term *dolomite* can be confusing because it refers to both the mineral dolomite as well as the rock containing the mineral.

Similar to the process of manufacturing cement, limestone is calcined to produce calcium oxide (CaO), commonly referred to as lime. Depending on the original chemical content, heating the stone in a kiln to a temperature between 900°F and 1,700°F (482°C and 927°C) produces calcination, which is essentially the removal of the carbon dioxide (CO_2) from the carbonate (CO_3) to form an oxide. This process of thermal decomposition produces a product that weighs 44%–52% less than the original stone. This change in mass is known as loss on ignition and is essentially the volume of CO_2 removed. Unfortunately, the carbon dioxide emitted from the cement- and lime-making processes is a major source of carbon emissions into the atmosphere.

CaO, which is known as quicklime, is caustic, alkaline, and somewhat unstable and can be difficult to handle in certain industrial applications. A large percentage of lime production is slaked or rehydrated by adding water to the quicklime to overcome the handling issues. When quicklime is slaked it becomes the chemical compound known as calcium hydroxide ($Ca(OH)_2$). Lime manufacture is the next highest use of limestone after aggregates and cement manufacture. Lime and hydrated lime are used industrially in basic oxygen furnace steel manufacturing, in pollution control (flue gas desulfurization), in plaster, as food additives, in paper manufacturing, and in the petroleum industry.

Ground limestone and dolomite are used in many agricultural applications. Finely ground carbonate rock is spread directly onto farm fields to neutralize soil acidity, which is generally a result of using nitrogen, phosphorus, and potassium (NPK) fertilizers. The neutralizing power of agricultural lime (aglime) is expressed in calcium carbonate efficiency, which is a function of both chemistry and the fineness of the aglime. Quicklime and hydrated lime are sometimes used as soil-stabilizing agents to reduce the moisture content in certain moisture-retaining "fat" clays. Pulverized limestone is also an additive in many animal feeds to increase the calcium in the livestock's diet. It is also fed to chickens to increase the thickness and durability of their eggshells.

Both lime and ground limestone are used in environmental projects to neutralize excess acid content in rivers and lakes, to neutralize acid mine drainage from coal mines, as water purification products in municipal water treatment, and as flue gas desulfurization to scrub stack gases in coal-powered electric plants. Very large aggregates (riprap, gabion stone, and jetty stone) are used for erosion control in rivers, lakes, ocean jetties, and ports. Limestone/dolomite is utilized as a fluxing agent in the steel manufacturing process and also in glass manufacture. Pulverized limestone has been used extensively in the underground coal mining as "mine dust" applied to the floor and walls in underground mines to dilute the highly combustible coal dust in the air.

Pulverized limestone is a highly specialized portion of the limestone industry. Pulverized limestone, generally the high-calcium (95%+ $CaCO_3$) variety, is ground to very fine consistencies and used as industrial fillers. These mineral fillers need to be inert (nonreactive) and of the proper consistency required for the finished product. Many products, but not all, utilizing mineral fillers are color sensitive and require high-brightness (reflectivity) white stone. Mineral fillers need to be high purity, less than 0.5% moisture, with low oil absorption, low iron content, very fine to ultrafine gradations, and meet the individual customer's specifications. Most industrial fillers have median particle diameters in the range of 3–10 micrometers and in some very specialized applications utilize 1–3 micrometer materials. Rigorous quality control testing with specialized instruments is required to assure complete customer satisfaction.

FIGURE 2.6 Side-by-side air-swept roller mill systems

Industrial fillers are exactly what their name implies: less expensive bulking agents that are nonreactive with the critical elements of the product being manufactured. Mineral fillers are used in plastics, rubber products, paint, carpet backing, wallboard mud, clay targets, asphalt roof shingles, adhesives, caulk, putty, gaskets, acoustical tile, wire and cable insulation, PVC pipe products, automotive body panels and moldings, floor mats, floor tile, ceramic and glass production, appliance parts, and a myriad of other industrial and home use products. Very high-purity fillers are used in food and pharmaceutical applications such as chewing gum, toothpaste, antacid tablets, and prescription tablet production and need to be produced in specialized sanitary environments.

Pulverizing a limestone product can increase the value on a per ton basis 10- to 20-fold as compared to crushed stone, but the production effort in reducing the product to the 3–10 micrometer size range is far from inexpensive. Various processes are utilized to produce industrial fillers including ball mills, roller mills, hammer mills, and rod mills for dry processing, where the stone is dried prior to grinding (Figure 2.6). Some ultrafine products are produced using wet grinding methods through autogenous and semiautogenous wet mills and sized and dried after grinding. Dry processing is probably the most common system using air-swept roller mills. These mill systems consist of a feeder, flash dryer, roller mill, system fan, air-classifying cyclones, and silos for the finished products (Figure 2.7). The feed rate and product gradations are controlled by the differential pressure within the enclosed system.

The finished product is generally conveyed to the storage silos by dilute-phase air-conveying systems. Producing industrial fillers is very energy intensive, utilizing considerable electric power and natural gas for the drying process. The finished products are shipped to customers by several methods; 50-pound (23-kilogram) bags are still in common use today as well as mini bulk bags containing 2,000–3,000 pounds (907–1,360 kilograms). The majority of products are shipped to customers as bulk products via pneumatic tanker trucks and rail cars.

Uses of industrial sands include glass manufacture, proppant sand for oil well fracking, molding sands for the foundry casting industry, and specialized sand for filtration uses.

FIGURE 2.7 New state-of-the-art pulverizing plant

Mineral aggregates, such as sand, gravel, and crushed stone, are essential to the maintenance of the American lifestyle. Crushed stone products are used for construction purposes (in cement, concrete, asphalt, concrete block, and road base); environmental purposes (erosion control, flue gas desulfurization, and filtration); mineral fillers (in shingles, plastics, glass, caulking, and rubber products); chemical and pharmaceutical uses (antacids, toothpaste, etc.); and agriculture products (aglime, fertilizer, and animal feed supplements). Several catchy phrases have been utilized over the years to define the critical role of aggregates to the American economy: rocks build America; if it can't be grown, it has to be mined; natural building blocks for quality of life; rocks, the foundation of our economy. In their own way, these simple catchphrases stress the importance of aggregates to us all.

CHAPTER 3

Geology of Natural Aggregates

The geology of natural aggregates is a specialty area within the field of industrial minerals geology. To understand this specialty, one must learn the basics of the aggregates industry—processing methods, markets, transportation needs, customer specifications, industry standards—and apply all of these factors to the geology of the area under study. It is not as simple as locating a deposit of sand and gravel or granite on a geologic map and going into the aggregates business. The old adage of "build it and they [customers] will come" has failed many an aggregate producer when in the wrong location. Locating and profitably exploiting an aggregate deposit is a combination of a minable deposit of quality aggregates (good geology), in a market area with the right customer base, located on accessible transportation routes, and having the approval of the neighboring community.

Like any science, geology has many subspecialties: petrology, mineralogy, structural geology, environmental geology, sedimentology, geomorphology, paleontology, and glacial geology, to name a few. Upon entering the work force, geologists move into areas of specialized industries: coal, petroleum, base metals, precious metals, environmental consulting, industrial minerals, and so on. Aggregates are considered to be within the realm of industrial minerals, which run the gamut from abrasives, to ceramics, to fluxes, to foundry sands, as well as glass raw materials, pigments, refractories, and construction materials. The word *geology* was derived from two Greek words: *geo*, meaning earth, and *logos*, meaning study. "Earth study" encompasses a lot of territory, thus the large number of specialties.

This chapter is not intended to be a geology textbook nor a detailed summary of all the potential geological deposits of aggregate in the United States, but a summary of the basics needed to get started developing an understanding of aggregates geology. Throughout this book, the terms *aggregates* and *aggregate* are used interchangeably. There are several definitions of aggregates depending on the source. ASTM D879 defines aggregates as "a granular material of mineral composition such as sand, gravel, shell, slag, or crushed stone used with a cementing medium to form mortars or concrete, or alone as base courses, railroad ballast and so forth." Aggregates is defined in the *Aggregates Handbook* as "any combination of crushed stone, sand or gravel in their natural or processed state" (NSSGA 2013). Numerous other definitions exist but they are all very similar in that natural aggregates are some combination of sand, gravel, and crushed stone used for various construction and industrial purposes.

The properties that quality aggregates need to possess in order to be incorporated into durable, long-lasting, construction projects are as follows: individual gradation of both coarse and fine aggregates, combined aggregates gradation, maximum particle size, particle shape, bulk density, specific gravity, absorption, surface texture, strength, soundness (freeze–thaw durability), pore structure, thermal expansion, deleterious materials, and organic impurities.

The gradation of natural aggregates is essentially manufactured in a processing plant by using multi-decked, mechanical screening devices with different size screens on each screen deck. Natural sand and gravel deposits have a gradation spectrum in situ, but it rarely, if ever, match the specifications of a customer without some mechanical manipulation. Crushed stone deposits require drilling, blasting, and crushing prior to the mechanical screening to meet specification sizes. The ultimate gradation specification depends on the final use of the material, whether it is for portland cement (PC) concrete, bituminous concrete, road base, railroad ballast, or riprap. The grain size distribution (gradation) is important in PC concrete to control strength, cement content, water content, and air voids, as well as handling and finishing properties of the mix. Bituminous concrete gradation specifications vary depending on mix design and application. Road base mixtures require additional fines and crushed material to control compaction for ultimate density, stability, and compressive strength.

Maximum particle size is another specification that varies depending on usage application. Larger particles have less surface area than fine particles and affect the water-to-cement ratio and ultimately the strength of the finished product. Top sizes between ½ and 1 inch (12.7 and 25.4 millimeters) make good high-strength PC concrete. Aggregates larger than 1 inch (25.4 millimeters) tend to make bituminous concrete unstable, difficult to compact to maximum density, and produce too many voids in the mix. Most road bases specifications require a maximum top size of 1.5–2 inches (38.1–50.8 millimeters) because larger particles are difficult to compact. Particle shape has a significant effect on strength and density in most mixes. Flat and elongate particles (slivers) tend to break under compaction stress, and round particles in bituminous mixtures act like ball bearings and can cause rutting under heavy loads. Cubical particles are the most serviceable by providing interlocking grains that bind together tightly for additional strength and density.

In my experience, the specific gravity of an aggregate is probably the best single indicator of quality. Specific gravity is the ratio of the mass of a specific volume of aggregates to the mass of an equal volume of water. An aggregate with a specific gravity of 2.7 is 2.7 times heavier than water, which weighs 62.4 pounds per cubic foot (1,000 kilograms per cubic meter). Aggregates with low specific gravities generally contain deleterious materials like argillaceous content (clay minerals), coal, lignite, or other materials with a lower specific gravity. Specific gravity is utilized in the design of both PC concrete and bituminous concrete mixtures. Changing to an aggregate with a different specific gravity than used in the original design mixture will likely result in an underperforming mix. Generally, in specifying aggregate, a high specific gravity and low absorption are the best indicators of quality. Specific gravity is sometimes also referred to as relative density.

The pore structure within an aggregate particle affects the porosity, permeability, and absorption of the aggregate. Porosity is essentially the amount of pores (void) space within a given aggregate. Some rock types can have porosities up to 75% of the total volume of the rock (pumice, porous sandstones). Rocks with low porosity levels make better-quality aggregates and by definition have higher specific gravity. Permeability is the ability of the pore structure to transmit fluid internally within the rock. The permeability is a function of the interconnection of the pores. The pore structure of an aggregate particle can be impermeable, isolated pores with no connection, or permeable, having interconnected pores. Absorption is related to both the amount of porosity and permeability of a sample. Absorption is the amount (%) of liquid (usually water) that a dry sample of aggregate can absorb in relation to its dry weight. Highly absorptive aggregate requires more binder (PC cement, asphaltic cement, etc.), which is the most costly portion of any mix design. Aggregates with absorptions of 1% or less are preferable. High permeability and porosity are desired by the petroleum industry when pumping crude oil from below the surface but are the bane of aggregate producers.

Unit weight of a gradation of aggregates is the density of the aggregate mixture including the voids in the mixture. Unit weight is calculated by dividing the dry weight of aggregates required to fill a container of known size by the known volume of the container. The percentage of voids can then be calculated utilizing the ASTM C29 test method. The void space in an aggregate gradation is the result of a combination of particle shape, gradation, and degree of compaction. Most mix designs require minimizing the voids in the mixture to minimize the amount of binder or cement required to maintain a dense, stable mix.

The strength of aggregates is the ability of the material to resist mechanical breakdown. Aggregate particles in various mixtures are subjected to mechanical stresses like compression in road base, abrasion during processing (crushing and screening), mixing stresses (ready-mix trucks, asphalt dryers), impact by truck tires on pavement surfaces, particle-to-particle abrasion (railroad ballast), and so on. The strength of an aggregate is dependent on the level of cementation of individual grains, porosity, and frequency of internal fractures or zones of weakness. Aggregates are routinely tested for mechanical strength using various tests like Los Angeles abrasion or Micro-Deval, where aggregate particles are subjected to abrasion by steel balls rotating in a drum.

The soundness of aggregate is its relative ability to resist weathering: cycles of freezing and thawing, repeated wetting and drying, and temperature extremes. These factors are relevant to the climate where the aggregate mixture is being utilized. In states like Florida where repeated cycles of freezing and thawing are rare, this becomes less of a design factor. Freeze–thaw durability is especially significant is areas with high levels of precipitation where the temperature fluctuates above and below freezing numerous times during a typical year. Climates where the winter freeze lasts several months with little or no fluctuation are, like warmer climates, less subject to these durability issues. Highly absorptive rocks and rocks containing high percentages of clay are very susceptible to weathering and make poor aggregates, especially in climates susceptible to freeze–thaw cycles. More information on aggregate properties, specifications, and test methods is covered in Chapter 7 on quality control.

Numerous different entities and agencies specify the requirements for aggregates utilized on various construction jobs. These include ASTM International, AASHTO, individual state departments of transportation (Ohio Department of Transportation, Kentucky Department of Transportation, Indiana Department of Transportation, etc.), the Federal Highway Administration, the Federal Aviation Administration, architects on specific jobs, and ultimately the final customer. Unless otherwise specified, most construction projects will defer to their state's department of transportation specifications.

UNCONSOLIDATED SAND AND GRAVEL

As previously stated in Chapter 2, 1.53 billion metric tons of crushed stone and 970 million metric tons of sand and gravel for a total of 2.53 billion metric tons of aggregates were produced in the United States in 2019. Sand and gravel makes up approximately 40% of the aggregates used in the United States annually. Sand and gravel occurs naturally in every state in the country but not necessarily in economically exploitable deposits. Sand and gravel deposits have one commonality in that they are formed by the action of erosion, transportation, abrasion, and deposition by water. The most common sources of sand and gravel are alluvial, fluvial, lacustrine, glacial, eolian, deltaic, and coastal deposits.

All rivers and streams are associated with sand and gravel deposits with the extent, depth, and gradation of the deposit dependent on the size of the stream, the stream gradient, the source of the sediments being transported, and the time period involved. Young streams draining steeply sloping topography tend to form narrow V-shaped valleys. These narrow, steep valleys produce fast-moving streams that predominately erode the underlying bedrock and deposit very little. Older, more mature streams are located in wide, flat valleys and have less gradient and slower currents. These more mature streams have a tendency to meander and erode material from the outside of the bend where the current is fastest and deposit material on the inside of the bend where the current wanes. The inside of the meanders form into deposits of sand and gravel. As base levels of streams change over time, these stream deposits can form elevated terraces along the course of the river.

The gradation of these terrace deposits can vary considerably, with coarse gravels deposited during times of high streamflow and finer materials (sand and silt) left behind when the flow ebbs. These more mature streams meander continuously, changing course in a constant movement across the valley floor, sometimes abandoning the stream course entirely, leaving elongated lakes or sloughs behind known as oxbow lakes. The stagnant water in these abandoned channels will eventually become eutrophic and leave silt, clay, and organic deposits behind to ultimately mingle with the coarser granular materials. Terrace deposits along large mature rivers are a good source of commercial sand and gravel depending on the type of bedrock the streams flowed over (source rock). Over time sediments in streams can migrate hundreds of miles. Unsound gravels from a source rock hundreds of miles upstream can sometimes be found embedded within sources of good, durable aggregate material.

Anywhere that a mature stream or river has flown is a promising location for a commercial sand and gravel deposit. Not all river valleys will produce commercially viable deposits, depending on the source rock, volumes of sand and gravel, thickness of the deposits, gradation of the material deposited, organic content, proximity of the deposit to markets, and so forth. Sand and gravel deposits are inherently heterogeneous, with massive variations of rock types, gradations, clay inclusions, and ultimately quality. These sediments vary considerably from deposit to deposit as well as within an individual deposit. I once drilled a test hole in a producing gravel pit (110-foot [33.5-meter] thick) that was within 50 feet (15.24 meters) of the dragline excavation. The test hole contained seven distinct clay layers totaling over 35 feet (10.7 meters) thick. The dragline, 50 feet (15.24 meters) away, was producing pure, clean, well-sorted sand and gravel containing no clay.

The northernmost portion of the United States has been extensively affected by continental glaciers. Figure 3.1 is a map of North America (produced by the Illinois Geological Survey) showing the extent of continental glaciation. A significant portion of 25 states was affected by various stages of this glaciation. Continental glaciers were over 1 mile (1.6 kilometer)–thick masses of ice slowly moving across the terrain as they continually accreted more snow and ice. These moving ice masses were powerful enough to gouge the Great Lakes out of the underlying bedrock. This gouging action plucked up pieces of the underlying rock and carried them hundreds of miles from their original location. There were five separate periods of continental glaciation, each interrupted by warm interglacial periods when the ice melted. These warm

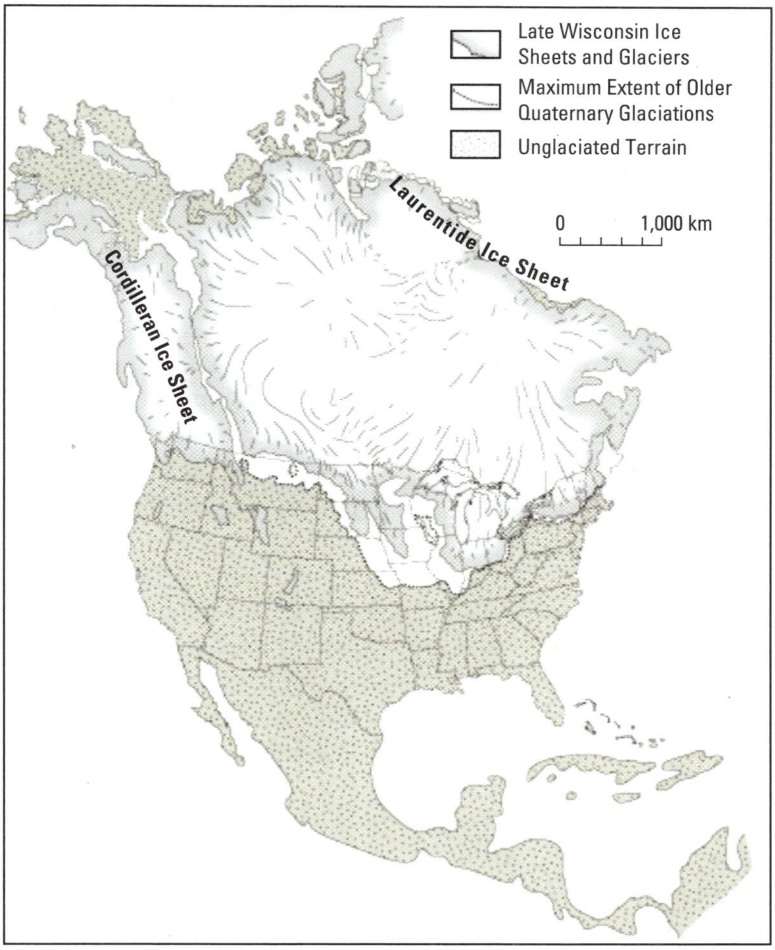

Source: Killey 2000. © 2000 University of Illinois Board of Trustees. Used with permission of the Illinois State Geological Survey.

FIGURE 3.1 Extent of continental glaciation

interglacial periods each lasted thousands of years. The several continental glaciation events during the Pleistocene period lasted about 2.5 million years, ending approximately 12,000 to 15,000 years ago. Although global warming due to humans' use of fossil fuels is evident today, who's to say that the last 12,000 years have not been part of another interglacial period and that the Earth may once again in the future witness massive sheets of ice approaching from the polar latitudes?

During the southward migration, the ice powerfully altered the underlying topography, removing the tops of mountains, rerouting the course of mighty rivers, and smoothing the landforms. The moving masses of ice and rock fragments formed several types of deposits that today are significant sources of commercial sand and gravel. Kames were formed as ice contact deposits where the ice melted leaving mounds of sand and gravel wedged along other landforms. Eskers are a result of meandering, subglacial stream courses within the melting ice that eventually filled up with sand and gravel and collapsed into meandering mounds of potential aggregate.

FIGURE 3.2 Melting Alaskan glacier

Probably the most significant source of commercial sand and gravel resulting from glaciation are glacial outwash deposits. As the climate warmed, the huge volumes of water produced by the melting of the ice flowed away from the receding glaciers. These massive streams laden with sediment followed the existing topography and buried numerous large, preglacial river valleys with outwash. Some of these buried valleys contain deposits of sand and gravel with thicknesses exceeding 300 feet (91 meters). These buried valleys have been the source of large volumes of commercial sand and gravel aggregates over the years, mostly in the midwestern and north-central United States.

Glacial till is an ice contact deposit commonly found layered within outwash deposits. It is formed at the base of the ice where it contacts the underlying bedrock. The grinding action of the moving ice pulverizes the rock into silt and clay mixed with larger granular materials. These sandy, clayey silts form hard, dense layers of deleterious materials and can actually exhibit flow structures where the till flowed under pressure, leaving behind unpredictable deleterious layers within the glacial outwash. These till layers within outwash are indicative of multiple advances and recession of the overlying glacial ice.

Mountain glaciers, once very common in the Rocky Mountains, are on the verge of extinction because of global warming patterns, but are still fairly common in Alaska (Figure 3.2). Mountain glaciers produce the same basic types of sand and gravel deposits as continental glaciation; kames, eskers, glacial outwash (Figure 3.3), and buried valleys. These more rugged mountainous areas are generally fairly far removed from major population hubs where the larger markets for aggregates are centered. Many smaller producers exploit these mountainous deposits, a by-product of which can be minor quantities of alluvial gold in the right locations.

Alluvial fan deposits are another major source of sand and gravel in the more arid portions of the country, generally in the western states (Figure 3.4). Basin areas located between and adjacent to mountain ranges are generally broad, flat valleys where sediments accumulate. Steep, fast-moving streams draining the mountains of snowmelt produce great volumes of water and sediment, especially during the spring runoff months. These turbulent, flooding streams erode the steep valley walls and drain into the adjoining flat basins. The abrupt change

FIGURE 3.3 Catamaran-style clamshell dredge mining glacial outwash in southern Ohio

FIGURE 3.4 Small alluvial fans at the base of a butte in eastern Oregon

of stream gradient and velocity of the water upon reaching the flat basin reduces the carrying capacity of the water and large volumes of sediment are quickly deposited. As the water spreads out and soaks into the floor of the dry basin, the larger, coarser rock fragments fall from the water first, near the base of the mountains, with the finer material flowing further onto the basin floor.

This type of deposit tends to fan out at the base of the stream, with the coarser particles piling up higher and higher against the flank of the mountain. As the finer material flows farther away from the end of the stream, the landform widens and becomes thinner with increased distance from the mountainside. A natural gradation is formed from the "fine" toe of the fan to the "coarse" top of the fan. Over millennia these fan-shaped deposits can coalesce and may reach hundreds of feet thick and hundreds of acres in lateral extent.

Other sources of sand and gravel exist, but those mentioned above are the principal sources of commercial exploitation. Deltaic deposits (river deltas) generally consist of clay and silts with very few coarse constituents, although stream deltas formed in mountain lakes can contain coarser materials. Lacustrine or lake deposits are most often associated with fine silts and clays with little or no coarse material. Some commercial sand and gravel mining occurs within the coastal plains in the oceanic marine environment, but these deposits require specialized equipment capable of mining at greater underwater depths and are therefore more costly to produce. Coastal plain deposits may become more viable in the future as more conventional terrestrial deposits near larger coastal cities are depleted.

Eolian (windblown) deposits of sand exist in some areas of the country. These deposits include sand dunes in desert areas, and even in some glaciated areas, aeolian sands sometimes form overburden deposits on top of glacial outwash. Windblown sands usually have a "frosted" surface texture caused by the abrasion of the sand particles impacting one another as they are blown from place to place. This frosted texture has been shown to be problematic because it affects the adhesion of various types of cement to the surface of the individual sand grains.

CRUSHED STONE (BEDROCK) DEPOSITS

It would be almost impossible to try and describe all the types of bedrock deposits that have and can be utilized as mineral aggregates. Bedrock deposits require the following properties to be a source of durable aggregate: freeze–thaw durability, abrasion resistance, proper pore structure, low water absorption, resistance to thermal expansion, high specific gravity, and a lack of deleterious material and organic impurities. Unlike sand and gravel, which has a natural gradation produced by the action of flowing water, the gradation of crushed stone is totally in the control of the aggregate producer. Bedrock has to be drilled and blasted, crushed, screened, and sometimes washed to provide products able to meet standard gradation specifications. The three basic types of rocks, which are classified based on their genesis, are sedimentary, igneous, and metamorphic.

Sedimentary Rocks

Sedimentary rocks, as the name implies, are derived from the cohesion of sediments into solid masses of rock. The two basic subtypes of sedimentary rocks are clastic, which are mechanically deposited, and chemically or biochemically deposited sedimentary rocks. Clastic rocks are generally classified by their texture and grain size. Conglomerates and breccia are a combination of sand, gravel, and cobble-sized grains (6+ inches [15.2 centimeters] through 200 mesh) that have become cemented together. Sandstone is composed of cemented sand-sized grains (minus 4 through 200 mesh), while siltstone consists of cemented silt grains (minus 200 through 325 mesh). Shale is essentially a combination of clay-sized (minus 325 mesh) minerals consolidated into a solid rock mass by agglomeration. Unlike silt grains, clay minerals tend to be platelet shaped (flat and elongated), containing a negative charge on one end and a positive charge on the other. They are originally attracted and held together electrostatically.

A compositional continuum exists between the clastic rock types. Few if any clastic rock types are pure sandstone or pure siltstone; they are more likely to be silty sandstone or clayey siltstone. Clastic sedimentary rocks make up only a small percentage of the crushed stone sold as aggregate. In 2019, only 3% of crushed stone was produced from sandstone and quartzite (USGS, n.d.-b). The value of clastic rock as aggregates is a function of the quality of the cementitious material that holds the grains together. Sandstones, conglomerates, and siltstones are generally glued together by either calcareous (calcium carbonate) or siliceous (containing silica) cement. For the most part, but not always, the siliceous cement is more likely to form a hard, dense bond with minimal porosity. Sandstone can be friable and porous if the cementing material is insufficient to bond the individual grains together tightly.

FIGURE 3.5 Large calcite crystals (twined) at Jeffersonville Limestone, Indiana

Siltstone is seldom used as aggregate because it tends to have low specific gravity, is often porous, is potentially poorly cemented, and tends to be argillaceous (contains clay). Because of the extreme variations in grain size and cementation, conglomerate is rarely used as aggregate material. The fissile nature of shale, even if it were well cemented, eliminates its use as aggregate. When crushed or broken, shale forms elongated particles that are deleterious and it lacks durability because it rapidly weathers into its original clay constituents. Processing some types of shale by heating it in a kiln to a sintering temperature forms gas bubbles within the shale and fuses the individual grains together to form a usable lightweight aggregate. This is an expensive, energy-intensive process with a somewhat limited market.

Chemically precipitated sedimentary rocks, on the other hand, made up 69% of the crushed stone sold in the United States in 2019 according to the U.S. Geological Survey (USGS, n.d.-b). This equates to 1.06 billion metric tons of aggregates from sedimentary origins. Limestone and dolomite, both forms of carbonate (CO_3) rock, are deposited in a similar manner and are very similar chemically. Limestone in its purest form is made up of the mineral calcite (Figure 3.5) and chemically consists of calcium carbonate ($CaCO_3$). Dolomite is composed of the mineral of the same name (dolomite) but has a chemical composition of $CaMg(CO_3)_2$. Examples of primary dolomite are somewhat rare. Most dolomite is considered to be a secondary mineral thought to be formed by magnesium-rich seawater reacting with the calcium carbonate in limestone involving replacement of some of the calcium ions by magnesium. Because magnesium ions are smaller than calcium ions, dolomite has a tendency to be more porous than limestone because the secondary mineral replacement shrinks the grain size, leaving voids in the rock.

Chemically precipitated carbonates are formed as a combination of dead organisms that extract carbonates from seawater (fossils) and lime mud generally precipitated from seawater by various forms of bacteria. This combination settles on the sea floor and eventually lithifies by accretion with additional overlying sediments. Many limestones show signs of seabed

FIGURE 3.6 Examples of fossil coral commonly found in limestone

organisms, worms, clams, marine plants, and so on, reworking the sediments prior to solidification in a process known as bioturbation. Evaporation can also play a role in the process, with carbonates forming in vast tidal flats where the seawater fluctuates under the influence of the tides. It is not uncommon to see marine fossils of clams, snails, calcified plants, corals (Figure 3.6), and other marine organisms in limestone. Fossils tend to be somewhat less evident in dolomites and seem to have been partially obliterated by the mineral replacement process.

Being formed in an ocean basin environment, these chemically deposited rocks were influenced by sediments from rivers and streams flowing into the basins bringing along sand, silts, clay minerals, and organic matter. Severe storms like hurricanes, changes in water temperature, and excessive wave action churning up the sediment helped combine other elements into the mix. Not all limestone and dolomite is suitable for use as aggregate. Some can be porous, friable, or soft, contain excessive organic matter, be interbedded with shale, or contain deleterious amounts of undesirable chemical precipitates like chert. Carbonate rocks are abundant in many parts of the United States and can be found in every state but must meet the required specifications for sound, durable aggregate upon final processing (Table 3.1).

Other chemically precipitated types of sedimentary rocks exist but do not make good aggregate materials. These are generally precipitated by organisms such as diatoms, radiolarians, and sponges from the siliceous constituents of seawater and have a tendency to react adversely with the alkali chemicals in PC (the alkali-silica reaction).

Igneous Rocks

The term *igneous* has its origins from the Greek word meaning "fire." According to the USGS (n.d.-d), "igneous rocks form when hot, molten rock crystallizes and solidifies. The melt originates deep within the Earth near active plate boundaries or hot spots, then rises toward the surface. Igneous rocks are divided into two groups, intrusive or extrusive, depending upon where the molten rock solidifies." Intrusive or "plutonic" rocks result when molten rock, also known

TABLE 3.1 Simple classification of sedimentary rocks

Chemical Precipitation		Clastic (Mechanical)	
Rock Type	**Description**	**Rock Type**	**Description (Grain Size)**
Limestone*	$CaCO_3$	Conglomerate	+6 inches (15.24 centimeters) to 200 mesh
Dolomite*	$CaMg(CO_3)_2$	Breccia	Angular conglomerate
Caliche	Calcareous soil	Sandstone*	Minus 8 to 200 mesh (if well cemented)
Coquina	Cemented shells	Siltstone	Minus 200 to 325 mesh
Marl	Calcareous shale	Shale	Minus 325 mesh platelets
Chert	Siliceous precipitate	—	—
Chalcedony	Siliceous precipitate	—	—
Opal	Siliceous precipitate	—	—
Flint	Siliceous precipitate	—	—
Agate	Siliceous precipitate	—	—
Coal	Organic hydrocarbon	—	—
Phosphate	P_2O_5	—	—
Salts	Evaporites	—	—

*Rock type is commonly used in crushed stone

as magma, is trapped within the Earth. Because it is held deep within the Earth, the magma cools very slowly and the resulting minerals are fairly coarse-grained. The final portion to solidify forms what is known as a pegmatite, which can contain crystals of enormous size. Intrusive igneous rocks become exposed on the Earth's surface as a result of erosion and tectonic uplift forces. The core of most mountain ranges consists of intrusive rocks, whose resulting upward pressure caused the overlying rocks to be faulted, folded, and thrust upward into mountains.

Extrusive or "volcanic" rocks form when the magma is extruded from within the Earth via active volcanos and fissures in the Earth's surface. The extrusive magma when reaching the Earth's surface is then known as lava (Figure 3.7). When exposed to the cooler atmosphere of the Earth's surface, the lava cools very quickly (relative to plutonic rock) and forms a very fine-grained, sometimes glassy texture that often contains gas bubbles (vesicular).

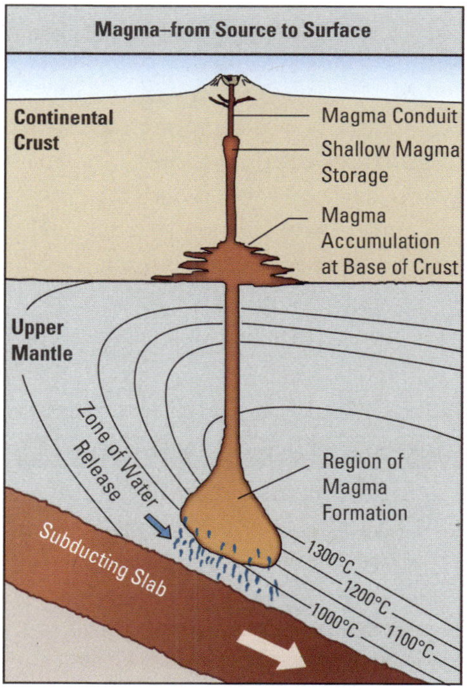

Source: USGS 2013. Graphic by Lisa Faust.

FIGURE 3.7 Molten rock beneath the Earth's surface

Igneous rocks are not only classified by grain size (Figure 3.8) or texture but also by mineralogy and consequently color. Light-colored igneous rocks are known to be acidic or felsic. Examples of felsic igneous rocks are granite, rhyolite, granodiorite, diorite, and andesite. Basic igneous rocks are known as mafic and are low in silica, high in magnesium and iron, and dark in color. Mafic rocks include gabbro, diabase, and basalt; ultramafic rocks include dunite.

Mafic rocks are generally grouped into what is collectively known as "traprock" within the aggregates industry. Certain extrusive igneous rocks with high silica content cool quickly enough that they form into a volcanic glass (Figure 3.9).

Other types of extrusive igneous rocks are formed by explosive volcanic eruptions. As excessive gas pressure builds up in a volcano, an explosive eruption can occur forcing tons of volcanic ash into the air. As ash flows, a ground-hugging cloud of hot ash and steam generally rhyolitic in composition, or the ash can be ejected high into the atmosphere and circle the Earth for years, ultimately affecting the entire planet's climate, can occur. These volcanic ashes can be deposited as pumice or cinders, or consolidated into a formation known as tuff.

Most igneous rocks make strong, durable, and very hard aggregate materials, with some exceptions (Table 3.2). The materials extruded in explosive events like pumice, tuff, and cinders have very low specific gravity and high porosity and are therefore unacceptable as aggregates. Some of the more acidic forms of igneous rocks can have adverse chemical

FIGURE 3.8 Texture of gray granite in central Idaho

FIGURE 3.9 Volcanic glass (obsidian) in western Wyoming

TABLE 3.2 Classification of igneous rocks by silica content

Felsic (Light Color)	Intermediate	Mafic (Dark Color)	Ultramafic	Explosive Deposits	Texture
>63% SiO_2	52%–63% SiO_2	45%–52% SiO_2	<45% SiO_2		
Granite* Granodiorite*	Syenite Diorite* Diabase	Gabbro*	Peridotite*	—	Coarse
Microgranite* Microgranodiorite*	Microsyenite Microdiorite*	Dolerite	Lamproite	—	Medium
Rhyolite Dacite	Trachyte Andesite	Basalt* (Traprock)	Komatite/ Picrite	Ash Pumice Tuff	Fine
Obsidian	—	—	Tachylyte	—	Glass

* Common sources of aggregates

reactions (e.g., alkali-silica reactivity) with cement. Many of the fine-grained igneous rocks associated with lava flows become too porous because of the inclusion of vesicular gas bubbles and the volcanic glasses like obsidian shatter into conchoidal glass shards if crushed.

Metamorphic Rocks

By definition, metamorphosis means change in form. Metamorphic rocks are rocks that formed through the alteration of other existing rock types by the mechanics of extreme heat (>150°C [302°F]) and pressure (>1,000 bar [14,500 psi]) caused by forces within the Earth. My wife had an Introduction to Geology professor that called metamorphic rocks "ruined rocks." Factors that play a role in "ruining" the original rock types include intrusive bodies of hot magma, shearing/slipping movement along faults, hot mineral-rich hydrothermal fluids, tectonic forces (the collision of continental plates), and pressure caused by the weight of the overlying rock mass. Many metamorphic rocks exhibit foliation, which is an alignment and reorientation of mineral grains generally perpendicular to the greatest pressure exerted on the rock. Metamorphosis can exhibit itself as the enlargement of existing crystals within a rock type (e.g., limestone into marble). At times, metamorphic rocks exhibit "banding" or layering of similar minerals (mafic minerals in layers, and felsic material in other layers) as in the rock type gneiss.

Certain minerals are a result of metamorphism and are not found in any other rocks. These so-called index minerals include andalusite, garnet (Figure 3.10), kyanite, sillimanite, and staurolite. Fine-grained minerals can be modified into a coarse-grained rock type through the application of heat and pressure. New minerals can form via contact between two different minerals where chemical changes can occur. Ten to fifteen percent of the surface of the Earth is underlain by metamorphic rocks, but they make up less than 6% of crushed stone aggregate in the United States. The term *protolith* is used to describe the original rock form prior to alteration (e.g., limestone is the protolith for marble). Degrees of metamorphosis can occur with variations in temperature and pressure. For example, the protolith shale with low heat and pressure will change into slate (very fine-grained); with more intense heat it becomes the rock phyllite (medium-grained); add more intensity and it turns to schist (medium/coarse-grained); in areas of high grade, metamorphosis it can become a gneiss (coarse-grained).

Courtesy of USGS

FIGURE 3.10 Garnets (index mineral) in schist

There are two generally accepted forms of metamorphism. In contact metamorphism, an intrusive body of magma exerts heat and pressure stresses on the existing bedrock. The amount of change in the protolith is dependent on the distance from the hot magma source. Regional metamorphosis happens over a much larger scale in areas where tectonic forces can affect entire mountain ranges and shifting continental plates abruptly meet. Metamorphic rocks can

TABLE 3.3 Classification of metamorphic rocks

Rock Type	Grain Size	Texture	Protolith
Slate	Very fine	Foliated	Shale, siltstone
Phyllite	Fine	Foliated	Slate
Schist	Medium	Foliated	Phyllite
Gneiss*	Medium/coarse	Foliated	Schist, granite, volcanics
Marble†	Medium/coarse	Nonfoliated	Limestone, dolomite
Quartzite†	Medium/coarse	Nonfoliated	Quartz sandstone
Anthracite	Fine	Nonfoliated	Bituminous coal

*Possible source of crushed stone.
†Good source of crushed stone.

be found in New England, the Lake Superior area, the Adirondack Mountains, the Piedmont Plateau, and throughout the Appalachians in the eastern United States. In the western United States, metamorphic rocks are scattered throughout the Rocky Mountains, the Great Basin, and the Cascade and Sierra Nevada mountain ranges (Table 3.3).

Marble (protolith limestone) and quartzite (protolith sandstone) generally are hard, durable, nonporous, and dense, and make good crushed stone aggregate. The more foliated forms of metamorphic rock—slate, phyllite, and schist—because of the parallel realignment of the mineral grains (mostly mica) become laminated and tend to break into elongated particles and the micaceous content can adversely affect the strength. These rock types generally make poor sources of crushed stone. Gneiss can at times be durable enough to make crushed stone but may also show weakness along bands of mineral segregation.

CHAPTER 4

Exploration and Reserve Analysis

Over the years, many methods have been used to explore for aggregate deposits. When I first entered the industry in the 1970s, many aggregate producers were small, multigenerational, family-held operations serving localized markets. Many started as farmers who found that their 40-acre (16.2-hectare) field in the river bottom was riddled with gravel and cobbles, flooded annually, and produced stunted crops or that they had an outcrop of bedrock on their property near the base of a hill. When their pickup truck hit a "pothole" in the farm lane, they knew where to dig to find some granular material to fill the hole and smooth out the lane. They eventually started selling aggregates to their neighbors to repair their own farm lanes and before they knew it were selling to the county highway department, then the state highway garage, and the local concrete plant, and a new business blossomed from there. Little formal exploration was accomplished and what was done was poorly documented. This led in many cases to the ultimate failure of their business when the reserves petered out unexpectedly, but also produced some very successful operations.

Some of these producers were smart enough to move across the fence line with a tractor-mounted backhoe and dig a test pit or two to see what was beneath the next field. I've seen many handwritten note papers describing the contents of test pits with descriptions like "good gravel," or "mostly sand," or "muck" with nebulous or no locations (e.g., 200 feet [61 meters] west of the big walnut tree). Certainly not very scientific and generally useless unless you were present at the time. Industrial mineral geologists were few and far between in the industry until some larger engineering and construction companies saw the need to find sources of aggregates to exploit for themselves. With an influx of capital from larger companies came the need to justify investment in equipment, which is very hard to do without proven reserves that have been accurately documented.

Any grassroots exploration program begins with a search of published documentation. All states have a geological survey that is a repository of years of geological research data on the state in question. Most states require that the drilling logs of water wells be recorded with the state division of water or county department of health, and many states have the location and logging data for petroleum exploration wells on file. Each state has a website containing volumes of information on the geology of the state. Many states now have interactive state geological maps where you can click on a site with your mouse and get basic geological information instantly. Searchable lists of publications on various geological subjects are common, where you can purchase or sometimes download pertinent information and reports. Listed on the websites will be contacts for state geologists with specific areas of expertise (environmental, petroleum, industrial minerals) who can be an invaluable resource and are always happy to assist. Maps of existing mineral extraction operations in the state can be a valuable resource to identify trends and locations where the resource in question is exploited. Some states publish annual statistics on mining production and sales on a producer-by-producer basis including their locations. These are a great research tool to see what the competition is up to.

Different types of geological maps exist depending on the area under investigation. Bedrock geological maps are a general representation of what types and ages of bedrock exist beneath the surface in any given area. The level of detail of the maps is a function of the amount of time and labor spent in field mapping, the number of drill holes available for correlation, and the amount of funding available for these mapping projects over the years. This will vary from state to state and organization to organization. As a general rule, states with large natural resource industries such as mining or petroleum will have the most detailed geological information available. Bedrock geology maps essentially remove the surficial deposits and give an indication of what bedrock lies beneath. Figure 4.1 shows the bedrock geology of Ohio in fairly general terms. This can be somewhat misleading in areas where substantial thicknesses of unconsolidated deposits overlie the bedrock.

When searching for unconsolidated materials like sand and gravel, a map of the bedrock geology is of little use because the landscape has been stripped of the surficial deposits by the cartographers. Many states also publish maps of the surficial, quaternary, or glacial geology overlying the bedrock. On the map of the glacial geology of Ohio in Figure 4.2, the areas shown as kames and eskers or outwash would be good areas to target in a search for minable sand and gravel deposits. These statewide maps are published on very large scales: 1:2,000,000. Some states publish geological maps on 7.5-minute quadrangles at a scale of 1:24,000, which is a much more useful scale for detailed work.

In areas where the surficial deposits are significantly thick, maps indicating the thickness of these deposits are very handy when attempting to determine the depth to bedrock or thickness of the overlying sediments. Maps showing the topographic surface on top of bedrock are sometimes available and subtracting the elevation of the bedrock surface from the existing topographic elevation will yield an approximate thickness of the sediments. Figure 4.3 indicates the thickness of glacial sediments in Ohio using the color shading technique. This map shows areas where up to 726 feet (221 meters) of sediment was deposited by the topographical smoothing effect of continental glaciation.

The U.S. Geological Survey (USGS) has a website with extensive research studies, papers, reports of investigations, bulletins, and so on, that also operates regional offices in most states. An interactive geologic map of the United States is available at USGS (n.d.-a). Another source of information is the county soil surveys published by the U.S. Department of Agriculture and the Soil Conservation Service (SCS). Many of these are published online and every county in the country has an SCS office that can assist. Usually an extensive table within the soil reports lists engineering properties of the soils as well as the parent materials from which each soil type was formed. There are also often detailed maps of the county on an aerial photographic base showing the location of all the various soil types.

Once you've learned the big picture from previously published studies and zeroed in on likely areas of interest, the local county assessor's office will be able to provide maps of property ownership in the area. The majority of counties now have property ownership databases available online where you can scroll the entire county and learn who owns what while still in your pajamas, having your morning coffee.

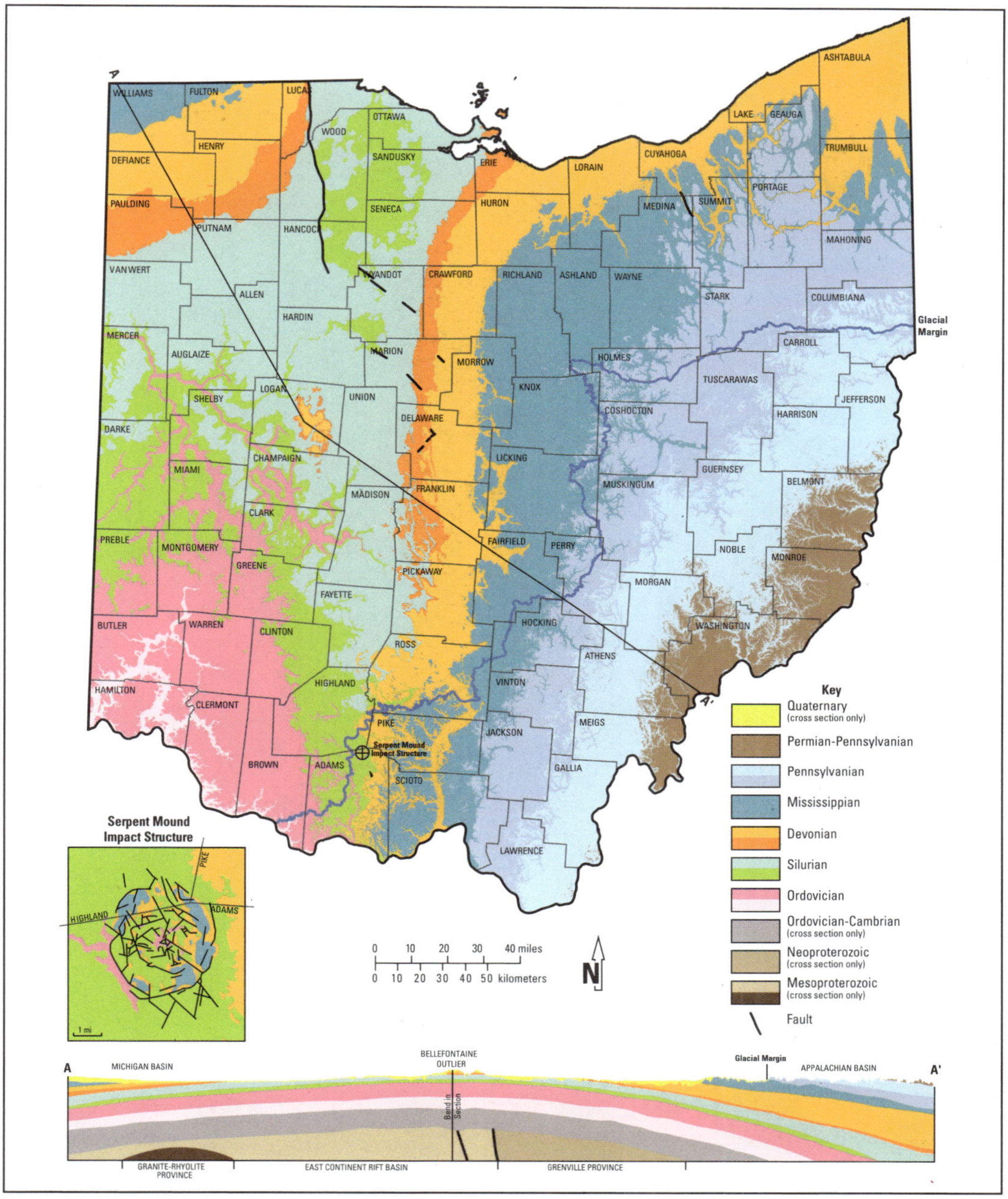

Source: Ohio Division of Geological Survey 2006

FIGURE 4.1 Bedrock geology map of Ohio

This is the stage of the project where it's time to meet with your marketing and sales team. Once you've determined the potential location for viable aggregate deposits, you'll need to determine the best location from the standpoint of the market, customer base, and ease of transportation to the market. Which identified potential deposits are located close to an interstate highway system, a highway interchange, a major railroad, or a navigable waterway? Is your competition closer to or further from the market? Transportation costs are a major factor

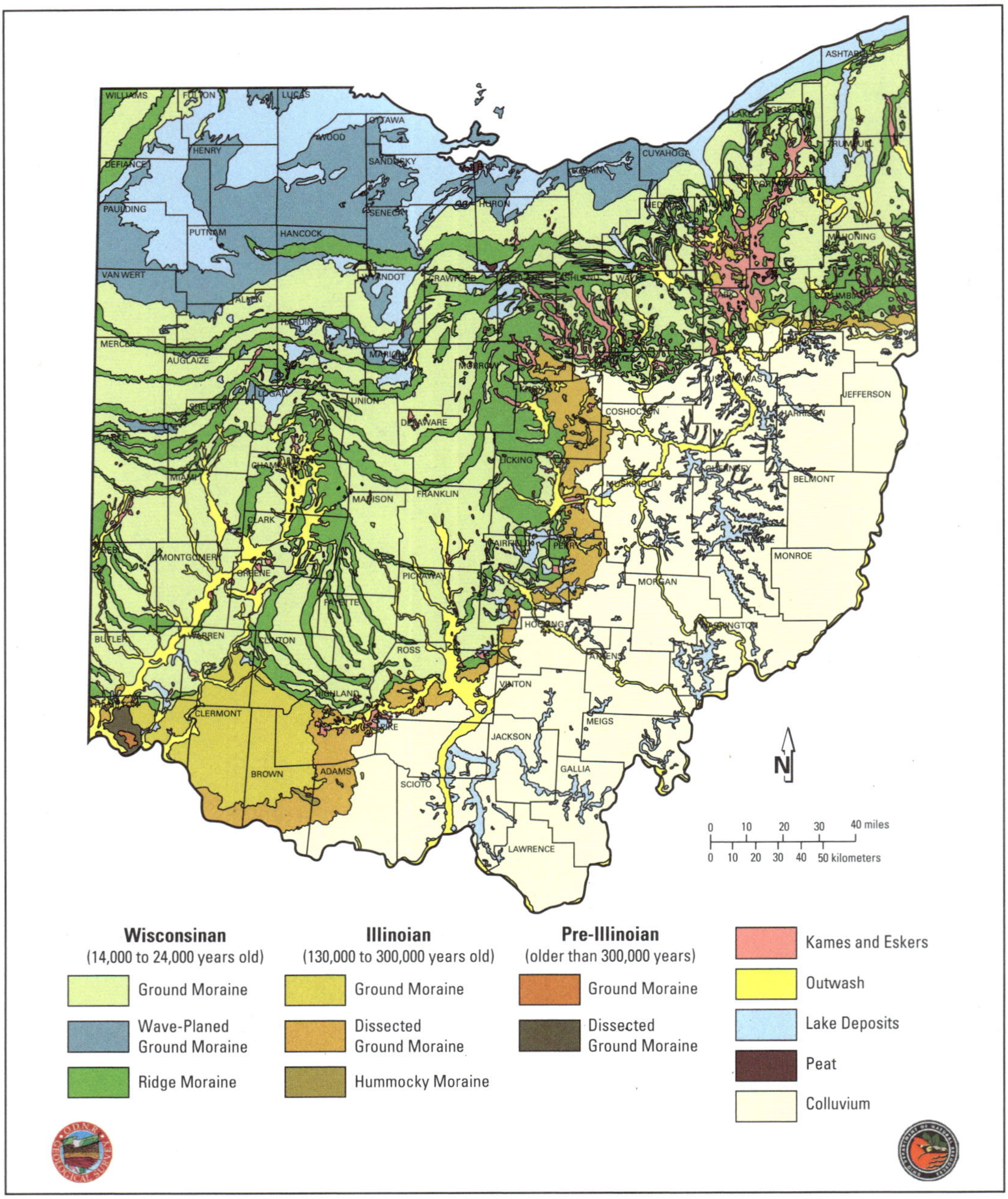

Source: Ohio Division of Geological Survey 2005

FIGURE 4.2 Glacial map of Ohio

(and cost) when moving heavy, bulk commodities to customers. Is the potential location at a transportation advantage or disadvantage compared to your competitor's operations?

There is more to this process than just finding and proving a deposit of aggregate material. If the deposit cannot be developed, sold, and transported competitively in the regional market, then the value is limited to some nebulous, future reserve. Rail transport is more economical than over-the-road truck transport, and water transport is the most economical, especially

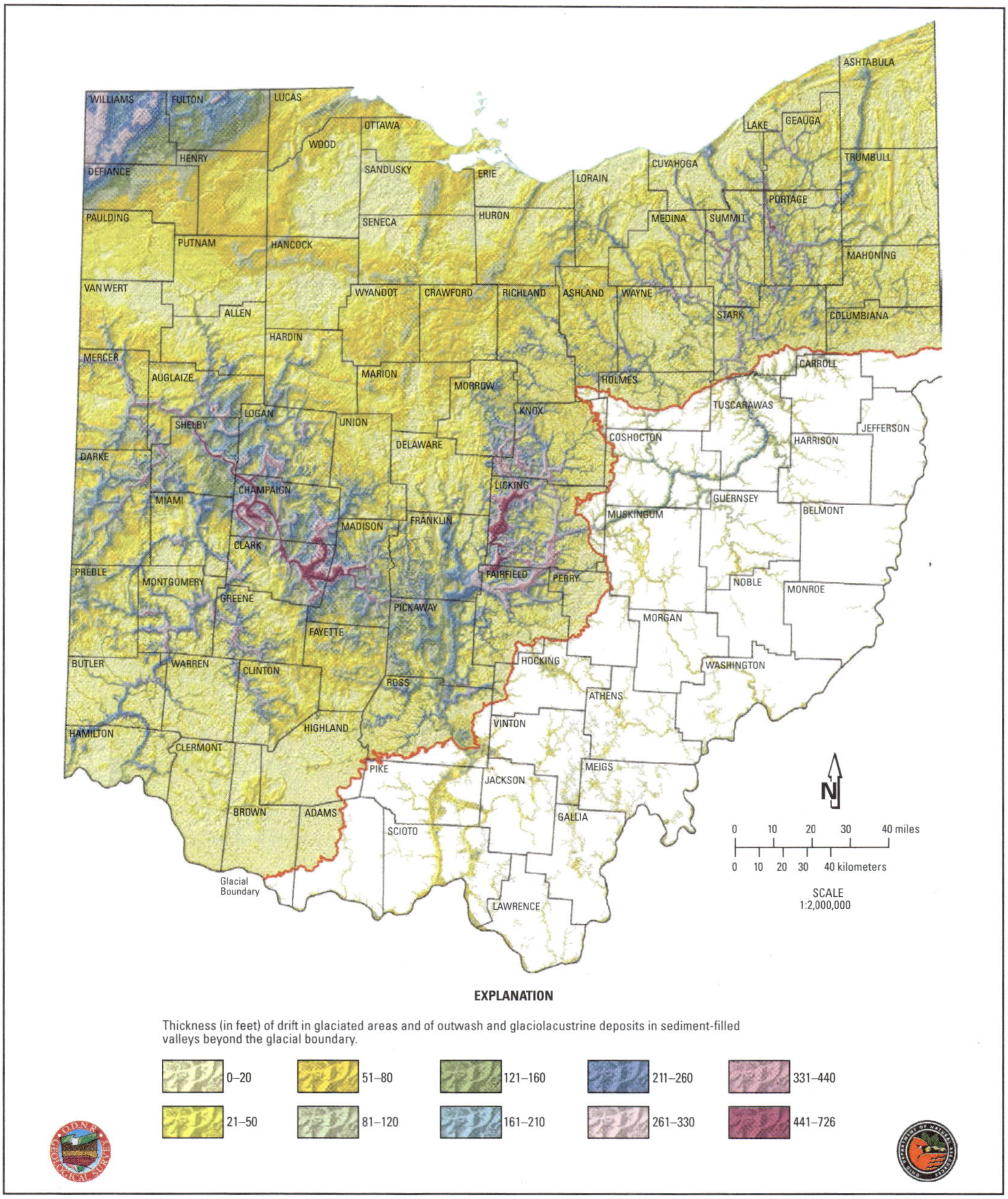

Source: Ohio Division of Geological Survey 2004

FIGURE 4.3 Drift thickness map of Ohio

over long distances. Both rail and barge transport require rehandling costs and additional infrastructure and personnel on the receiving end of the transport, the cost of which needs to be considered in the evaluation. The additional infrastructure requirements include real estate, barges and/or rail cars, conveyors, rail car/barge unloading equipment, dust control, permits, and wheel loaders to load customer trucks at the receiving/sales yard.

Prior to beginning a specific property reserve analysis, it is always a good idea to complete your negotiations with the owner of the property. The price should be agreed before the first test hole is begun and a legal agreement signed. This can be done via a letter of intent, an option to purchase, or an option to lease. If the price is not nailed down prior to a successful exploration program, the property owner can decide at a later date that the property is more valuable than initially thought and increase the asking price. The decision to buy versus lease is up to the individual company and is dependent on corporate structure, accounting methods, tax considerations, capital availability, and ultimately the desire of the property owners.

A mineral royalty lease can be negotiated on a specific cost per ton mined or a percentage of the selling price. Any savvy property owner will want a royalty escalation clause based on inflation when negotiating a long-term royalty agreement. A percentage of the selling price provides for inflation fluctuations and allows your company, as the lessee, the ability to budget a known percentage on a renewing basis. There are advantages to leasing for a property owner (lessor) as well. A mineral lease royalty agreement gives the lessor a steady source of income throughout the life of the deposit (normally 20–50 years) while retaining ownership of the property. A royalty agreement can be advantageous to the lessor from a tax standpoint by spreading out the profits over many years and possibly avoiding some capital gains tax issues. A properly reclaimed aggregate deposit will also retain an asset value after completion of mining.

When a specific property is acquired, it's imperative to have a detailed topographic map of the property, preferably on an orthophoto quadrangle base showing aerial photography. This detailed site map should be on a usable scale in the range of 1:1,200 to 1:4,800, depending on the size of the property. To the digitally savvy among us, this can be easily accomplished using aerial photographic drones and computer-aided design (CAD) or geospatial database software, but a hard copy of the map should be kept on site for reference and notations when doing field work. This property map needs to include details such as topography, roadways, easements, streams, buildings, property lines, wooded areas, surface and buried utilities, section lines, survey markers, areas of environmental concern, and any other pertinent details.

It is very important to retain accurate locations and elevations of all test holes, test pits, rock outcrops, and sample locations. Test data without accurate location data are essentially useless and a very costly waste. In today's world, numerous cell phone apps are available that provide acceptable GPS coordinates and elevations. If possible, survey in the locations and elevations with GPS survey instruments for better accuracy. Always make sure your data are secure and backed up for future reference with GPS, CAD, and geospatial data programs. A professor at field camp (between my junior and senior year) stressed: "Always complete your field notes and record your data at the end of each day; you could get hit by a truck on the way home." When I started in the profession, we were trained to make maps using a plane table and alidade, and to survey with a transit and rod or theodolite. I envy the technology (GPS, CAD) that is available these days and fantasize about what I could have accomplished in the 1970s and 1980s if it had been readily available then.

Determining the number of test borings required for a specific property is always a tricky question. If you ask this question of the industrial mineral geologist in the field and then the CFO of your company, you will probably get radically different answers. Exploration drilling can be very expensive ranging from $15 per linear foot ($48 per meter) of drilling up to $60 ($192) depending on number and depth of holes, diameter of holes, amount of sampling required, time of year, and other factors; therefore, the question becomes partially an economic decision. Quality testing of drill cores or bulk samples obtained from the drilling project is an additional cost and could easily exceed the cost of drilling. I once gave a presentation to a state aggregates association on the costs of exploration projects versus the cost of a surprise during the mining and processing operations. In the long haul, the cost of doubling your exploration budget was nothing compared to the increased operating costs of finding a surprise production interruption, not to mention the potential for lost revenue.

I have encountered more than one reserve surprise in my career. I once toured an underground copper mine in the upper peninsula of Michigan where the ore zone that was being mined sloped downward at about 15°–20°. At some time in the distant past the ore had mysteriously run out, which resulted in the mine production being nothing but brecciated material with no remaining copper ore grade. After a period of hand wringing, economic turmoil, massive employee layoffs, and additional geological exploration and interpretation, it was finally determined that the breccia zone was a result of a large normal fault, the movement of which had dropped the ore zone downward by 2,300 feet (701 meters) on the opposite side of the fault (Figure 4.4). That was a very big and very costly surprise! The industrious miners eventually excavated a series of switchback decline tunnels through the fault zone, picked up the ore zone once again, and resumed mining.

During the era when I was consulting, I was called by a producer in Pennsylvania who was rapidly running out of his limestone deposit and was at a loss to know how to proceed. His quarry was very long and narrow and as he attempted to widen the quarry in one direction he failed to find any limestone, and as he moved perpendicularly in the opposite direction was faced with an ever thickening overburden problem (Figure 4.5). I had no immediate familiarity with the geology of the area but was observing the roadside geology on the long drive to the site. It was apparent after touring the quarry (and observing the roadside geology) that the operator had been mining along the strike of a steeply dipping limestone formation that was trending downward at about 30°. The alternatives were clear; either to strip and waste the

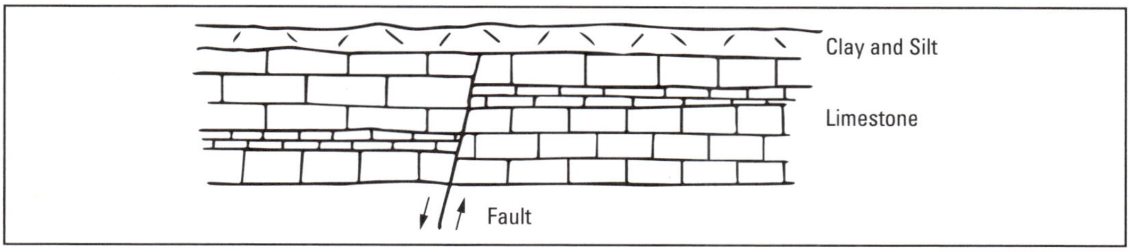

Source: Lamar 1967

FIGURE 4.4 Sketch showing movement along a normal fault

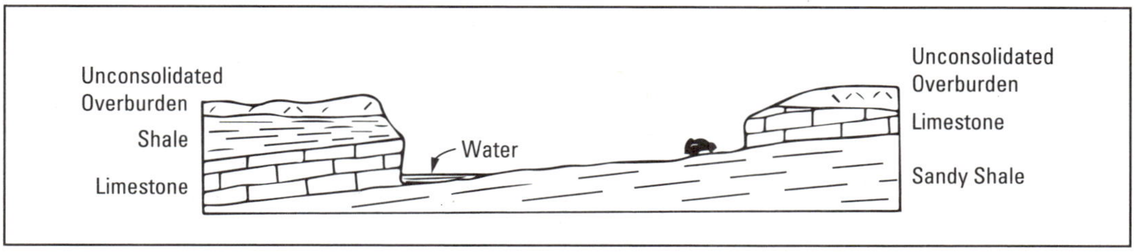

Source: Lamar 1967

FIGURE 4.5 Sketch showing the effects of dipping beds in a quarry

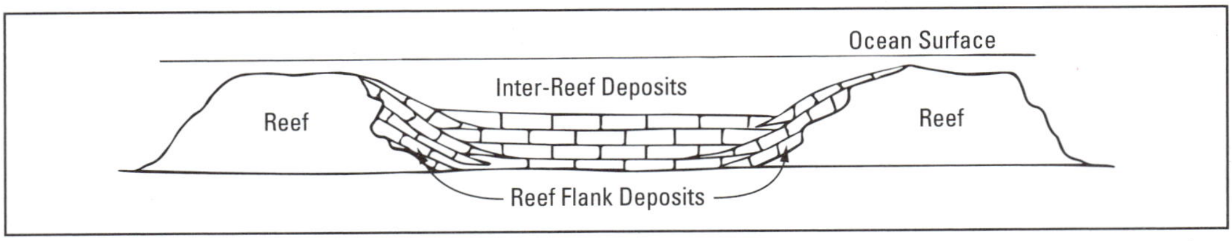

Source: Lamar 1967

FIGURE 4.6 Example of facies changes within a reef-bearing limestone formation

excess overburden and deepen the quarry significantly along the dip, or purchase some additional property adjacent to the strike trend of the quarry.

On another consulting project, an underground producer of high-calcium limestone (for lime and other chemical purposes) was running into serious quality problems. When advancing headers in a certain direction, the high-calcium limestone deposit was increasingly contaminated with significant amounts of deleterious materials resulting in too much magnesium and silica in the finished product. After touring the mine and later a surface quarry in the same area, it became apparent that the deposit being mined was a high-calcium limestone reef facies. The new headers were intersecting the reef flank and inter-reef deposits consisting of silty dolomite and shale (Figure 4.6). Additional drilling and chemical testing were needed to define the limits of the high-calcium reef rock and the location of the reef flanks versus the inter-reef deposits.

Most exploration projects should be done in phases with the pertinent decisions being made based on the boring logs and quality tests completed in the previous phase. Unconsolidated deposits of sand and gravel are generally more heterogeneous than bedrock deposits and thus require more drilling. However, facies changes within sedimentary rock and unexpected weathering along joints and faults can cause surprises in crushed stone operations as well. The first thing I do when determining drill-hole locations on a site is to pinpoint one boring on the highest elevation on the property. In flat-lying deposits this will generally give you a good idea of the entire section that will be encountered during the exploration process, including overburden, the production zone, and ultimately the total thickness of the deposit. This initial reference boring needs to be logged very carefully and may later be used as a reference in additional boring logs (e.g., 42–55.7 feet [12.8–17 meters], same as Ledge C in Boring #1).

The purpose of the initial phase of an exploration project is to determine the existence of the target mineral zone and give a general indication of quantity and quality. Drill holes can be relatively few and widely spaced at this initial stage. If these widely spaced test borings are consistent in lithology, thickness, material quality, weathering, and formation contact elevations, you can be fairly confident and develop a Phase 1 reserve estimate. The more inconsistency found in the boring logs and/or the test results, the more data that will be required in the subsequent phases of drilling. Phase 1 should provide you with a level of confidence as to whether to proceed with further data gathering. If this phase is marginal or disappointing, it may be time to cut your losses and look elsewhere. If Phase 1 is encouraging, drill another phase until you reach the level of confidence required to move forward with the permitting process. At this phase, it is just as important to find and identify the negative characteristics of a deposit as it is to verify the quality of the target. Identify the potential surprises before they are encountered in the mining process.

Some of the more sophisticated mine planning software available can generate random test boring locations. These are computer-generated, statistically random locations that are designed to remove any and all bias from the decision on where to locate the test sites. (My earlier assertion of locating one boring at the highest elevation is an example of my personal bias.) These random locations can, at times, put two test borings within several feet of one another and leave large acreages virtually untested. There are very sophisticated programs that can calculate reserves from three-dimensional boring information using kriging and other statistical methods. These software packages are expensive and most were developed to analyze very complex ore grades containing minute values of gold, copper, nickel, silver, and so on or multiple chemical parameters. Whether these software packages are applicable to your potential aggregates deposit is a function of cost, need, employee training, confidence level, and desired return on investment.

Regardless of your confidence level in your exploration results, there will ultimately be multiple phases of drilling and testing on most properties. Operational issues may arise requiring additional boring information (e.g., a dredge encountering a hard layer of glacial till not previously identified or excessive weathering along a joint trend causing quality issues). Future reserve questions may also arise like what's below the shale layer in the quarry floor? Or is there more sand and gravel beneath the till layer and how can we remove it? Or the adjoining property owner wants to sell his property, should we purchase it?

After drilling, logging, and sampling multiple test borings, it is often valuable to produce cross sections or isometric projections, on a usable scale, of the deposit correlating the various layers within each test boring or core. Examples of the layers to compare would be top and bottom of the overburden or soil layer above the deposit, any obvious shale layers within a limestone deposit, areas of high schistosity in metamorphic rock, changes in facies within sedimentary layers, marker horizons within the deposit, and bottom of mining horizon. This can be done manually or there are numerous computer programs that do an excellent job of producing three-dimensional sections with exquisite graphics. In complex sand and gravel deposits, correlation may be difficult because of the rapidity of changes within these heterogeneous materials. In the past, we utilized a color-coding system for cross sections in these situations; green (indicating money) for coarse sand and gravel, yellow (caution) for sand, and red (stop) for clay or till layers. This made for some very colorful sections but easily identified good versus bad areas within the deposit.

EXPLORATION FOR SAND AND GRAVEL

One might think that sand and gravel, being unconsolidated, would be easy to sample and analyze in the field. That is not necessarily the case. Shallow deposits can be sampled utilizing test pits dug with a backhoe or excavator. This type of machinery has very limited depth capacity (20–25 feet [6–7.6 meters]) depending on the length of the machinery's boom and stick. The unconsolidated nature of the material limits the use of test pits because the more material removed from the pit and the deeper the excavation, the more the remaining material collapses into the hole. Test pits can be excavated in multiple layers but the excavations become extremely large, cumbersome, and time consuming. If, when excavating a test pit, groundwater is encountered near the surface, further depth penetration becomes problematic.

Obtaining representative samples of unconsolidated deposits can be difficult. Auger drills have been extensively utilized in this process, but like test pits they have limitations. An auger is a screwlike length of pipe, with flights similar to the threads on screws, that is advanced into the ground by a truck-mounted drill. The soil material that is being drilled is ejected by the auger on to the surface (much like wood shavings from a wood screw). Logging and describing the material as it is ejected onto the surface was once a preferred method of test drilling sand and gravel deposits. The limitations of this method become apparent when encountering clay layers that mix with the granular materials as they are ejected. It can be difficult to determine where one layer ends and another begins. Drilling below the water table also can affect the results; at times soupy, muddy material will be ejected and at times nothing discharges around the augers. Empirical methods of observing the performance of the drill penetration and rotation have been developed, but with only limited success. Having incorrect or assumed information is worse than having no information at all.

(A)

(B)

Courtesy of Sialtec Europe

FIGURE 4.7 (A) Auger, and (B) hollow-stem auger

Hollow-stem augers were an advancement and are still used extensively in soil sampling for foundation analysis. Larger-diameter augers with an internal diameter (ID) opening of 4 inches (101.6 millimeters) allow for a split spoon sampling tool to be inserted through the center of the auger and pounded into the ground with a 300-pound (135-kilogram) weight and a sample taken ahead of the drill string (Figure 4.7). The number of blows with the weight per 6-inch (152.4-millimeter) interval of advancement is recorded and used to estimate density. These augers work well in fine granular deposits, but the ID of the split spoon sampler is 1.125 inches (28.5 millimeters). Any gravel or cobbles larger than this will be pushed to the side, broken, or plug the sampler, resulting in no sample recovery.

Cable-tool-type water well drills work well in sampling coarser granular deposits but are cumbersomely slow and costly. Drilling a 100 foot (30.5 meter)–deep hole with a cable tool rig can take as long as 4 or 5 days, which is a huge waste of a geologist's time. A cable tool rig uses a large weight on the end of a cable to pound a casing pipe (usually 6-inch [152.4-millimeter] ID) into the ground. Another tool called a bailer on a separate cable is used to remove the material from inside the pipe, which can then be sampled and retained for analysis. After 10–20 feet (3–6 meters) of pipe is advanced and bailed out, another section of pipe is then welded on top of the first and the process is repeated until the final depth is achieved.

Rotary drilling has been used successfully to sample sand and gravel deposits. Rotary drills using drilling mud (a combination of bentonite and mica) can work, but only in relatively fine deposits. The bentonite mud mixture expands and coats the walls of the hole and holds it open long enough to sample the deposit with a larger split spoon sampler. This method is limited and when areas of coarse gravel are encountered the mud is insufficient to prevent the collapse of the hole. Reverse-circulation rotary drills (either water or air) have been successful in sampling this type of deposit. Reverse circulation pushes air or water under pressure into the hole on the outside of the drill string and forces the sample of granular material up through the inside (annulus) of the drill pipe and catches the sample in a cyclone. The size of the sampled material is limited by the ID of the drill pipe.

Rotosonic drilling, which has matured over the last 15 years, has revolutionized drilling and the accuracy of sampling of granular sand and gravel deposits (Figure 4.8). Rotosonic drilling utilizes sonic vibrations along with rotary drilling to advance a double casing into the deposit. A high-frequency resonance energy is induced into the drill string by oscillation of the drill head and is combined with the drill rotation. Vibration frequencies of up to 150 hertz can be developed by some systems. This combination (sonic resonance and rotation) has a

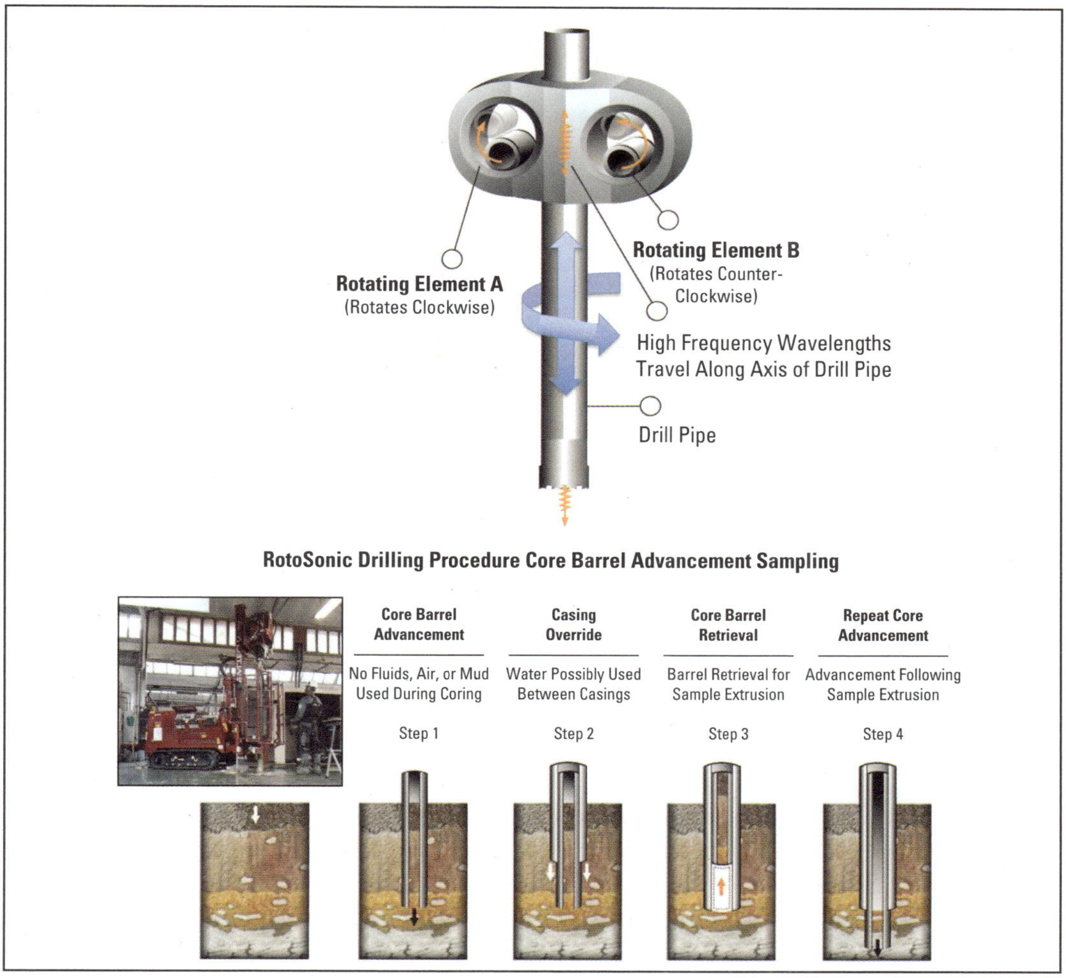

Original source unknown

FIGURE 4.8 Rotosonic drilling

liquefaction effect on the materials surrounding the drill bit, reducing friction and increasing penetration of the drill. The inner casing is advanced, filling up the inner annulus of the pipe casing with a continuous sample or core of material. The outer casing is then advanced around the inner casing to prevent collapse of the hole. The inner casing is then removed and the sample is disgorged by sonic vibration into a clear plastic sleeve, leaving a continuous sample for observation and/or further analysis. Rotosonic is a very fast and clean drilling method using minimal water and no drilling mud or air. It is ideal for environmental drilling projects and can actually core solid rock under some circumstances.

Detailed logs of all holes drilled need to be retained permanently for any drilling completed, whether the property is purchased, leased, or eliminated from consideration. This information could become valuable at some future date even if you do not acquire this particular property today. You never know if adjacent properties may become available in the future. Logs should be completed by a geologist or competent technician trained to identify the specifics of the deposit. A competent person should be on site at all times the drill is in operation to supervise the progress. Never rely on a drilling crew to lay out samples while you are away doing some other task. An innocent mistake by an unsupervised drill crew can be the cause of a costly mistake in the future. Drillers should always keep a separate drilling log for future comparison (Figure 4.9).

The following is the minimum information that needs to be included on a drilling log: date, property name or owner, boring number, location coordinates, collar elevation, total depth of the hole, bottom elevation, static water level (SWL), name of driller or drilling contractor, name of person logging hole, and a detailed description of the material encountered and sampled. When I number test boring logs, I generally start with the year, an abbreviation for the property or owner, and finally the boring number (e.g., 2018-BB-1: boring number 1 drilled in 2018 on Billy Bob's farm). In future years it can be handy to be able to differentiate on a map the chronology of the drill holes and phases. The description should include depth of the material encountered during drilling including color, texture, grain size, moisture (wet or dry), clay content, plasticity, density, obvious deleterious material, top size of gravel, and so on. The technician logging the boring should become familiar with the drilling equipment, noting changes in penetration rate (an indication of density or gravel size), rotation rate (indicating plasticity or presence of clay layers), refusal of penetration (bedrock, boulder), and so forth. A good understanding of the Unified Soil Classification System is essential for describing unconsolidated deposits (Figure 4.10).

Always record the depth within the boring where water is encountered. Make sure to note in the log at what depth the material is dry, moist, wet, or saturated. Upon completion of the boring, the depth to water from the surface inside the boring should be measured using a weighted tape. If possible and if the boring has to sit overnight, remeasure the depth to water and note how long the hole sat idle between depth readings. SWL can be critical information when determining reserves. Above-water reserves and below-water reserves require different types of mining equipment and thus affect the cost of recovering the material. Above-water reserves can be dry banked, as the old-timers used to say, meaning excavated by a front-end loader or shovel at a fairly nominal cost. Underwater reserves will require additional equipment such as a dragline or dredge.

Surface geophysical methods are also available for exploring for sand and gravel deposits. These include seismic reflection, seismic refraction, direct current (DC) resistivity,

Company Name

Drilling Log

Property Name: ______________________ Test Hole #: __________

Location Coordinates: ______________________ Collar Elevation: __________

Total Depth: __________ Bottom Elevation: __________ Static Water Level: __________

Date Drilled: __________ Driller: __________ Logged by: __________

Comments: ______________________

____

Depth		Elevation	Sample	Description
From	To	(Top)	#	
0				

Page _____ of _____

FIGURE 4.9 Typical drilling log

ground-penetrating radar, and electromagnetic methods. Lucius et al. (2007) gives a good description of each of these methods. All these systems have limitations in characterizing sand and gravel deposits and must not be considered a substitute for drilling and sampling. The conclusion section of Lucius et al. (2007) even states: "geo-physical surveys can provide information on the areal extent and thickness of deposits, overburden thickness, and depth to water table. However, resolution decreases with depth for all surface geophysical methods. In addition, estimates of layer thickness can be different from actual thickness because of uncertainties in the measurement process and model ambiguity."

My personal experience using DC resistivity has been very disappointing over the years. In the mid-1970s, while struggling to justify drilling costs, I convinced my employer to purchase a four-electrode DC resistivity system. I had read a publication touting the benefits of this

UNIFIED SOIL CLASSIFICATION AND SYMBOL CHART		
COARSE-GRAINED SOILS (More than 50% of material is larger than No. 200 sieve size.)		
GRAVELS More than 50% of coarse fraction larger than No. 4 sieve size	Clean Gravels (Less than 5% fines)	
	GW	Well-graded gravels, gravel-sand mixtures, little or no fines
	GP	Poorly-graded gravels, gravel-sand mixtures, little or no fines
	Gravels with fines (More than 12% fines)	
	GM	Silty gravels, gravel-sand-silt mixtures
	GC	Clayey gravels, gravel-sand-clay mixtures
SANDS 50% or more of coarse fraction smaller than No. 4 sieve size	Clean Sands (Less than 5% fines)	
	SW	Well-graded sands, gravelly sands, little or no fines
	SP	Poorly graded sands, gravelly sands, little or no fines
	Sands with fines (More than 12% fines)	
	SM	Silty sands, sand-silt mixtures
	SC	Clayey sands, sand-clay mixtures
FINE-GRAINED SOILS (50% or more of material is smaller than No. 200 sieve size.)		
SILTS AND CLAYS Liquid limit less than 50%	ML	Inorganic silts and very fine sands, rock flour, silty of clayey fine sands or clayey silts with slight plasticity
	CL	Inorganic clays of low to medium plasticity, gravelly clays, sandy clays, silty clays, lean clays
	OL	Organic silts and organic silty clays of low plasticity
SILTS AND CLAYS Liquid limit 50% or greater	MH	Inorganic silts, micaceous or diatomaceous fine sandy or silty soils, elastic silts
	CH	Inorganic clays of high plasticity, fat clays
	OH	Organic clays of medium to high plasticity, organic silts
HIGHLY ORGANIC SOILS	PT	Peat and other highly organic soils

LABORATORY CLASSIFICATION CRITERIA		
GW	$C_u = \frac{D_{60}}{D_{10}}$ greater than 4; $C_c = \frac{D_{30}}{D_{10} \times D_{60}}$ between 1 and 3	
GP	Not meeting all gradation requirements for GW	
GM	Atterberg limits below "A" line or P.I. less than 4	Above "A" line with P.I. between 4 and 7 are borderline cases requiring use of dual symbols
GC	Atterberg limits above "A" line with P.I. greater than 7	
SW	$C_u = \frac{D_{60}}{D_{10}}$ greater than 4; $C_c = \frac{D_{30}}{D_{10} \times D_{60}}$ between 1 and 3	
SP	Not meeting all gradation requirements for GW	
SM	Atterberg limits below "A" line or P.I. less than 4	Limits plotting in shaded zone with P.I. between 4 and 7 are borderline cases requiring use of dual symbols.
SC	Atterberg limits above "A" line with P.I. greater than 7	

Determine percentages of sand and gravel from grain-size curve. Depending on percentage of fines (fraction smaller than No. 200 sieve size), coarse-grained soils are classified as follows:

Less than 5 percent .... GW, GP, SW, SP
More than 12 percent .... GM, GC, SM, SC
5 to 12 percent .... Borderline cases requiring dual symbols

PLASTICITY CHART

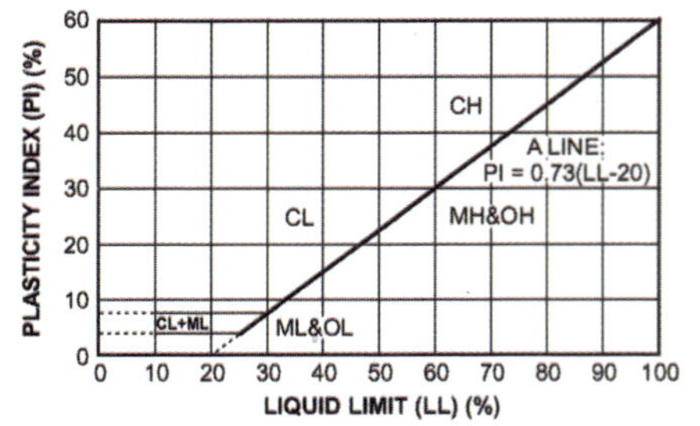

Courtesy of Virginia DOT

FIGURE 4.10 Unified Soil Classification System

type of system in sand and gravel exploration. The spacing of the electrodes was supposed to be analogous to the depth of the deposit. Electrode spacing was changed at regular intervals and readings recorded and graphing methods used to predict changes with depth. We experimented with survey line arrays directly above active gravel pit faces where we could observe the open face of the excavations and compare the observations to the resistivity readings. We then went to unexcavated areas and attempted to extrapolate the observed resistivity values to virgin ground and found very little correlation between the two.

Fast-forward to 2010 and another published article about the benefits of resistivity in sand and gravel exploration. The availability of resistivity arrays with as many as 100 electrodes coupled with computer interpretation of the data convinced me to try this system again in an extremely difficult deposit containing multiple layers of glacial till. I hired a geophysical consultant recommended by the state geologist who guaranteed me success in defining the deposit using resistivity along with computer interpretation. Four lines were identified on the property and multiple-channel, multiple-electrode arrays utilized to produce very colorful cross sections with each color indicating a different material: clay, silt, fine sand, coarse sand, medium gravel, and coarse gravel.

Two weeks after I received the cross sections and paid an invoice for several thousand dollars, I brought a rotosonic drill onto the property and drilled four test borings directly along one electrode array. What the resistivity data indicated to be clean, well-sorted sand and gravel was actually about 80% hard, dense glacial till consisting of silty clay. When I confronted the geophysicist who had done the survey with the boring logs, he shrugged and said "I don't understand how that could have happened." I regretted having already paid his invoice.

Based on my personal experience, I am skeptical about the application of surface geophysics to sand and gravel exploration. Ground-penetrating radar is very limited in depth capacity, and resistivity in my experience does not work. I have no personal experience as to the applicability of electromagnetic methods. I do think that seismic methods could very possibly be used to locate the overburden/bedrock interface in crushed stone operations. In the end there is no good substitute for actual drilling and sampling.

EXPLORATION FOR CRUSHED STONE

Bedrock deposits can be sampled if exposed faces or outcrops exist on the surface. This type of sampling should be done by an experienced geologist familiar with the rock type being sampled. Road cuts expose the local bedrock in many locations but certain safety precautions must be exercised along highways and the local highway patrol may not approve of these procedures. I once took the superintendent of a quarry along to look at an outcropping of limestone along an Indiana highway. He climbed the outcrop ahead of me to get a closer look and quickly retreated when he encountered a copperhead baring its teeth. Beware of snakes! The location and elevation of any bulk sampling of outcrops must be recorded and photographic evidence should be a protocol. In certain areas outcrops appear intermittently with areas of covered, vegetated slopes interspersed. What lies below the covered sections of the slopes? In sedimentary rock the covered slopes generally indicate an area of less competent, more easily weathered, or eroded rock.

Surface samples taken from rock outcrops for quality testing purposes have been weathered and are thus likely to be inferior in quality to unweathered samples of the same bedrock deposit. Unless the outcrop or road cut was formed very recently, quality testing may be of little value. Detailed observation of how the rock formation has been affected by years of weathering can shed light on the potential quality of the deposit. Certain types of deposits can actually "case harden" because of weathering and exhibit hardness properties that are not typical of the actual bedrock. Some older glacial outwash sand and gravel deposits case harden and form an outer rim of cemented conglomerate that can be several feet thick.

Rotary drilling is one method for sampling crushed stone deposits. Rotary air or rotary mud machines produce small chips of stone known as cuttings that can be recovered from the exhaust air via a cyclone or strained from the drilling mud recirculating from the holes. These cuttings are very small chips and not always easy to identify. Softer materials like shale can be lost in circulation and brittle materials like chert are likely to shatter. In evaluating aggregate deposits, it is as important to know what deleterious materials are in the rock as it is to know what good properties exist. Rotary drilling can be utilized for a quick and dirty analysis, but can be troublesome when trying to determine quality. The only true way to evaluate a bedrock deposit is to core drill using diamond coring tools (Figure 4.11).

FIGURE 4.11 Freshly cut rock core

In the core drilling process, an outer tube or outer barrel attached to a drill bit encrusted with industrial diamonds is rotated into the rock. The diamonds on the outside and bottom of the bit act as an abrasive and allow the core barrel to advance cutting into the rock. Contained within the outer barrel is an inner tube or inner core barrel that is held firmly within the outer barrel by the stop ring. Water or drilling mud is circulated to cool and lubricate the bit and remove any fine cuttings from around the drill string. When a core run, generally 10 feet (3 meters) in length, is completed, rotation is stopped and the drill string is detached from the drill. At this point a wireline attached to an overshot is lowered via a winch into the annulus of the outer barrel. The overshot couples with the locking coupling on top of the inner core barrel. This process releases the locking device and the inner barrel is hoisted to the surface by the wireline. The core of rock within the inner core barrel is recovered and the inner barrel reinserted into the bottom of the outer barrel, where it locks in place. The drill string is reattached to the machine and drilling resumed until another 10-foot (3-meter) length of core can be recovered. This process is repeated as many times as necessary until the required depth is achieved.

Triple-barrel coring systems are sometimes utilized to recover highly fractured bedrock but are uncommon in crushed stone exploration. The most common core sizes used are in the range of 2 inches (50.8 millimeters) in diameter with the designation of NX or NQ. Figure 4.12 is a diagram of a typical double-barrel wire line core system. Table 4.1 shows common core diameters in both inches and millimeters.

The recovered core should be carefully placed in specially designed core boxes with particular attention given to marking the top and bottom of the core, the boring number, and the depth of the sample. The core is measured and a notation as to the percentage of core recovered is noted in the boring log utilizing the following formula:

$$\frac{\text{length of core sample}}{\text{length of drill run}} \times 100 = \%\ \text{recovery}$$

It is important to record core recovery percentage to track any voids in the rock or zones of weakness or excessively weathered areas that would not allow for total recovery. The core can be logged as it is recovered from the drill or moved in the core boxes to a secure location for further detailed analysis. A core splitter is a specially designed device utilized to split the core in half to enable a better description of the rock sample. The process of coring with abrasive

Courtesy of Sialtec Europe

FIGURE 4.12 Core drilling system

diamonds tends to polish the outer core surface, which can obscure many details within the rock sample. For the purpose of getting accurate geologic descriptions, it is advisable to routinely split core samples. In special cases, rock core samples need to be lab tested for rock mechanical properties (e.g., compressive strength, tensile strength, modulus of rupture) and this core should remain intact (not split).

Rock quality designation (RQD) is an empirical test of the quality of rock core indicting the degree of fracture or jointing of the rock mass being drilled. To calculate RQD percentage, the length of each piece of recovered core that is longer than 4 inches (100 millimeters) is measured and these individual lengths are added together. This figure is then divided by the entire length of the core run. The formula is

$$\frac{\text{sum of (length of sound pieces } >4 \text{ inches [100 millimeters] each)}}{\text{total length of core run} \times 100} = \text{RQD \%}$$

As an example, a 10 foot (3 meter)–long core run has eight pieces longer than 4 inches (101.6 millimeters); the individual pieces have lengths as follows: 5.5 inches (139.7 millimeters) + 16 inches (406.4 millimeters) + 5 inches (127 millimeters) + 12 inches (304.8 millimeters) + 30.5 inches (775 millimeters) + 19.75 inches (502 millimeters) + 4.25 inches (107.9 millimeters) + 5 inches (127 millimeters) = 98 inches (2,489.2 millimeters). When put into the formula, you get the following RQD percentage: (98 inches [2,489.2 millimeters] /120 inches [3,048 millimeters]) × 100 = 81.7%. RQD data are especially useful when attempting to determine the weathered nature of a body of rock or when engineering determinations are required (e.g., tunneling, underground mining). This example would be considered good, hard rock. Table 4.2 equates RQD to the core quality of a single core run.

Additional information needs to be included on a core drilling log than is necessary on a boring log for unconsolidated material like sand and

TABLE 4.1 Common core sizes (after Christensen Dia-Min Tools Inc.)

Size	Diameter of Core in. (mm)	Diameter of Borehole in. (mm)
EX, EXM	0.846 (21.5)	1.484 (37.7)
EWD3	0.835 (21.2)	1.484 (37.7)
AX	1.185 (30.1)	1.890 (48.0)
AWD4, AWD3	1.138 (28.9)	1.890 (48.0)
AWM	1.185 (30.1)	1.890 (48.0)
AQ Wireline, AV	1.067 (27.1)	1.890 (48.0)
BX	1.654 (42.0)	2.358 (59.9)
BWD4, BWD3	1.614 (41.0)	2.358 (59.9)
BXB Wireline, BWC3	1.433 (36.4)	2.358 (59.9)
BQ Wireline, BV	1.433 (36.4)	2.358 (59.9)
NX	2.154 (54.7)	2.980 (75.7)
NWD4, NWD3	2.059 (52.3)	2.980 (75.7)
NXB Wireline, NWC3	1.874 (47.6)	2.980 (75.7)
NQ Wireline, NV	1.874 (47.6)	2.980 (75.7)
HWD4, HXB Wireline, HWD3	2.406 (61.1)	3.650 (92.7)
HQ Wireline	2.500 (63.5)	3.791 (96.3)
CP, PQ Wireline	3.346 (85.0)	4.827 (122.6)

Courtesy of Boyles Bros. Dimantina

TABLE 4.2 Rock quality designation of core

RQD (%)	Core Quality
91–100	Excellent (fresh rock)
76–90	Good (hard rock)
51–75	Fair (moderately weathered)
25–50	Poor (weathered rock)
<25	Very poor (completely weathered)

gravel. The same basic information needs to be recorded as to location, elevation, water level, property ownership, and so on. Core recovery percentages should be included as well as RQD data. An example of a typical core drilling log is shown in Figure 4.13. Some geologists prefer the addition of a graphic log added to their core logs. The USGS has a publication with standard symbols for all lithologic units and geological mapping symbols (USGS 2006). Most geological and mine planning software packages contain these symbols as well.

When logging borings, whether in unconsolidated sediments or bedrock, concentrate on the primary purpose of the individual drill hole, which is the identification of quality aggregate deposits with the intent to develop a reserve analysis. Always keep in mind that this record is designed to be permanent, may outlast your career with the company, and that other employees will need to understand and interpret the pertinent details. Economic decisions and large capital expenditures will be based on the accuracy and detail contained in your drilling logs. The *Engineering Geology Field Manual, Second Edition* (Bureau of Reclamation 2001) is an excellent field reference, but bear in mind that it is geared toward engineering geology and construction.

After completion of the boring, rock cores should be treated as an asset. Rock cores are not just a box of useless rocks shoved into the corner or thrown out upon completion of the project. Cores are a financial asset based on the cost involved of the drilling, the geological description/analysis (core log), the quality analysis, and the value of the reserves they represent. Rock cores are also a scientific and informational data asset, if the core has been properly

Company Name

Core Drilling Log

Property Name: ______________________________ Test Hole #: ______________

Location Coordinates: ______________________________ Collar Elevation: ______________

Total Depth: __________ Bottom Elevation: __________ Static Water Level: ______________

Date Drilled: __________ Driller: ______________ Logged by: ______________

Comments: ____

Depth		Elev.	Core	%		RQD
From	To	(Top)	Run #	Recovery	Description	%
0						

Page _____ of _____

FIGURE 4.13 Typical core drilling log

logged, accurately located, and meticulously documented. If at all possible, based on financial ability and storage space, half of a split core (preferably the index core) should be retained and stored in an indoor space (protecting it from the effects of weathering) for future reference (Figure 4.14). On many occasions I have wished I had an actual core sample to look over rather than just a core log when faced with an operational surprise or quality problem. As the geologist or technician, never assume you will be available if a future question arises about a drilling project. People are promoted, retire, or move on to different jobs, but with proper documentation, data storage, and a library of rock core, the next generation will remain well informed.

The surface geophysical exploration methods covered under unconsolidated deposits may apply under some circumstances in the exploration for bedrock deposits. Seismic refraction or reflection would probably be the most applicable. This method should be capable of distinguishing between unconsolidated material and solid bedrock and may be a good tool to identify the thickness and volume of overburden on top of a potential quarry deposit. Certain borehole geophysical methods are excellent tools for evaluating subsurface deposits. Borehole geophysics entails lowering an electronic probe into a drill hole and utilizing recording instruments to record various parameters at different depths. Borehole geophysics are utilized extensively in the petroleum industry. Natural gamma-ray detection is probably the most valuable downhole tool to aggregates exploration. Gamma rays naturally occur in clay minerals and can be detected by natural gamma detectors. Argillaceous (clay-bearing) minerals are deleterious in aggregates. Shale is bedrock that consists of clay minerals and is easily detected as well. Borehole geophysical instruments are extremely sophisticated and very expensive, which is not conducive to keeping them on hand for an occasional exploration project, but they can be rented by the day, week, or month from several different sources.

FIGURE 4.14 Split core of high-calcium limestone from the Ste. Genevieve Formation in North-Central Kentucky

Exploration for natural mineral aggregate deposits is an important part of developing economically viable mines, pits, and quarries. The exploration and evaluation should be completed by an experienced, professional industrial mineral geologist. Small companies may not be able to afford to keep an experienced geologist on staff on a full-time basis. If not, numerous consulting firms are available that specialize in providing services to the aggregates industry. I would advise investigating a consultant's credentials thoroughly, including references from other aggregate producers, prior to contracting their services. I have seen a number of reserve analyses over the years completed by local testing and engineering firms that produced very poor results because of a lack of experienced personnel doing the work. One would not hire a dentist to perform thoracic surgery.

RESERVE ANALYSIS AND CALCULATION

Numerous geological, mine planning, and CAD software programs are available that are capable of analyzing three-dimensional geological data and producing estimates of mineral aggregate reserves. Without access to such software, there are other ways to calculate reserves the old-fashioned way. Solid geometry is the first method that comes to mind. Length × width × thickness = volume, which can easily be converted to tonnage when the density of the aggregates material is known. This is an easy calculation for flat-lying deposits with a relatively consistent thickness.

As a general rule, there are 2,500 short tons of sand and gravel per acre-foot (20,271 short tons per hectare-meter), and 3,500 short tons of solid rock per acre-foot (28,483 short tons per hectare-meter). These figures are fairly conservative estimates. This obviously varies with the actual density of individual rock formations and should be recalculated for final reserve estimates based on actual data, but the rule-of-thumb estimates can be utilized for relatively quick calculations. A 75-acre (30.35 hectare) field containing glacial outwash sand and gravel having an average thickness of 42 feet (12.8 meters) would yield approximately 7.87 million gross tons (75 acres [30.35 hectares] × 42 feet [12.8 meters] × 2,500 short tons per acre-foot [20,271 short tons per hectare-meter] = 7,875,000 short tons). Similarly, 152 acres (61.5 hectares) underlain by a 54 foot (16.4 meter)–thick ledge of limestone would contain about 28.7 million gross tons of limestone (152 acres [61.5 hectares] × 54 feet [16.4 meters] × 3,500 short tons per acre-foot [28,483 short tons per hectare-meter] = 28,728,000 short tons).

TABLE 4.3 Specific gravity ranges for various rock types

Rock Type	Specific Gravity
Andesite	2.4–2.8
Basalt	2.7–3.2
Dolomite	2.7–2.8
Gabbro	2.9–3.1
Granite	2.6–2.7
Limestone	2.7–2.8
Marble	2.6–2.9
Sandstone	2.0–3.2
Traprock	2.7–3.1

The reasons these general rules are useful are the similarities in specific gravity of various rock types (Table 4.3). A rock with a specific gravity of 2.7 has a density of 168.5 pounds per cubic foot (2,700 kilograms per cubic meter) (2.7 × 62.4 pounds per cubic foot [1,000 kilograms per cubic meter] [density of water])

$$168.5\frac{\text{lb}}{\text{ft}^3}\left(2{,}700\frac{\text{kg}}{\text{m}^3}\right) \times 27\frac{\text{ft}^3}{\text{yd}^3}(1) = \frac{4{,}549.5\frac{\text{lb}}{\text{yd}^3}\left(2{,}700\frac{\text{kg}}{\text{m}^3}\right)}{2{,}000\frac{\text{lb}}{\text{st}}\left(907.2\frac{\text{kg}}{\text{st}}\right)}$$

$$= 2.27\frac{\text{st}}{\text{yd}^3}\left(2.97\frac{\text{st}}{\text{m}^3}\right)$$

$$2.27\frac{\text{st}}{\text{yd}^3}\left(2.97\frac{\text{st}}{\text{m}^3}\right) \times 1{,}613\frac{\text{yd}^3}{\text{acre-ft}}\left(152\frac{\text{m}^3}{\text{ha-m}}\right) = 3{,}669\frac{\text{st}}{\text{acre-ft}}\left(451.4\frac{\text{st}}{\text{m}^3}\right)$$

Thus, 3,500 short tons per acre-foot is a conservative rule-of-thumb estimate (compared to the 3,669 short tons shown above) and is an easy calculation if the acreage and thickness of deposit are known. A similar calculation can be done for sand and gravel. When estimating unconsolidated overburden quantities, the 1,613 cubic yards per acre-foot (152 cubic meters per hectare-meter) is a useful number as well. The number of acres (hectares) of overburden to be removed times the thickness in feet (meters) times 1,613 (152) will give a good approximation of the in-place volume in cubic yards (cubic meters) that needs to be stripped.

Another method for calculating reserves in deposits that vary considerably in thickness is to develop an isopach (thickness contour) map of the mineral body. An isopach map is the same as a topographic map but instead of producing contours based on the same elevation, the contours represent different thicknesses of the reserves. Once this type of map is constructed, the areas between the different thickness contours can be measured and reserves calculated for each different area and then added together for a total reserve number. A series of cross sections can be drawn through the property correlating test boring data and the areas of each individual cross section's mineral reserve determined and summed for a total reserve. Universal spacing, random spacing, triangular spacing, perpendicular at midpoint between holes, and polygon blocks are various other methods of dividing the areas surrounding test borings and using the thickness at each boring to represent a defined area, and then adding these areas for a total reserve.

No matter how the reserve estimate is developed, it is a good idea to be conservative with your numbers. No one likes a surprise like running out of reserves 3 years earlier than anticipated. Many significant financial decisions are made based on the geologist's reserve estimate. Your first reserve estimate will be a **gross reserve** figure showing the entire amount available in the property. Various federal, state, and local regulations will dictate setbacks from property lines, roadways, utility easements, pipelines, and so on. These requirements need to be assessed, measured, and removed from the estimates of gross reserves. Upon excavation of unconsolidated deposits, the material will fall to its natural angle of repose (usually about 37°) because of gravity and erosion forces. These slope losses need to be calculated also and factored into the evaluation as reserve losses. In thick alluvial fan and dry bank glacial deposits, slope losses can be significant. Reclamation plans need to be considered in reserve analyses, as well as processing plant, scale house, roadway, and office locations. A **minable reserve** number has been determined when these reserve eliminations have been subtracted.

After determining minable reserves, processing losses must be taken into consideration. Neither sand and gravel nor crushed stone are ever 100% salable. Sand and gravel deposits often contain considerable percentages of silt and clay (minus 200 mesh material) that is screened and washed out of the products during processing. Gradations of the pit run sand and gravel taken from test borings should be analyzed for percentage of material passing the 200 sieve and this is considered waste. In some deposits this can amount to 15%–20% of the total. The crushing of coarse gravel and crushed stone produces an additional loss of 5%–7% in dust, which is collected in bag houses or washed into settling ponds. The operations personnel should be consulted for estimates of processing loss or waste before final reserve estimates. In some markets certain products may not be salable; for example, areas with a glut of sand in the regional deposits. These too may have to be considered a processing loss. Once these processing and nonsalable percentages are removed from the minable reserves, we have the bottom-line reserve numbers that we are looking for: **salable reserves**.

If you are employed by a publicly traded company, there are protocols based on guidelines from the U.S. Securities and Exchange Commission and other international agencies (if located outside of the United States) covering the methods of publicizing mineral reserve numbers for public reporting for investment purposes. The three categories of mineral resources are inferred, indicated, and measured; the two categories of mineral reserves are probable and proven. These categories are defined by the amount of geological, material quality, and market certainty. *The SME Guide for Reporting Exploration Results, Mineral Resources, and Mineral Reserves* (2017) does an excellent job of defining these terms. A specific section discusses industrial minerals reserve reporting.

Sections 68 through 71 of the Guide states: "With respect to industrial minerals, stone and aggregate, the Modifying Factors may be significantly more critical than geoscientific knowledge in determining Mineral Reserves. Such factors as quality, transportation, cost to markets, location and quality of competing deposits and ability to compete with such deposits to access the market, are important and should be carefully considered before declaring Mineral Resources and Mineral Reserves." Section 70 additionally states: "As a general rule, a Mineral Reserve cannot be declared unless there are reasonable expectations that all permits, ancillary rights and authorizations required for mining can be obtained and a viable market identified.

For some minerals such as sand, gravel and aggregates, permitting requirements may be such that reasonable expectations can only be defined by comparison with competing reserves. When a deposit is scheduled to be mined at a date some distance in the future, declaration of a Mineral Reserve implies reasonable expectation at the time of reporting that the necessary markets and permits could be obtained when needed."

Once again, we see the relevance and importance of several market factors: transportation cost to market, location and quality of competing deposits, access to market, and the ability to be competitive. As noted earlier, many factors besides controlling a viable aggregate deposit are part of the determination of its value. Permits including zoning, environmental, mining, and reclamation are also necessary and fall under the category of community approval. Permitting requirements are covered in more detail in Chapter 6.

CHAPTER 5

Location and Transportation

Just like any other real estate decision, location is the key to a successful aggregates mining operation. In 2019, 2.53 billion metric tons of aggregate was produced and sold in the United States. With an estimated population of just over 329 million people, those figures equate to 7.7 metric tons of aggregates sold for each individual American. These figures will vary in individual markets, with large metropolitan areas requiring a greater volume than rural areas of the country because of the amount and type of construction projects. More people = more construction = more aggregates sales. This is evident by the list of states that produce the most aggregate. In 2016 the top 10 states in crushed stone production listed in descending order were Texas, Pennsylvania, Florida, Missouri, Ohio, North Carolina, Virginia, Georgia, Illinois, and Kentucky. For the same year the top 10 states in sand and gravel production were California, Texas, Minnesota, Michigan, Arizona, Washington, Colorado, Ohio, Utah, and New York. These are all states with large populations with the exception of Kentucky, which happens to export large volumes of aggregate out of state via several navigable waterways.

Once, shortly after starting a new job, my employer asked me to look into a piece of property they had previously purchased as an aggregate reserve that was located in a fairly rural area of the Midwest. He asked if I could come up with an estimate of what I thought the annual sales might be from this location. I first did some research on the population of the five-county area surrounding the site. Then, knowing the locations of the competitors in the area, I developed a map showing their locations in relation to the site in question. I calculated the size of the market by multiplying the population by the average consumption per capita of aggregates in that particular state. On the map I then drew a 20-mile (32-kilometer) radius around our site and the other producers to see where the market areas would intersect. Based on this analysis, I estimated that the expected sales would be a maximum of 175,000 short tons per year. This was not a particularly appealing volume of annual sales in a commodity business like the aggregates industry.

During this process, early in my career at this particular company, I discovered the wrath of the company vice president (VP) who had originally purchased the site, which he defined as having an "excellent potential market." Soon thereafter we moved equipment onto the site, at the insistence of the VP, and began production to "test the market" (Figure 5.1). We operated at this location intermittently for the next 4 years and never exceeded 160,000 short tons of sales or made a nickel of profit. Needless to say, there was never a fifth year of operation. But I was smart enough not to gloat and remind the VP of my initial market projections. When and if this area increases in population or the competitive situation changes, it may become an economically profitable reserve. I'm reiterating this story as an example of a location that was less than ideal for a sand and gravel mining operation. Luckily my employer had pockets deep

FIGURE 5.1 Portable plant testing the market

enough to weather a few years of minor economic loss to test the market; a poorly capitalized company may have gone out of business.

I am not, nor do I claim to be, an expert in marketing, but I do have firsthand knowledge of the aggregates industry. A serious aggregates production facility requires a considerable capital investment in mining and processing equipment, not to mention the cost of real estate, site development, permitting, local government approval (zoning), and utilities. A serious marketing analysis is a strategic necessity when attempting to locate a new source of aggregate. This type of analysis must include the location and estimated sales volumes of each competitor in the market, as well as estimates of the total available sales in the marketplace.

Potential increases in the market, either long or short term, need to factor into any marketing investigation. Is Amazon anticipating locating a regional distribution center in the area? Is the metropolitan airport predicting a terminal expansion and addition of more runway capacity? What is the potential for a new interstate bypass around the city? The state departments of transportation (DOTs) publish budgetary and anticipated spending on regional highways. County highway departments have annual budgets and regional improvement plans. There are publications that can assist in answering some of these questions, one of which is the Momentum Index published by Dodge Construction Network, which includes regional information on upcoming construction projects. Bid solicitations for very large construction projects are sometimes published in newspapers or construction periodicals.

Population trends are available on numerous websites, such as state and federal departments of commerce, local chambers of commerce, or county, city, and state planning agencies. The most recent federal census occurred in 2020 and it contains a wealth of new information on population trends. As I stated earlier, more people is directly related to more aggregates sales. Is the target market area gaining population like Boise, Idaho, which has been called the fastest growing area in the country? Or is the market area within a part of the Rust Belt, which has been losing population at a slow but steady rate over the last few decades? These trends in people movement will affect the future market for aggregates, either up or down, in many regions of the country.

Get to know as much as possible about your competition! Is their location closer to or further away from major transportation routes? How many truck miles is the competition

FIGURE 5.2 Trucks, the most common transport mode

from major customers (ready-mix plants, asphalt plants, highway department stockpiles, etc.)? Where are the larger aggregate consumers presently buying their crushed stone? It's easy enough to park down the road from a large asphalt or ready-mix plant and see how many trucks belonging to what company are delivering them their raw materials. Get to know the customer base prior to entering the market. Most states have associations of asphalt producers, ready-mix producers, and general contractors, all of which have websites listing their members and their locations (e.g., Asphalt Pavement Association of Indiana, Associated General Contractors of Ohio). Meet with the state DOT officials and county engineers and discuss their upcoming highway programs, specifications, and budgets.

Aggregates are a heavy, bulk commodity and transportation costs become a significant part of the cost to the customer. The cost of a ton of aggregates delivered by truck to the jobsite can easily double or sometimes triple depending on the distance from the quarry to the job (Figure 5.2). Trucking costs vary considerably from market to market depending on labor rates, fuel costs, traffic congestion, highway access, and DOT truck weight regulations. It is always in the customer's best interest to be located close to a source of good-quality aggregate to minimize transportation costs. In many markets, aggregate producers tend to lease plant sites on their property to ready-mix and asphalt producers. This mitigates a portion of the transportation costs and concentrates similar industrial activities in the same area. Many local planning agencies like this approach of concentrating trucking-intensive industries in the same area.

Getting an idea of the reserves picture as it relates to your competitors is a very valuable piece of information. This is generally not a simple process because reserve data are always kept highly confidential. Simple observation, familiarity with the regional geology, and knowing the real estate holdings of the competition can provide valuable information about their reserves. A little bit of sleuthing is involved at this stage. Programs like Google Earth open the world of aerial observation to everyone (Figure 5.3). Although not updated every year, it's amazing how much information you can glean from a website like this. If Google Earth is not up to date and you need more current information, you can hire an aerial photographic service to fly the region and provide scaled photos of the site. This, of course, is a much more expensive option.

FIGURE 5.3 Google Earth image of a large limestone mining operation in Indiana with two asphalt plants and two ready-mix plants on site

Drones can also be an excellent source of aerial reconnaissance. Most counties in the United States provide the ability to go online and search for the ownership of all the property in the county. Once you've searched the competition's property holdings, compare it to Google Earth and determine what percentage has been mined versus what remains in a virgin state. This can be somewhat problematic if the competitor leases property and it is shown in another property owner's name on the county website. But you can reasonably assume if an adjoining property is obviously being mined that these are reserves for the competitor.

If you have a general knowledge of the regional geology, you can extrapolate what geological formation the competitors are mining and approximately how thick the deposit is. This is a good time to consult geological maps of the area if available. Are multiple benches evident or are they mining one ledge only? From the scale of the photo can you estimate how high the benches are? What kind and capacity of excavation equipment are they using to recover the sand and gravel deposits? Is it a clamshell dredge that can excavate several hundred feet or a dragline with a short boom capable of digging only 40 feet (12 meters)? Observations from local roadways can also be helpful. How far or how often does the dredge move in 1 week's time? Do they have a substantial area prestripped of overburden? For the last 30 years of my career, I always kept a set of good binoculars in my vehicle for just such an opportunity (or was it for bird-watching?). These methods will not tell you exact reserve numbers but can be a good indicator of what the competitors are doing. Get to know the competition. Watch the newspapers and legal notices concerning zoning hearings where the competition may be attempting to expand their reserves. It's always interesting to attend a competitor's zoning hearing as an observer. Machiavelli in his book *The Prince* said: "Keep your friends close but keep your enemies [in this case competitors] closer" (Machiavelli 1984).

Some states publish statistics on mining operations that include their locations and sometimes the annual production figures. Most of this type of information is going to be found at the state geological survey websites. A good example of this is the Report on Ohio Mineral Industries (Ohio Division of Geological Survey 2018). Figure 5.4 is an example of the type of data available in this report for the year 2018. It includes mine reference numbers that relate to an interactive map, production tons, sales tons, dollar value, geologic unit mined, and principal products produced. Reporting these data is a legal requirement in the state of Ohio. Not all states require this level of reporting or publish this much data.

This type of data is a treasure trove when putting together a marketing analysis for a new aggregate operation. Combined with estimates of the total market based on population statistics, it is easy to determine each competitor's percentage of market share. Take, for example, the first listing in Figure 5.4: Arch Materials' Batavia mine, located in Clermont County, Ohio. The state mine number Ct-A corresponds to an exact location on the map in the 2018 Report on Ohio Mineral Industries. No Mineral Resources Management (MRM) permit number is listed because this is an underground mine and only surface mining is regulated by Ohio MRM. (There is a proposal pending in the Ohio state legislature to add underground mining to the jurisdiction of the Ohio MRM.) This mine produced 1,975,000 metric tons and sold 1,893,350 metric tons in 2018. The total value of sales is listed as $17,605,000. If you do the math, the average selling price was $9.30 per ton. Arch Materials mines the Black River Group of the Ordovician period and produces riprap and concrete and asphalt aggregates for commercial construction.

Arch Materials serves the Cincinnati regional market area. World Population Review (2022) lists the metropolitan population of the Cincinnati metropolitan statistical area as 2.24 million people. At 7.7 metric tons/person, this equates to a 17.2 million metric ton total market. A metric ton weighs 2,204.6 pounds (1,000 kilograms), while a short ton weighs 2,000 pounds (907 kilograms); therefore, a short ton is equal to 0.907 metric tons. If we convert 17.2 million metric tons to short tons, we get a total market for the area of 15.6 million short tons. If Arch Materials produces 1.975 million short tons at this location, its share of the Cincinnati market is therefore 12.7%.

To take this analysis further, we could go to the Clermont County, Ohio, website and determine the amount of property owned by or leased to Arch Materials. Some research into the geology of the Black River Formation in this part of Ohio would tell us that this is a very good, durable, structurally sound deposit of dolomitic limestone, dolomite, and in places high-calcium limestone. The Black River Formation is about 400-feet (122-meters) thick in this area. Knowing the acreage and the thickness of the deposit, we could extrapolate some approximate reserve numbers available on this property. This is not as simple as determining approximate reserves in a surface mining operation. Limestone and dolomite are mined underground by the room-and pillar mining method, where anywhere from 30%–40% of the reserve is left behind as pillars to support the weight of the overlying rock. Room heights can range from 40 to 90 feet (12 to 27 meters) depending on the mechanical properties of the stone. In mining a thickness of 400 feet (122 meters), multiple mining layers would have to be developed with sill pillars left behind between layers. A total extraction rate of somewhere in the range of 50% would probably be reasonable.

This is just a quick synopsis of how much can be done in a marketing analysis with a little bit of publicly published information. Scrutinizing the information published in a state

Company	Mine Name	County	State Mine Numbe	Mineral Resources Management Permit Numbe	Production (tons)	Sales (tons)	Dollar Valu	Geologic Unit (age)	Principle Use(s)	Address
Arch Materials, LLC	Batavia Mine	Clermont	Ct-A	---	1,975,000	1,893,350	$17,605,000	Black River Group (Ordovician)	riprap, portland cement concrete, asphalt, commercial building	
Barrett Paving Materials, Inc.	Ludlow Quarry #394	Miami	Mi-BPM	IM-1169	331,000	346,000	$3,460,000	Brassfield Formation (Silurian)	unspecified	
Barrett Paving Materials, Inc.	Miami River Quarry #400	Shelby	Shy-17	IM-389	657,000	666,000	$6,660,000	Lockport Dolomite (Silurian), Brassfield Limestone (Silurian)	unspecified	
The Belden Brick Company, LLC	Wallick Pit	Tuscarawas	Ts-1841	IM-45	13,016	13,016	$221,272	Brookville limestone (Pennsylvanian)	unspecified	
The Belden Brick Company, LLC	Shanesville Pit	Tuscarawas	Ts-1939	IM-47	14,906	14,906	$245,949	Brookville limestone (Pennsylvanian)	riprap	
Bluffton Stone Co.	Bluffton Stone Co.	Allen	An-1	IM-485	323,237	315,710	$3,112,898	Tymochtee Dolomite (Silurian)	riprap, portland cement concrete, road construction/resurfacing, asphalt, commercial building	
C. F. Poeppelman, Inc.	C. F. Poeppelman, Inc.	Miami	Mi-10	IM-335	421,658	382,921	$2,910,263	Lockport Dolomite (Silurian)	riprap, portland cement concrete, road construction/resurfacing, commercial building	4755 N. State Route 721
Cardinal Aggregates, Inc.	Cardinal Aggregates	Wood	Wd-C	IM-2231	633,921	627,333	$4,517,000	Lockport Dolomite (Silurian)	riprap, road construction/resurfacing, asphalt, commercial building	8026 Fremont Pike
Carmeuse Lime, Inc.	Woodville - Plant #10	Sandusky	Sy-3	IM-282	886,918	936,774	$6,198,422	Lockport & Salina Dolomites (Silurian)	riprap, portland cement concrete, road construction/resurfacing, asphalt, commercial building, unspecified	1967 County Rd. 42
Carmeuse Lime, Inc.	Maple Grove	Seneca	Sa-2	IM-352	788,775	788,775	$4,141,069	Lockport Dolomite (Silurian)		1967 County Rd. 42
Christman Quarries	Christman Quarry	Monroe	Me-C	IM-1088	48,339	48,339	$447,135	Fishpot limestone (Pennsylvanian)	road construction/resurfacing	47278 Swazey Rd.
Con-Ag, Inc.	Con-Ag, Inc.	Auglaize	Ae-31	IM-744	440,911	443,504	$3,383,239	Cedarville Dolomite (Silurian)	riprap, portland cement concrete, road construction/resurfacing, commercial building	16672 County Rd. 66A N
Cumberland Limestone, LLC	Cumberland Limestone, LLC	Guernsey	Gy-309	IM-1030	122,534	300,508	$3,974,206	Fishpot limestone (Pennsylvanian), Unidentified (Pennsylvanian)	road construction/resurfacing	53681 Spencer Rd.
Custar Stone Co., Inc.	Custar Stone Co., Inc.	Wood	Wd-5	IM-473	439,603	404,786	$4,047,850	Dundee Limestone (Devonian), Detroit River Group (Devonian)	riprap, portland cement concrete, road construction/resurfacing, asphalt, extenders/fillers	9072 County Rd. 424
Duff Quarry, Inc.	Duff Quarry, Inc.	Logan	Lgn-15	IM-453	569,039	569,039	$2,162,348	Tymochtee Dolomite (Silurian)	riprap, portland cement concrete, road construction/resurfacing, commercial building	9042 State Route 117
East Fairfield Coal Co.	Withers-Petersburg	Mahoning	Mg-631	IM-1103	15,780	79,550	$1,065,805	Vanport limestone (Pennsylvanian)	unspecified	10900 South Ave.
Egypt Valley Stone, LLC	Shugert	Belmont	Bt-E	IM-2299	696,457	696,457	$7,756,128	Fishpot limestone (Pennsylvanian)	riprap	
Enon Sand & Gravel, LLC	Enon Sand and Gravel	Clark	Ck-501	IM-340	44	44	$440	Cedarville Dolomite (Silurian)	road construction/resurfacing	11641 Mosteller Rd.
Erie Blacktop, Inc.	Erie Blacktop Quarry	Erie	Ee-27	IM-608	378,499	378,499	$3,639,646	Columbus & Delaware Limestones (Devonian)	riprap, road construction/resurfacing	4507 Tiffin Ave.
Fairborn Cement Co., LLC	Miller, Schieder, Batdorf, Stewart	Greene	Ge-8	IM-364	787,570	787,570	$4,238,493	Brassfield Formation (Silurian)	portland cement concrete, road construction/resurfacing	3250 Linebaugh Rd.

Source: Ohio Division of Geological Survey 2018

FIGURE 5.4 Screenshot from 2018 *Report on Ohio Mineral Industries*

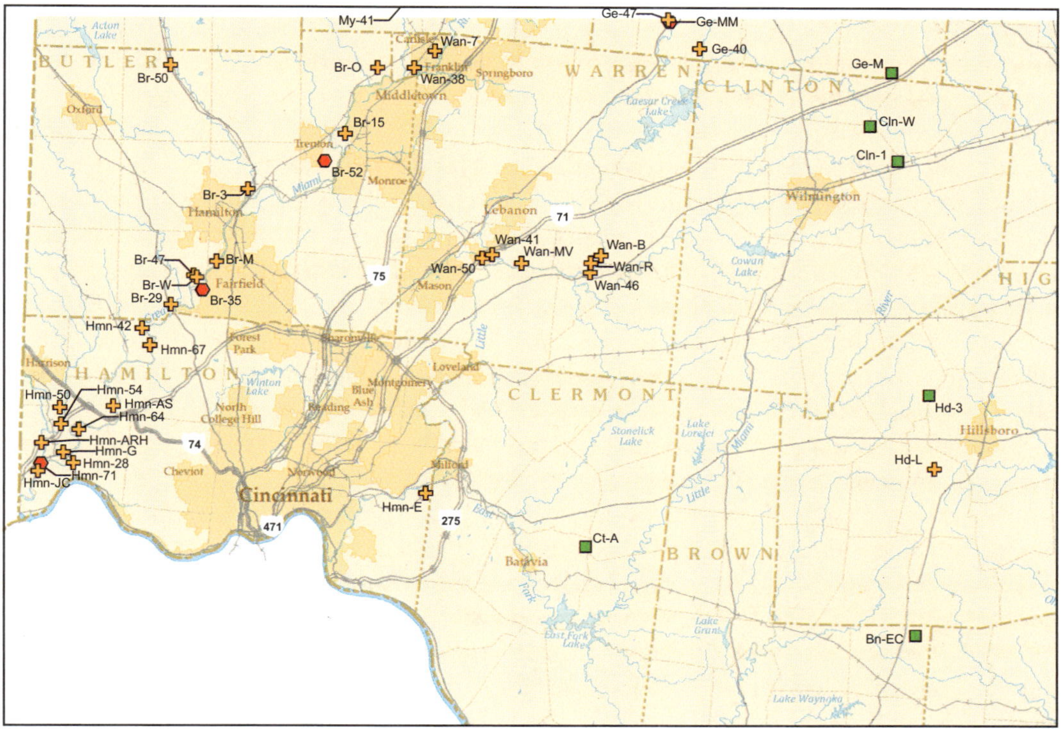

Source: Ohio Division of Geological Survey 2018

FIGURE 5.5 Portion of the map in the 2018 *Report on Ohio Mineral Industries*

mineral industries report, coupled with information from the state geological survey, some population statistics, a quick search of property ownership on a county website, a few basic assumptions, and some rudimentary calculations can yield valuable market research. This one operation in the Cincinnati market area services about 13% of the local aggregates market and probably has sufficient reserves to last for a century or more at present rates of production.

Figure 5.5 is a portion of the map in the 2018 Report on Ohio Mineral Industries showing southwestern Ohio including the Cincinnati area. The map symbols include green squares that indicate limestone and dolomite producers, orange crosses indicating sand and gravel producers, and red hexagons representing sales of multiple commodities. To get a total picture of the market, an analysis of each producer represented on the map would be necessary. Cincinnati is located in the southwest corner of the state and the aggregate market is influenced by producers in both northern Kentucky and southeastern Indiana, which would also have to be taken into account for a total picture. The Cincinnati standard metropolitan statistical area (SMSA) includes a fair amount of population living in northern Kentucky as well.

To exacerbate the market situation in an area like the Cincinnati SMSA, the city is located on the banks of the Ohio River. Several river terminals in the market area import aggregate materials via the Ohio River transportation corridor. Most of this imported limestone aggregate comes from quarries located on the Ohio River in north-central Kentucky and south-central Indiana. Statistics published by the U.S. Army Corps of Engineers (USACE) on river shipments can shed light on quantities of aggregates coming into the market area via the Ohio River.

TRANSPORTATION

Trucks are the most common method of transporting aggregates, especially within a 50-mile (80-kilometer) radius of a pit or quarry. Dump trucks come in numerous sizes and configurations and their carrying capacity depends on the number and configuration of the truck's axles. Small single-axle dump trucks are ideal for delivering small loads to residential or landscaping contractors. Larger multiaxle trucks carry larger loads for delivery to construction sites and industrial customers (asphalt plants, concrete plants, etc.). Interstate highways have a maximum gross weight limit of 80,000 pounds (36,287 kilograms) without requiring special permits. This includes the weight of the vehicle plus its load. The federal bridge law (23 CFR §658.17) determines the number of axles, the load weight per axle, and the axle spacing allowed to cross interstate bridges. Individual state DOTs determine and enforce gross weight limits and axle loads for state highways. Depending on individual state DOT regulations and the number of axles, a single dump truck can transport 22–27 short tons of aggregate. In some states a dump truck is permitted to pull a dump trailer called a pup.

Tractor-trailer-style dump trucks come in various configurations: rear dump, bottom dump, and side dump based on the construction application. Some states allow the use of double trailers pulled by one tractor, which can almost double the hauling capacity. Use of multiple trailers increases the delivery capacity, reduces transportation costs, and is more common in western states where long-distance hauls are more often required. Dump trucks with pups and tractor trailers (single or double) are far less maneuverable than a straight dump truck alone. This can limit the usage of these types of multiple trailers in locations with limited turning areas, small receiving dump hoppers, and so on. Large multitrailer rigs with bottom dump capacity are very efficient on large highway projects, airport runways, and large parking areas, while side dump trailers can be very efficient riprap haulers. Many factors affect the costs, but trucking aggregates generally ranges from about $0.15 to $0.35 per short ton-mile. A 25-short ton load on a 45-mile (56-kilometer) one-way haul at $0.30/short ton will add $337.50 in transportation costs. Assuming an $8.50/short ton cost for material, the stone would cost only $212.50 but the total delivered cost for the load would be $550.00. This is an example of the transportation costs exceeding the cost of the material.

Most aggregate trucking companies, whether company owned or contracted, try very hard to keep their trucks loaded at all times. Truck dispatchers will attempt to route trucks from plant to plant and customer to customer to minimize empty mileage. Backhauls, finding a return load, can substantially increase profitability in the trucking industry. Operating cost of a dump truck can range from $85 to $150 per hour, including labor, fuel, maintenance, tires, insurance, and depreciation. Delivery time includes not only drive time, but also time spent in traffic congestion, at traffic lights, loading time, weighing time, unloading time, and time on the jobsite. An empty return run from a job back to the quarry is very expensive, especially when operating a large fleet of delivery trucks.

Shipping methods, other than by truck, can be very effective in reducing transportation costs of aggregates and reaching much longer distances. Locating an aggregate processing facility on a rail line or navigable waterway can be advantageous depending on what markets can be served via the alternative transportation route. Expanding from a local to a regional or multistate marketing area is possible when shipping by rail or barge. One railcar can move anywhere from 100 to 125 short tons of aggregate, and a single barge up to 1,500–1,800 short tons.

According to the National Waterways Foundation, one tow with 15 barges can move the equivalent tonnage as 216 railcars or 1,050 tractor trailers (National Waterways Foundation 2017). The short ton-miles traveled per gallon of fuel for a barge is 647, for a train it is 477, and for a truck it is 145. The economies of scale should be obvious from these numbers. Environmental issues must be considered as well. The tons of carbon dioxide (CO_2) emitted to the atmosphere for barge transportation per 1 million short ton-mile (1,609,344 kilogram-mile) is 15.6, 21.2 for rail, and 154.2 for trucks. There are more than 12,000 miles (19,312 kilometers) of commercially navigable waterways servicing 38 states in the United States and many more miles of rail.

The three classifications of railroads based on their revenue and mileage of track are Classes I–III. Class I railroads are the large interstate systems like CSX, BNSF, and Union Pacific that carry large volumes of both national and international freight from coast to coast. Class II are smaller, more regional multistate systems and Class III are what is known as short-line railroads that run more specific localized routes that interconnect with the Class I and II systems. Shipping a few rails cars on the large Class I system can be problematic because of the large amount of long-distance freight volume the Class I systems handle. In a commodity business, the rules of the game dictate that the larger the number of railcars shipped, the better the freight rate. Most aggregates are shipped regionally by Class II or short-line railroads. Moving from one rail system to another generates additional switching fees, which can be cost-prohibitive. Unit trains (50–100 cars) carrying nothing but aggregates from Point A to Point B are the preferred method of the railroads, but smaller shipments can be negotiated, especially with the regional or short-line railroads.

Rail loading facilities at the quarry or pit site can be as simple as a large front-end loader or can be much more sophisticated. Large, conveyor-fed hoppers straddling the track make it relatively easy to feed aggregate materials via gravity into the cars awaiting beneath. Systems utilizing direct conveyor feed are also fairly common. I have toured automated coal tipples that automatically load a unit train of coal (150 cars), weigh each car, strike off the top of the load, and spray a dust control agent on the top, all while the train is moving at a steady 5 miles per hour (8 kilometers per hour). The operator needs to exercise diligence in the loading process so as not to overload the railcars beyond the load limit (LDLMT) marked on the outside of the car. Some rail yards are equipped with automatic scales, which will immediately sideline any overloaded cars. Railcars are generally equipped with radio-frequency identification (RFID) cards used to identify the cars along with their load limits and such when entering rail yards. A sidelined, overloaded car requires the car owner to show up at the rail yard and remove the excess weight, which is not an easy task with a load of aggregates.

Railcars capable of handling bulk materials like aggregates come in several designs. Rapid-discharge hopper cars are preferable if high-capacity hoppers are available beneath the tracks at the receiving load-out station. Rapid-discharge cars have a series of double doors across the bottom of the car and steep sides designed to induce rapid flow of material. Standard hopper cars have fewer and smaller discharge doors and flatter sides and tend to unload more slowly. Gondola cars are easier to load because they have lower sides, but they have no bottom-discharge gates and require an excavator or clamshell crane to unload at the delivery end. Specially designed ballast cars that are engineered to deposit ballast along the tracks at a controlled flow rate are mostly utilized for rail maintenance. Bottom-discharge cars can be operated manually using levers or pneumatically with compressed air supplied either from the locomotive or remote air compressors. Car shakers may be necessary to help induce faster unloading depending on the gradation, plasticity of the material, and potential for freezing temperatures. Frozen material can easily occur during early spring and late fall, so the operator needs to keep a keen eye on weather forecasts. A day of rain followed by freezing overnight temperatures can freeze a load of aggregates into a large block, effectively sidelining the railcar for an entire winter season.

Multiple parallel siding tracks with several switches or a long looping track system will be necessary on the loading and receiving end of the rail shipment. Because of traffic concerns, mainline rails cannot be blocked while the railcars are being loaded or unloaded. Depending on the size of the shipment and availability of sidings, the train may need to be broken into multiple segments and the cars separated from the train. A long, circular, looping track can eliminate downtime associated with train breakup issues by continuously moving a unit train over the hopper discharge system. When loading the cars at the point of shipment, it is imperative to keep cars with the same products together to eliminate multiple conveyor changeovers at the discharge site when unloading. The fastest and most efficient rail load-out system is a drive-over hopper design where each car is indexed then discharged into a hopper beneath the tracks with the capacity to hold the volume of 1.5–2 railcars. A feeder under the hopper will continuously feed aggregates onto a discharge conveyor that sequences to a series of stacking conveyors (for separating individual products) or a tripper conveyor system. Substantial steel and/or concrete structures are required to support the weight of the train, the track, and the material load. Portable rail unloading conveyors that slide under the track after removing two ties are utilized in areas with no permanent unloading infrastructure. A switching locomotive, portable Trackmobile, or car winching system may be another necessary capital expenditure if the railroad is unwilling to assist in unloading cars with their engine power.

Hopper cars can be rented or leased from the railroad or other owners, but there are limits on turnaround time and cars that are not perpetually in use subject the lessee to demurrage charges. Outright purchase of railcars is also an option but will require obtaining a four-digit alphanumeric designation from Railinc, a division of the Association of American Railroads. Railinc identification numbers for private car owners all end with the letter X and are used for ownership identity purposes (Figure 5.6). All railcars are required to be stenciled with the ownership ID number, a car number, the LDLMT, tare weight (LTWT), and total weight (TOTWT) permitted, in pounds. All railcars are subject at any time to safety and maintenance inspections by federal railway officials and railroad inspectors, and can be sidelined for any apparent maintenance or safety issue. It is not uncommon to receive a return load of empty cars and find a car or two missing from the train because of some observed maintenance or safety issue. Cars going through a rail yard can be sidelined by an inspector and will not be returned to service until the maintenance or safety issue has been repaired. A build date is associated with each car and there are federal regulations on how long a car can be operated between scheduled maintenance, brake rebuilds, and total longevity of the car, usually 30 years. Whether to lease or purchase rail cars is an internal corporate decision based on capital availability, debt load, cash flow, accounting methods, and so on.

When evaluating the alternative transportation options, it must be kept in mind that, unlike quarry stockpile areas, distribution sites on the receiving end of the transportation system will be required with rail or barge transport, as well as specialized load-out equipment at the quarry. These additional rehandling costs need to be factored into any evaluation. Both rail

FIGURE 5.6 Aggregate car with Railinc ID numbers

FIGURE 5.7 Three-hopper rail unloading pit under construction

and barge loading require specialized equipment at the point of origin. Unloading facilities at the point of sale will need additional capital expenditures for unloading equipment, conveyor lines and stackers, roadway access, real estate, utilities, permitting, and environmental compliance (Figure 5.7). Additional personnel will be needed to load and unload the barges or rail cars and reload products into the customer trucks; clerical personnel will also be required to operate the scaling, ticketing, and accounting functions. All of this comes with additional cost. The economies of scale of rail and barge transport will lower transportation costs, but some of these savings will be offset by the additional costs at the point of sale. The cost per ton-mile to transport bulk material via a railroad ranges from about $0.05 to $0.12 per short ton-mile, not including the cost of loading, unloading, stockpiling, and ticketing.

Moving aggregates to market via water is by far the most economical transportation mode. Barge traffic on the inland waterway system moves massive volumes of aggregates annually (Figure 5.8). The USACE monitors and controls the locks, dams, and navigational features on the inland waterway system with enforcement supplied by the U.S. Coast Guard (USCG). The two principal types of barges that move aggregates are hopper barges and flat deck barges, each having a capacity of 1,500–1,800 short tons. Barges are lashed together and pushed by a tow boat, some of which can have as much power as 25,000 horsepower (18,387 kilowatts).

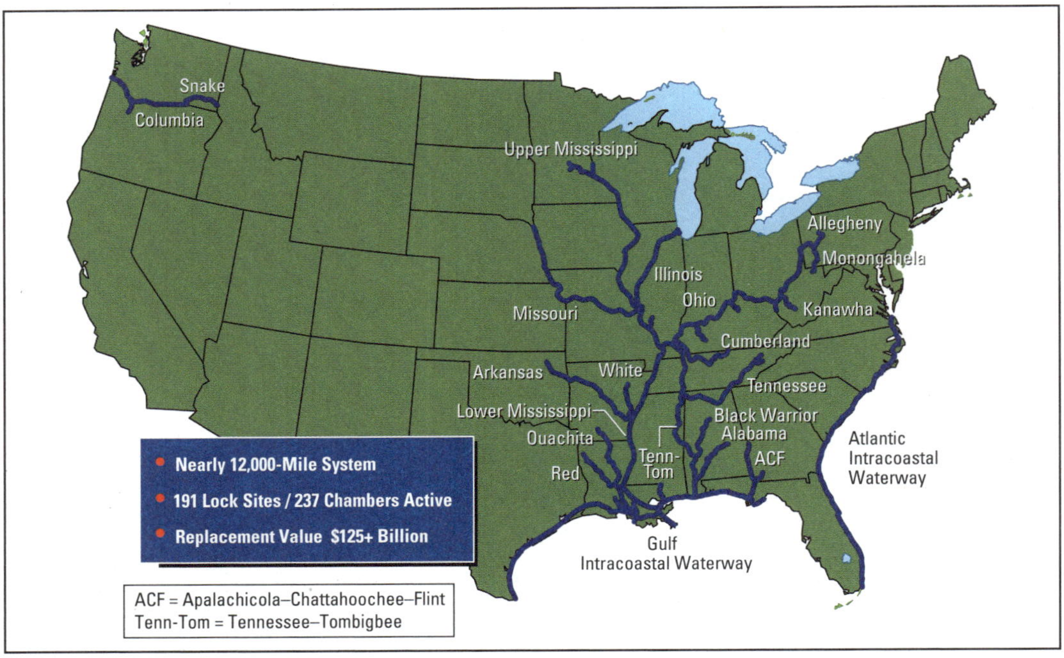

Source: USACE 2000

FIGURE 5.8 U.S. inland and intracoastal waterway system

Tows can be as small as one to four barges on smaller rivers and up to as many as 40 barges on the lower Mississippi River. Most of the major rivers can easily handle tows of about 15 barges (22,500–27,000 short tons). All natural rivers follow meandering, circuitous routes based on the topography traversed, which can substantially add to the miles traveled in the ton-mile cost calculation.

Barge loading facilities can be as simple as a dock on which a dump truck can back up to the water and dump its load directly into a barge. Conveyor system discharge is probably the most common method of loading barges. A conveyor extending out over the water feeds material into a barge as it is pulled forward by a small tow boat or a winch. Loading facilities require a permit from the USACE and most also have a mooring permit for a specified number of barges. The USCG patrols all the inland waterways and is responsible for safety and environmental compliance at loading and receiving docks.

Unloading and receiving facilities need significant infrastructure as do barge loading operations. Whether delivering to a customer site along the river or to a sales yard, a method of unloading the barges is required. This usually consists of either an excavator (some areas require a long-reach machine) or a crane equipped with a clamshell bucket. These machines can be stationary, attached to a river cell, or the mobile type operating on a flat deck barge moored to the shore. The material is removed from the barge and fed directly into a truck or a hopper that feeds a conveyor system branching off to different stockpiles. Much the same as rail cars, barges can either be rented/leased or owned outright by the producer. And, similar to rail cars, demurrage can be costly if the barges are leased and not turned around in short order.

Approximately 630 million tons of freight is shipped on the inland waterway system in the United States on an annual basis. The Mississippi River and its tributaries along with the

Gulf Intracoastal Waterway connect ports, such as Mobile, Alabama; New Orleans; Baton Rouge, Louisiana; Houston; and Corpus Christi, Texas; with major inland ports, including Memphis, Tennessee; St. Louis; Chicago; Minneapolis; Cincinnati; and Pittsburgh. According to the USACE, aggregates account for about 13% of all freight moved annually on the inland waterways or about 82 million tons (USACE 2000). As a side note, a portion of the asphalt cement used by hot mix asphalt plants is shipped via barge and is included in the petroleum products category in Figure 5.9. The average cost per short ton-mile for shipping by barge on the inland waterway system is $0.025–$0.045.

Oceangoing ships are an even cheaper alternative method of transporting aggregates. Much of the Gulf, south Atlantic, and Florida coasts are devoid of commercial deposits of aggregate. Transportation by truck is problematic because of traffic congestion in many of the larger metropolitan areas along the East and West Coasts like New York City, Boston, Philadelphia, San Francisco, and Los Angeles. Many of these metro areas depend on imported aggregates arriving via oceangoing freighters into the heart of these cities. Some of these oceangoing aggregates are produced within the United States, but most are imported from other countries. Large-capacity aggregates mining operations with connected deep-water port facilities export material to the United States from Mexico; Colombia; the Caribbean islands; Labrador, Nova Scotia, and British Columbia in Canada; and Scotland. Approximately 30 million metric tons of aggregates were imported into the United States in 2019 (Table 5.1).

Large self-unloading oceangoing bulk cargo ships with capacities as large as 85,000 metric tons are available to ship aggregates. These vessels have individual holds to separate various product types and an internal conveyor system under the holds that moves the bulk materials to a large on-deck conveyor capable of swiveling and discharging aggregate products into a port facility. For other types of large bulk material, oceangoing vessels can be unloaded via cranes with clamshell buckets into hopper and conveyor systems in port or into barges moored alongside that then shuttle the products to shore. These imported materials travel very long distances but at a minimal cost of $0.01–$0.02 per ton-mile. Very large capital outlays are

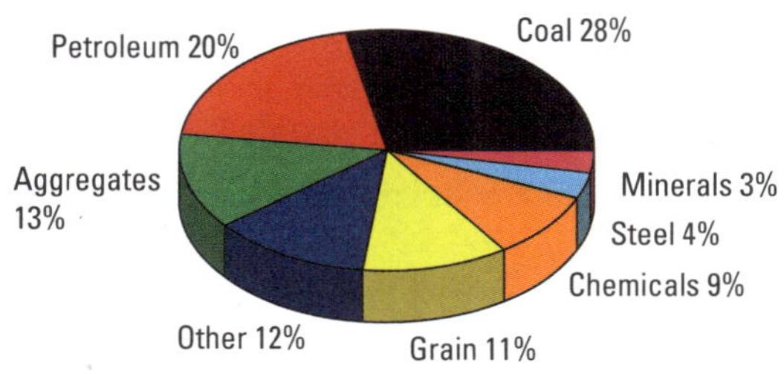

Source: USACE 2000

FIGURE 5.9 Principal commodities shipped on the inland waterway system

TABLE 5.1 Aggregates imports into the United States, 2015–2019

	2015	2016	2017	2018	2019
Crushed stone imports (million metric tons)	20	20	19	21	25
Sand and gravel imports (million metric tons)	4	3	7	6	5
Total aggregate imports (million metric tons)	24	23	26	27	30

required for the loading and unloading of deep-water oceangoing vessels and labor is required on both ends of the shipment. The ships themselves can be owned or leased for one voyage or for multiple voyages over a specified period of time. The shipping distances are long, but unlike the circuitous river routes they can be navigated with relatively straight bearings. This type of vessel is also commonly operated on the Great Lakes system servicing cities like Chicago; Milwaukee; Detroit; Toledo, Ohio; and Cleveland. Several large quarries in coastal areas of Michigan routinely ship to cities on the Great Lakes.

Transportation is a key element of where to locate a successful, profitable aggregate operation. Geology alone cannot be the determining factor. Sometimes it can't be helped that a competitor has a transportation advantage. Customers are located across a diverse marketing region that does not always correspond with where the quality deposits of aggregate can be found. If you want to dominate the local aggregate market, your company has to stand out among the competition. Price is always an issue and many contractors will buy from the lowest bidder (total cost of material plus transportation) unless you offer something more than the competition.

Providing the best customer service and selling the highest quality product at competitive pricing is the key to maintaining a high standing in the marketplace. What do I mean by customer service? Just because you're selling a commodity, don't ever treat your customers like one. Demand a high level of professionalism throughout your organization at all times; from upper management, from the sales/marketing force, from the quality control department, from the environmental people, from accounting, and from the production crew. This level of professionalism not only applies to your customers, but also to your vendors and suppliers, to government regulators, to neighbors, and to the general public.

Customer service means treating your customers with respect (whether you personally happen to like them or not). You will always have external customers (contractors, ready-mix plant operators, etc.), but you also have internal customers (your suppliers, other departments within your company, your fellow employees). The aggregates business has historically been famous for its rivalry between the sales and production departments. This is unacceptable; all departments within your organization should work together toward the same goal: the profitability of the company. All the departments within your company need to communicate regularly and honestly. As leaders, it is the job of management to be laser focused on the issue of everyone working together toward a common goal.

It is imperative that the sales department communicate to production what has been sold, when it is to be delivered, and at what rate. Sales should never anticipate that the production people will automatically have sufficient inventory of a specific product on hand at a specific time without first giving them notice of the sale. Processing plants run most efficiently in a series of different operating modes or combination of products, with distinct products maximized in specific modes. If the plant has been concentrating on coarse concrete aggregate because of extremely high demand, the inventory of base products could be lower than usual. Accurate inventories must be kept at all times and communicated from production to sales on a regular basis. I said earlier in this chapter that location is the key to a successful operation; in this case, communication is the key. It is not difficult to develop a spreadsheet that will calculate inventory on a daily basis as production is reported. Ending inventory + production – sales = starting inventory. If this is recorded on a daily basis in a computer database available to both the production and sales departments, everyone is kept informed. Inventories need to be audited and updated regularly by a survey method to assure accuracy. This will be discussed further in the chapter on operations management.

Never make a promise to a customer that you cannot keep; if you commit to 10 trucks running for 10 hours the following day delivering base product to a highway contractor, you have to make sure it happens, exactly as promised. Contractors are notorious for calling at the last minute and making immediate demands on your resources. You might be tempted to say that you received the order too late in the day and cannot fill the requirement tomorrow, but that will not be acceptable to a contractor that has a fleet of equipment and laborers ready to start work at 5:30 AM. Once you've let a customer down (internal or external), you have breached their trust. The customer's expectation becomes one of failure to follow through and when the next job or project comes up they will be looking elsewhere for service. Always be prepared for the unexpected!

Honesty and integrity are paramount in dealing with your internal and external customer base. Once you lie to a customer, a regulator, or a colleague, you've lost their respect and probably will never regain that relationship again. A small untruth may gain you some extra time or 15 cents extra per ton, but in the long haul will cost you a customer when it becomes evident. And they always become evident eventually. There is never a need for you to belittle your competitors; instead, outdo them with excellent service, professionalism, and quality product.

No aggregate producer is going to be able to service every customer with every product in the market. By virtue of different processing equipment and deposit variations, some customers' needs may not be a good fit for your operation. I remember a case where we installed a new density separation system at a sand and gravel processing plant with the sole purpose of removing lignite from the sand and making three gradations of sand: asphalt sand, concrete sand, and mason sand (Figure 5.10). Density separation provides the ability to make very precise cuts in sands and the product is then reblended to produce the final products. The project was very successful and produced clean, lignite-free sand products. This became a tremendous marketing advantage because our competitor's sand still contained a significant percentage of lignite, considered deleterious by the specifications.

FIGURE 5.10 Density separation plant under construction

Knowing that this equipment was capable of producing precise gradations, a salesman for the company sold several thousand tons of a very tight specification of filter sand to a local municipal water treatment facility at a very inviting price per ton. Were we able to produce the filter sand specification? Yes, we produced a beautiful product. But what the sales department failed to consider was the effect that making this product had on the other sand products at the plant. By pulling about 8% of this size of product out of the sand stream, we were no longer able to produce concrete and asphalt sand that met specifications. These were our bread-and-butter products and the increased selling price of the filter sand could no way make up for the loss of sales of the other products. Additionally, what were we to do with the 92% of the sand production that was out of spec?

Superior quality products are the key component of gaining and maintaining market share. This is where a well-designed quality control plan comes into play. All customers have product expectations, whether it is just a common fill product to level a low spot in their backyard or high-quality aggregate products to make into high-strength concrete. Customers pay a price for the aggregates they purchase and trust the material will perform to their expectations. Material specifications originate from the customer whether they design them themselves or rely on an architect, an engineer, or a number of agencies and organizations that publish end-use specifications for construction materials. The majority of aggregate specifications originate from ASTM International, AASHTO, the Federal Aviation Administration (FAA), USACE, the Federal Highway Administration, and the various state DOTs.

Specific jobs will dictate what set of specifications apply to the materials used. An airport runway will generally fall under the FAA specifications, while a harbor or dam project will most likely be governed by the USACE. Highway construction specifications will require materials to adhere to the state DOT specification book with reference to AASHTO. Most bituminous concrete and portland cement concrete producers adhere to ASTM or state DOT standards. General construction projects will normally require that purchased materials meet state DOT standards unless more specific architectural or engineering standards are required.

To stand out in the marketplace, a producer must always meet quality and gradation specifications. Anyone with a backhoe and property in a floodplain can bail out a pile of sand and gravel. But, without rigorous quality testing and a well-engineered processing plant, that person will most likely not be able to produce a quality aggregate product. There is nothing that will ruin a producer's reputation faster than delivering material to a project and have it rejected because it does not meet specifications. Not to mention the fact that the material has to be removed and trucked away. You might be at a transportation cost disadvantage to a competitor, but if the customer knows you can deliver a quality product on the customer's time schedule and the competition cannot, you'll get the job! The quality of your products and service are your reputation and your position in the marketplace. Customer focus + excellent service + professional employees + quality products = successful aggregates business.

CHAPTER 6
Permitting

The aggregates industry is one of the most heavily regulated in the United States. It would be impossible to list all the permitting requirements applicable to aggregates operations throughout the country because many are regulated at the individual state, county, and even local level. Some federal environmental regulations are enforced by state environmental protection agencies by mandate or contract. To summarize the expanse and scope of the permitting requirements, I attempt to describe them generically by type or area of permitting, for example, air permits, surface water permits, or mining permits. Unfortunately, many permitting requirements and regulations are redundant and enforced by multiple agencies.

In the first paragraph of Chapter 3, I made the following statement: "Locating and profitably exploiting an aggregate deposit is a combination of a minable deposit of quality aggregates (good geology), in a market area with the right customer base, located on accessible transportation routes, and having the approval of the neighboring community." You may have wondered what was meant by "approval of the neighboring community." That is the crux of the permitting issue, being a good neighbor or being forced to be a good neighbor via regulation. In the past, the mining industry, being made up of strong-willed, independent people, was not terribly sensitive to the needs or desires of their neighbors. These miners felt that if they owned the property, they had the right to do as they wished with it. This lack of sensitivity was the reason for the perpetuation of the regulations and permit requirements now in place.

LAND USE REGULATION AND PERMITTING

Let's start this chapter by looking at land use regulations. These are the grassroots regulations that are written and usually administered on a local level, whether it is a county, a township, or a municipality. These regulations vary by state and locality, but very little of the country is without some form of zoning regulation. Land use is generally tied to some type of regional comprehensive plan where a committee, commission, or group of citizens decide that certain areas should be utilized for specific purposes. Comprehensive plans are generally required to be updated periodically (every 5 or 10 years) and put out to a public hearing so the voters can evaluate and comment. These plans generally define specific use areas such as agricultural, office, commercial, industrial, and residential, with numerous subdivisions within each use category (such as single-family and multifamily residential and light or heavy industry). Many of these specific use areas were historical. When the first plan was written, the existing area that contained the central business district was defined as office and the shopping areas became commercial districts.

Once the comprehensive plan is established, a series of regulations are promulgated defining what uses are permitted within each area or district. This is what becomes known as the zoning code. Most zoning codes are comprehensive. They include regulations defining housing density, lot size and dimensions, setbacks from property lines for buildings, building construction standards, nuisance standards, permitted uses, nonpermitted uses, utility requirements, and so on. It is important to remember that most land use regulations vary considerably depending on locality. Case law as a result of various legal actions over the years will also affect the interpretations of these regulations. A good understanding of the local zoning code is essential prior to venturing into any zoning case. Many people see zoning regulations as an infringement of their civil rights (much like the old-time miners), but that is a subject for some other book.

Most zoning regulations contain a category called nonconforming use. This is generally a land use that was in existence prior to the promulgation of zoning regulations and does not conform to the standards for that specific district. Because this specific use predated the regulations, it could not legally be eliminated. To be considered a nonconforming use, the use must have been in existence prior to the enactment of the zoning regulation prohibiting the use, and the use must have been lawful when it commenced. Aggregate mining is a long-term land use where many years of reserves are required to justify the large capital expenditure requirements. Large tracts of land were purchased to enable producers to have a substantial reserve of minable material. It is not unusual to see mining operations categorized as nonconforming use. However, some courts have ruled that if the land was not actively being mined at the time the zoning was enacted, the nonconforming use designation is not applicable to the nonmined reserves. Nonconforming use is sometimes known as pre-existing use, or the property is said to be grandfathered.

Some zoning codes will include land use areas specifically defined for mining purposes. In some cases, this is based on current geological knowledge, or more than likely the location of existing mining operations defines such districts. Most zoning codes contain a provision for what is known as a conditional use or a conditional use permit. A conditional use is a use that can be authorized by the zoning authority in an area not specifically designated for that use, but with distinct conditions attached and agreed upon by the property owner. Most zoning codes will have a list of permitted conditional uses within each zoning district. It is not unusual to see mining listed as a permitted conditional use in agricultural, floodplain, and heavy industrial zoning districts.

A conditional use permit is obtained through an application process, usually in a format provided by the applicable zoning authority. Once the application is submitted, there will be a specified period of time where the application is reviewed for completeness. Normally an application fee is required at this time. The application is usually routed for comment to local administrative departments like the board of health, local transportation district, state department of transportation, local environmental authority, police and fire departments, regional planning, or county engineer's office. When the review process is completed, a public hearing will be scheduled in front the board of zoning appeals (BZA). This type of public hearing is generally considered adjudicatory, where witnesses are sworn in and subject to cross-examination. This is done in an effort to eliminate speculative testimony and to present only evidentiary fact.

In considering a request for a conditional use permit, the BZA will want the intended use to:

- Be adequately served by existing public facilities (police, fire, ambulance, schools, etc.)
- Not require additional cost to the taxpayers

- Have adequate transportation facilities for the anticipated traffic that will be generated
- Be in compliance with building codes and standards of the zoning district
- Not disturb existing and future uses in the area
- Be environmentally and visually compatible with the surrounding area
- Not be a strain on the existing utilities (power, water, sewer, etc.)
- Not be deleterious to the general welfare by producing smoke, dust, odor, fumes, and noise
- Not to destroy historic, natural, scenic, or environmentally sensitive areas
- Be designed and maintained in accordance with objectives of the zoning resolution

Following the public comment, both pro and con, the BZA will close the hearing, deliberate in public, and ultimately vote on whether to grant a conditional use permit. A simple majority is needed to grant or deny this ruling. During the deliberation period, the BZA will decide on any number of conditions that will be required of the applicant if the permit is issued. If a conditional use permit is granted by the BZA, it becomes a contract where the permit applicant must agree to abide by the conditions outlined. Depending on the state and local jurisdiction, the BZA's decision may be only a recommendation that needs to be approved at another level of authority like the county board of zoning commissioners or the board of county commissioners. This entire process is very time consuming, expensive, and legally intensive, and can take from 6 months to more than a year depending on the contentiousness of the issues. As in any legal process, the ruling can be appealed to a district court (or higher) by either the applicant or the opposition.

If the area in question is located in a zoning district that does not specifically allow mining or allow it as a conditional use, a zoning amendment can be requested. Zoning amendments are legislatively granted changes in the uses permitted within a particular zoning district. The process of acquiring a zoning change or amendment is similar to acquiring a conditional use permit but considerably more complicated and difficult to accomplish. In this case, the applicant is asking the local authority to change an existing law rather than just issue a permit with conditions that have already been allowed within a specific zoning district. Depending on the local jurisdiction, this process will probably require several public hearings before different boards and cost substantial monies in legal fees.

Many zoning codes also allow for another type of requested change known as a zoning variance. A variance is an administratively granted waiver of a specific restriction of objective standards like area or setback limitations, structure heights, or additional outbuildings. I remember a specific example of a required zoning variance that we needed at a sand and gravel operation. We were putting an addition onto an existing production facility that consisted of a density separation plant for sand production. The density tanks were long, tall structures, with cyclones and dewatering screens underneath that fed onto transfer conveyors. The final structure was about 105 feet (32 meters) tall when completed. The local zoning code stated that no structures taller than 85 feet (26 meters) were permitted in this district. This area was in the floodplain of the Ohio River and was zoned for mineral extraction and processing. It just so happened that a major power distribution line crossed the property very close to the proposed plant addition. I had our survey team determine the height of an already existing power distribution tower on the site. When I went to the hearing before the local BZA and showed them that the power distribution tower was 215 feet (65.5 meters) tall, they granted a variance with very little need for discussion (Figure 6.1).

FIGURE 6.1 A 215-foot (65.5-meter) power tower in a floodplain

Although this process seems somewhat straightforward on the surface, have no delusions: acquiring a conditional use permit or a zoning amendment is a complicated political process fraught with many dangers along the way. In filing an application for a conditional use permit or a zoning amendment, more information is better on the application. Most zoning application forms consist of one or two pages with generic questions like "Is the intended use compatible with the objectives of the zoning code?" Or "What additional services will be required for the proposed use?" Do not assume the bureaucrats reviewing your application have any knowledge whatsoever about the aggregates industry. Nor should you assume that the members of the BZA have any familiarity with an aggregates mining operation. Most will more than likely have heard all the negative issues surrounding quarries and gravel pits and have no idea of the importance and necessity of aggregates.

Add addendums to your application explaining in detail the nature of the proposed operation, the environmental standards that will be adhered to, the process of extraction, what other regulations apply to the operation, what safety standards will be applicable, how you plan to protect the value of the neighbor's property, and so on. Add a statement summarizing the importance of the industry, the economic impact, the potential tax revenue generated, and the savings (in transportation costs) to the taxpayers of having a local source of aggregates. I have completed several zoning applications that were 50+ pages long containing several figures and full-sized maps when the application form itself was only two pages of questions.

This tends to be an educational process where it is imperative the regulators are better informed than the public that will speak in opposition. If possible, prior to even submitting an application, attempt to meet with the county commissioners, township trustees, and/or BZA members to establish your credibility and let them know an application is coming. Most county or township websites will list the names and contact information for all the players. This can also be accomplished by attending chamber of commerce meetings, other public hearings, political rallies, or sit-down meetings at your request. For the same credibility reasons, it is good idea to meet with the adjoining property owners individually and discuss the proposed project well in advance of a public hearing. Don't let the immediate neighbors first hear about the project with a notice from the BZA. You may not convince them that they want a quarry as a neighbor, but at least you will know what their concerns are and what issues to address at the public hearing.

Generate a comprehensive mining and reclamation plan to present with the application and at the public hearing with stages showing where the quarry limits are anticipated to be in 5, 10, 15, and 20 years. Be sure to include a site barrier or landscaping plan to minimize the visual impact to adjoining homes, highways, and so forth. Indicate on the plan how surface drainage will be routed, the location of environmental mechanisms like settling ponds to treat wastewater, observation wells to monitor groundwater levels, and tree lines and earthen mounds to mitigate dust, noise, and visual effects. Include a transportation plan showing the major truck routes to and from the operation and how the entrance road is to be designed with line of sight and any necessary acceleration or deceleration lanes to be added to the highway (Figure 6.2). It is often difficult to decide on a specific reclamation plan when the future use of the property is 30 years in the future, but provide a proposed reclamation plan that is harmonious with the surrounding land uses.

A contentious zoning issue existed at one of our quarries. Reserves were running low and we had acquired rights to the property across a very large stream from the existing quarry that happened to be located in a different township and zoning jurisdiction. Fifteen years earlier, the previous owner of the quarry had attempted to rezone the same property and was denied the zoning change based on some outright hostile public hearings. The existing quarry had been in operation for about 30 years with well-known, established, and accepted truck routes. Access to the property across the stream would have been down a 2 mile (3.2 kilometer)–long lane lined by very nice rural residential mini-farms. This residential area was, of course, a major obstacle to the zoning change. We proposed leaving the processing plant and the truck route the same as it had been historically and to build a bridge across the stream for conveyor

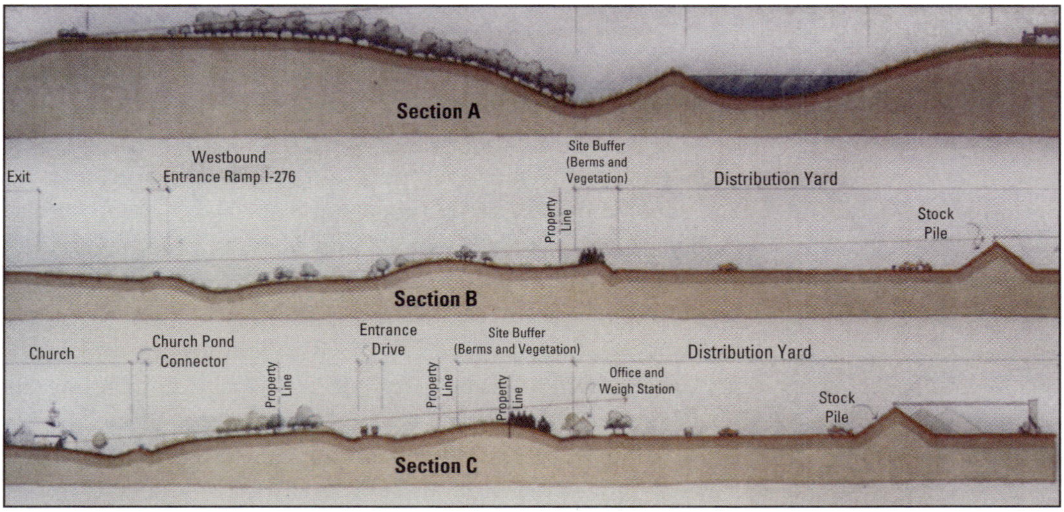

Original source unknown

FIGURE 6.2 Cross sections showing site barriers and visual line of sight

FIGURE 6.3 Conveyor access bridge under construction

and worker access to the new quarry site (Figure 6.3). After multiple public hearings, we were able to establish our credibility and convince the public to allow the operation. Because of the bridge span involved, this was not an inexpensive solution to the problem, but we extended our reserve base in a well-established, profitable market.

Always solicit other interested parties, like potential customers, county engineers, supportive neighbors, and local suppliers, to attend the public hearings and speak in support of the project. There will be many people willing to speak out against you at a public hearing, but unless you specifically ask for support, it will be conspicuously lacking. At most public hearings, the board will ask specifically to hear from those opposed and separately to hear from anyone in support of the project. It is not wise for your company to be the only citizens supporting the project at the public hearing. Be proactive in asking for support far enough in advance that your supporters can put the meeting on their calendar and don't be shy about reminding them of their commitment to assist at the meeting a few days beforehand. Most people are reluctant to speak in public and would rather conveniently forget about a meeting unless reminded. It is also helpful to have a number of your employees show up at the meeting wearing a shirt with a company logo to show support and give the public a sense of the people depending on the quarry for their livelihood.

Be familiar with all the issues that are likely to come up at the public hearing. Many will be evident after meeting with the neighbors. Questions concerning truck traffic, dust, noise, operating hours, environmental compliance, effects on property values, blasting, and surface and groundwater usage are all obvious issues. Address the potential issues in your presentation to the BZA prior to hearing them come from the opposition. If you don't have employees capable of answering technical questions, hire experts (consultants) that can address the queries. I know of aggregate companies that hire public relations firms to assist with zoning issues. Yes, this can all be very costly, but you cannot afford to be perceived as incapable of handling the technical questions or ignorant of the issues. Make sure to emphasize the regulatory scrutiny to which you will be subjected: U.S. Environmental Protection Agency (USEPA), state EPA, Mine Safety and Health Administration (MSHA), Homeland Security, U.S. Army Corps of Engineers (USACE), state mining reclamation inspectors, and so on. Help the BZA and the public to understand how tightly you will be regulated and scrutinized.

During a specific quarry zoning process, one of the hot-button issues was property values. Many of the immediate neighbors were extremely concerned about their properties being devalued by a quarry operation in the local vicinity. In an effort to pacify the neighbors' fears, we offered to sign contracts with them concerning their property valuations. We proposed having two appraisers, one specified by the company and the second by the owner, evaluate the property value and average the results. This was to be the original value of the property. We then offered to sign contracts with the owners inflating the property values annually based on the Consumer Price Index (CPI). If the owners decided to sell the real estate and could not get any offers equal to the original value inflated by the CPI, we would purchase the real estate at the inflated value. This offer was made to all property owners within 1 mile (1.6 kilometers) of the quarry. Several property owners signed the contract but others deferred. This contract offer nullified the property value issue and the zoning was ultimately accomplished. This became a standard offer in several subsequent zoning cases.

Any proposed mining operation, especially one located within a metropolitan market, is going to generate a lot of public interest, scrutiny, and media coverage. The American dream of home ownership is being directly threatened by an unknown entity (your quarry) attempting to enter the neighborhood. And with media coverage, that neighborhood can become an extremely large one. You can count on the fact that the public hearings will be contentious and at times the crowd can get very hostile. The perceived threat to someone's home is a very emotional issue. I've been at several public hearings where the meeting had to be moved to a larger venue like the high school auditorium because the local hearing room could not accommodate the crowd.

The public perception of mining in any form is extremely negative. A 2015 survey in the UK concluded that 75% of people would oppose a new quarry in the local area and only 10% would be likely to support it. In most surveys in the United States, pits and quarries rank only slightly behind landfills as the most despised land use imaginable. This is partly because of the public's ignorance of the need for, uses of, and importance of aggregates in their daily lives and partially because of the industry's past practices and some industry bad actors. The negative perception is also exacerbated by association with other mining industries like coal that have, in the past, left vast landscapes of unreclaimed land scarring certain parts of the country and contaminating streams with acid mine drainage. Thus, the need for gaining the approval of the neighboring community.

The term *NIMBY*, derived from "not in my backyard," has been applied to people that are antidevelopment. The modern world, where almost everyone has access to the internet, has provided a new level of scrutiny by the opposition at a hearing. Anyone with a computer can obtain vast amounts of information, both positive and negative, and quickly become an "expert." Any minor infraction of an environmental rule by the applicant, even if 15 years ago, can and will be brought to light with a little research at the state EPA website and be used against you at a hearing. The MSHA website lists all accidents, orders, and citations ever issued to a producer, which will bring additional scrutiny of your past safety record. Any news article pertaining to your company, parent company, the companies' owners, board members, or subsidiaries, whether positive or negative, is also fair game. Be prepared to address any past bad press and to utilize the good press to your advantage.

MINING RECLAMATION REGULATIONS AND PERMITS

In 1977, the federal government passed the Surface Mining Control and Reclamation Act, which established the Office of Surface Mining Reclamation and Enforcement. This legislation has provided reclamation for thousands of acres of abandoned coal mine lands and regulates the reclamation of active surface coal mines. No federal laws requiring reclamation of aggregates properties exist at this time. Many but not all states have enacted their own legislation to regulate the reclamation of industrial minerals mines. The majority of these laws are fairly similar in nature and initially require the operator of a surface mine to obtain a mining permit. These permits require detailed maps showing the location of the operation and what county, township, or municipality has jurisdiction. The majority of state permits are interdependent with the local land use regulations discussed previously, requiring proof of zoning approval for the specific operation.

Most mining permit applications will require an extensive application map (Figure 6.4), made to a specified scale, with details that include the following:

- Location coordinates or latitude and longitude
- North arrow, map scale
- Number of acres owned and to be affected
- Topography and surface features
- Property lines and easements
- Identification of adjoining property owners
- Buildings, roads, highways
- Drainage details including streams and ponds
- Utility lines, pipelines, transmission lines
- Geological information
    - Geological formation to be mined
    - Thickness of the deposit
    - Thickness and character of overburden
    - Depth to groundwater
- Cross sections showing
    - Premining topography
    - Postmining topography
    - Depth of proposed excavation
    - Limits of proposed excavation
- Sequence and direction of mining
- Areas of proposed overburden storage
- Proposed water impoundments
- Proposed sumps, settling ponds, and outfalls from dewatering

A map of the reclaimed property is required by some states as well. In addition to the detailed maps, an application form will be required asking for information on the mine operator like corporate structure, management, ownership, directors, statutory agent, and a listing of the applicant's other mining permits. The application will generally require a statement of intended future use of the property post-reclamation. Many applications ask what methods will be utilized to protect the adjoining property owners. Liability insurance requirements will be outlined in these documents and an insurance certificate required. Most laws require that

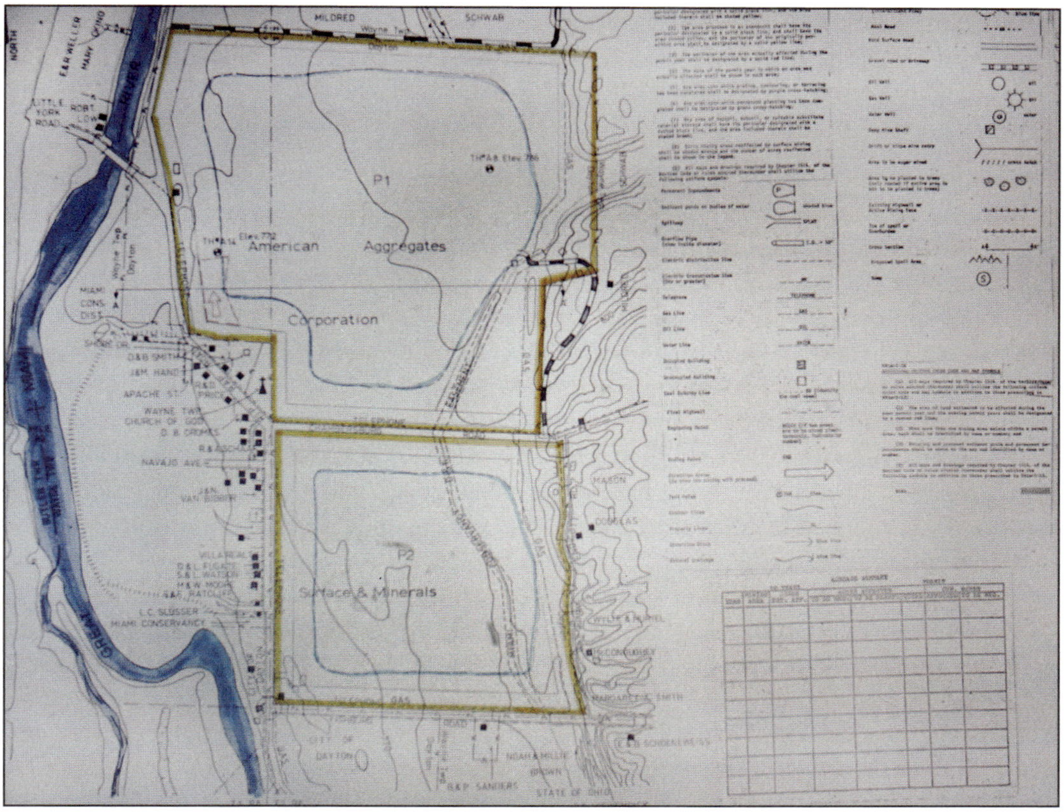

FIGURE 6.4 Typical surface mine permit map from Ohio

the acreage affected be bonded at a set rate per acre. Bonds can normally be supplied as a cash bond to be held by the state until reclamation is completed or a surety bond from a recognized bonding agency. After submission of the required fees, insurance certificates, bonds, and approval of the application, the mining process can begin. Depending on the exact requirements, this process will normally take several months.

Rules will be promulgated following the actual legislation that outlines the reclamation methods, environmental requirements, and what is deemed necessary to protect the adjacent properties. These rules will include some or all of the following:

- Setbacks (mining limits) from property lines, highways, easements, buildings, streams, and so on
- Limits on the placement and storage overburden
- Restrictions on groundwater removal
- Control of surface drainage
- Sediment and erosion control
- Management of slopes and highwalls
- Control of fugitive dust emissions
- Restrictions on flyrock and vibration from blasting
- Certification of blasters
- Storage of explosives

- Limits on the slopes of final reclaimed land
- Permissible above-water and below-water slope angles in water impoundments
- Height restrictions of highwalls in hard-rock quarries
- Requirements and growing time limits for replanting vegetation upon reclamation

Annual (progress) reports are required by many states and inspections are mandated once or twice yearly by inspectors employed for that purpose. Most mining permits will be limited to a specified period of time (10 years, 20 years) and will have a specified renewal process after that time. If the operation is sold to another producer, the permit must be legally transferred and approved by the state agency. There is usually a method to amend these permits if mining circumstances change or additional land is added. Upon completion of mining and reclamation, a final report and bond release will be required. Most permits allow for interim bond releases if an operator can prove a certain portion of the property has been adequately reclaimed and vegetation re-established. Each step of the process (application, annual report, permit renewal, amendment, bond release, and final report) will generally require additional fees paid to the regulator. Most state reclamation laws include the authority to permanently shut down an operation and revoke the bond for violations of the law.

Groundwater is a controversial issue in many areas of the country. In 2009, the state of Ohio amended its reclamation law to include language protecting adjoining property owners' right to groundwater. Prior to a mine operator dewatering a pit or quarry, a study modeling the effects of the groundwater withdrawal and subsequent cone of depression needs to be completed. The mine operator may complete and submit a groundwater model of the projected cone of depression or the Ohio Division of Water (ODW) will complete a model at the request of the Division of Mineral Resources Management (DMRM). A model submitted by an operator will in any event be reviewed by the ODW prior to acceptance. A computer model entitled MODFLOW developed by the U.S. Geological Survey (USGS) is the generally accepted modeling software.

If a property owner located within the modeled cone of depression experiences "diminution, contamination, or interruption of that water supply" (Ohio Revised Code Chapter 1514.13 [B] [2]) and complains to the operator or DMRM, the mine operator must, within 72 hours of receiving a written complaint, replace the water supply with a temporary supply of similar quality and quantity. There are a number of steps following the temporary supply where the ODW becomes involved to determine the actual cause of the problem. If the investigation determines the mine operator's dewatering is the cause, the mine operator is ultimately responsible to provide a permanent source of replacement water. This is just one example of some of the variations in mining and reclamation legislation that exist from state to state.

The state of Indiana has no mining and reclamation legislation that pertains to aggregate producers. About 20 years ago, the talk of pending legislation prompted the Indiana Mineral Aggregates Association (IMAA) to author its own guidelines for mining and reclamation. The IMAA founded a subsidiary group called the Environmental Stewardship Council (ESC) that is required to follow the IMAA guidelines for mining and reclamation. The association administers and provides personnel that annually inspect its members' operations for compliance. All producing members of the IMAA are required to join the ESC. The membership is encouraged to contact local zoning authorities and request that the ESC guidelines be incorporated into local zoning codes as requirements. The membership has had limited success with certain zoning authorities. This is an example of a proactive approach to reclamation that has, so far, eliminated one level of governmental management by imposing a form of self-regulation.

A comprehensive understanding of the mining and reclamation regulations that apply to your area of operation is an absolute necessity. These regulations will ultimately affect your operating costs as well as the public perception of you as a professional operator.

FLOODPLAIN PERMITS

A vast majority of commercial sand and gravel mining operations are located adjacent to major rivers and streams, which are the genesis of the sand and gravel deposits. Numerous hard-rock quarries are also located in low-lying valleys where overburden has been minimized by erosion. The subject of allowing development in a floodplain can be a confusing issue because the regulations originate at the federal level but are administered locally and not always by the same local agencies. The National Flood Insurance Program (NFIP) is administered by the Federal Emergency Management Agency (FEMA) but is coordinated and enforced by state and local governments. The NFIP is a voluntary program under which local communities and FEMA agree to enact and enforce floodplain management regulations in exchange for federally subsidized flood insurance. Subsidized flood insurance is only provided to property owners in areas where the local government has opted into the NFIP program.

The main objectives of the NFIP are that new buildings not be affected by flood damage and that any new development in a floodplain not increase the historical flooding levels. Participating local communities have the responsibility to regulate all development within identified flood hazard areas in compliance with minimum flood damage reduction criteria established by the NFIP. The federal government identifies flood hazard areas within the individual communities by developing flood hazard boundary maps, flood boundary and floodway maps, and the flood insurance rate map (FIRM). These maps identify the 100-year flood elevations, the 100-year floodplain boundary, and the 500-year floodplain boundary, and delineate the floodway. The FIRM also includes designated insurance rate zones. Areas identified as Zone A on these maps are considered to be flood hazard areas. FIRMs for areas participating in the NFIP are generally available online and at the local office of floodplain management. The area located within the 100-year flood profile is considered to be the floodplain and any development in a floodplain, such as aggregate operations, is therefore within the regulated area. The 100-year flood is a flood that can be expected to happen once in 100 years or has a 1% chance of happening in any individual year, and a 500-year flood has a 0.2% chance of happening in an annual cycle. Unfortunately, Mother Nature does not produce floods on a regular schedule. It is possible to have a 100-year rain event twice in the same week. Over the last decade, there has been a trend of more severe weather events, which some meteorologists believe to be a result of climate change and a harbinger of trends to come.

A floodway is defined as the area of the floodplain that needs to be left unimpeded to allow for the free movement of floodwater and generally contains the deepest, highest-velocity flood flow. The floodway fringe is a storage area for generally shallow, slow-moving flood water. Development is usually permitted in the floodway fringe as long as buildings are constructed above the designated 100-year flood elevation. The NFIP requires that a local official be assigned to implement and enforce local management regulations and administer a system to issue permits for development. In many cases, this is accomplished via local zoning regulations and zoning officials on a county administration level. A development is defined as any human-made alteration in the floodplain, which includes buildings or changes in the landscape such as filling or excavations. Any alteration that will increase the flood elevation or alter an identified flood hazard requires a permit.

HEC-2 is a software program developed by the USACE to model the hydrologic and hydraulic effects of alterations in a floodway. An operator applying for a development permit in a floodway will likely be required by the local floodplain administrator to complete a HEC-2 study to determine what effect their proposed processing plant, excavation, fill, roadways, levees, overburden stockpiles, and so on, will have on flood flows and levels. The study needs to prove that the cumulative development will cause a rise of less than 1 foot (0.3 meters) in the flood elevation upstream from the proposed development.

This may seem to be a daunting feat, but it generally is doable. Excavations actually increase the efficiency of the floodway by removing material from the floodplain. The material removal process normally exceeds the impediments caused by processing plants, fill, levee construction, and overburden piles. Acquiring a development permit does not eliminate the operator's risk of flood damage to processing equipment, mobile equipment, and so on, in the event of a flood. Plant designs should be engineered to protect critical equipment such as electrical transformers, motors, and electrical switchgear, raising them above the highest potential elevation of floodwaters. Management should be constantly aware of the weather forecasts, storm events, and potential flooding in anticipation of moving mobile equipment to high ground on short notice. Floating equipment like dredges, service boats, floating docks, and pump platforms need to be adequately secured to shore in the event of unexpected extreme increases in water levels.

An operator planning a new development in a stream valley should research the available FEMA maps to determine if the property is in a regulatory floodway. If so, it will be necessary to determine who the local floodway administrator is and meet to determine if a development permit is required. Upon receipt of a development permit application, the administrator will determine the need for a HEC-2 analysis. Various consulting companies have the ability to perform HEC-2 analyses.

WETLANDS ISSUES

"Wetlands are areas that are inundated or saturated by surface or ground water at a frequency and duration sufficient to support, and that under normal circumstances do support, a prevalence of vegetation typically adapted for life in saturated soil conditions. Wetlands generally include swamps, marshes, bogs, and similar areas" (USEPA 2022c). This is the accepted definition of a wetland by the USEPA and USACE, the principal federal agencies tasked with enforcing Section 404 of the Clean Water Act (CWA). Other federal agencies have limited regulatory authority over wetlands; the U.S. Fish and Wildlife Service and the U.S. Department of Agriculture Soil Conservation Service (SCS). Many states have been delegated the authority to administer the federal wetland regulations for the USEPA and/or have additional state regulations governing wetlands.

Wetlands are protected and regulated because they provide a vital role in our ecosystem (Figure 6.5). Wetlands are a filter source for absorbing excess nutrients, sediment, and pollution prior to reaching rivers and streams. They also slow and lower flood levels by absorbing excess waters that overflow from rivers in times of high flow. Thousands of aquatic and terrestrial species of plants and animals rely on wetland habitats for their very survival. Wetlands

FIGURE 6.5 Example of an obvious wetland

also provide a multitude of recreation and educational opportunities such as hunting, fishing, bird-watching, hiking, and boating.

To be determined a wetland, an area must have all of the following three characteristics: (1) the presence of water at or near the surface, (2) wetland vegetation, and (3) hydric (wetlands) soil types. Swamps, bogs, and marshes are fairly easily recognized by most people, but there are other, much more subtle wetland varieties that require an expert to recognize. Most bogs, swamps, and marshes are thought to be located in low-lying areas, but a common type of upland wetland also exists and to the untrained eye appears to be a normal forest environment. Over the years, prior to wetland regulations, many thousands of acres of wetlands were drained to facilitate development and enhance agriculture. Miles of underground drainage pipes exist that were installed by farmers over the last 100 years to dry up wet spots in their fields. These predrained fields still contain hydric soils but are generally exempt (grandfathered) from the permitting process.

Individual county soil surveys published by the SCS will have a listing of hydric soils and are a good place to start when attempting to define wetlands on your property or a parcel you wish to acquire. Unless you have an exceptional grasp of agronomy, hydrology, and botany, it is usually a good idea to hire a wetland expert if you think a wetland might exist on your property. An expert on wetland delineation will be able to identify the soil types, wetland vegetation, and hydrology necessary to determine and identify the quality and size of a wetland. They will also be knowledgeable in the regulations in your state, and have contacts within the USEPA and USACE to facilitate the Section 404 application process if you decide your activity will ultimately affect a wetland. Certain exemptions and nationwide permits exist that cover specific industries and situations and should be investigated prior to applying for an individual Section 404 permit. A nationwide permit could streamline the process if one is applicable.

A permit will be required if you intend to mine through marshy areas, construct levees, store overburden, build roads, or drain a marshy area. There are basically three legal options when encountering wetlands in a mining operation: avoidance, minimize the impact, or mitigate and monitor. The decision on how to handle a wetland on a mining property will be a financial one. Is it less costly to avoid the area and leave a portion of the mineral reserve in the ground or mine through the wetland and mitigate? Mitigation will take the form of restoring the wetland after impact, creating a new wetland elsewhere on your property, creating a wetland on another property within the same drainage basin, or purchasing wetland acreage from a wetland mitigation bank. Depending on the location and quality of the wetland affected, the regulatory agency will likely require compensatory mitigation at a ratio of from 1.5:1 up to 3:1 times the acreage affected. If you destroy 10 acres (4 hectares) of wetland, you will be required to create 15–30 acres (6–12 hectares) as mitigation. The USACE will also require a monitoring plan for a period of time to guarantee successful creation of a self-sustaining wetland ecosystem. Wetland creation is a tricky and costly undertaking; the responsibility can last for many years.

An isolated wetland is defined as a nontidal wetland that is not connected to or tributary to the navigable waters of the United States. The USACE assumed authority over isolated wetlands based on the fact that migratory birds utilized these wetlands when migrating. A 2001 U.S. Supreme Court ruling known as the SWANCC decision restricted the USACE from regulating isolated waters of the United States. Many individual states, however, immediately enacted new legislation to regulate isolated wetlands on a state level. A review of your state wetland regulations is generally prudent when encountering isolated wetlands on a mining property.

The Tulloch Rule is another significant piece of legislation affecting wetlands. The USEPA and USACE attempted to expand their authority in 1993 to regulate certain developers that were attempting to avoid Section 404 permits by digging drainage channels up to the edge of wetlands without technically affecting the wetland itself. The Tulloch Rule expanded Section 404 to include "any addition (of fill material) including any redeposit" into a wetland even in "de minimus" amounts (Genovese 2000). A U.S. Court of Appeals (COA) invalidated the Tulloch Rule in 1998. The COA ruled that under the CWA the USEPA and USACE have no authority to regulate depositing fill materials in wetlands if they resulted in only "incidental fallback" (*National Mining Association v. United States Army Corps of Engineers* [145 F.3d 1399, 330 U.S.App.D.C.329, 1998]). At this time, excavation and land clearing can be done in a very carefully controlled way to a wetland without a 404 permit if the operation results in only incidental fallback. This is known as the one-step method where the excavation is completed by an excavator loading directly into a truck with no spillback, no stockpiles, no windrows, and so on, within the boundary of the wetland.

Changes in the CWA are in progress as I write this section and I personally am not a wetlands expert. When dealing with these issues in the past, I have always hired a knowledgeable consultant. During the Obama administration, the CWA was widely modified, with a definition of *waters of the United States* broadened as well as the authority of the USEPA and USACE. Numerous lawsuits quickly followed and the revisions were held up in various federal courts for several years. The CWA was revised again by the Trump administration and this definition reverted to the original, covering only navigable waters and their tributaries. More changes to the CWA are in progress and there will surely be more precedent-setting lawsuits. The Biden administration will no doubt have its own ideas concerning the CWA.

There are potentially severe financial penalties for violating Section 404 of the CWA. Prior to purchasing a new property, it's a good idea to pursue a wetlands evaluation so you know up front what obstacles that you face. If you're going for zoning on a new property, rest assured your application will be reviewed by various environmental agencies and if wetlands exist the regulators will be asking how you plan to deal with the issue.

Photo by James Thomas. Cleveland State University. Michael Schwartz Library. Special Collections. Cleveland Memory Project.

FIGURE 6.6 Cuyahoga River fire on June 22, 1969

Avoidance, minimize the impact, or mitigate and monitor are the options you will face. Although not required when filing for a Section 404 permit, a knowledgeable, certified wetlands expert can be an invaluable resource.

SURFACE WATER PERMITS

The previous section summarized the application of Section 404 of the CWA; in this section we discuss Section 402 of the CWA, which focuses on the prevention of pollution of surface water (streams, lakes, and rivers). I vividly remember as a college student the day the Cuyahoga River caught on fire in downtown Cleveland in 1969 (Figure 6.6). This was a sobering reminder of the environmental abuse that humans have foisted on our planet. The publicity from this event had a major impact on the 1971 rewrite of the Federal Water Pollution Control Act (FWPCA). Cleveland, like many midwestern cities, was an intensely industrialized and highly polluted city that used the Cuyahoga River as a sewer to carry industrial wastes away and into Lake Erie. Signs along the shores of Lake Erie warned against eating fish from the lake due to toxic levels of mercury. Thankfully, these stories belong to the past and the Cuyahoga River and Lake Erie are restored to their preindustrial cleanliness and fishing is permitted once again. In the year 2000, the Cuyahoga Valley National Park was established in the upper reaches of the Cuyahoga Valley. Pitifully, this only happened as a result of strict environmental legislation on the federal level, the CWA.

The FWPCA was enacted in 1948 but had far too little impact until the nation's attitude toward the environment radically changed in the 1970s. The FWPCA was entirely rewritten in 1972 and amended in 1977 by the CWA and amended again by the Water Control Act of 1987. Collectively, these acts have become known as the CWA. The CWA does not regulate groundwater. The principal aim of the CWA is to assist states in their efforts to improve and maintain the quality of the nation's surface water. The act is administered by the USEPA with assistance of the USACE where issues of dredging and filling are concerned.

Much of the nation's drinking water is derived from surface water sources. Contaminated surface water can seriously affect health in humans. Neurological disorders, reproductive problems, and gastrointestinal sickness have all been associated with polluted surface water. Although some naturally occurring deleterious chemicals can be found in surface water in certain areas (uranium, radon, arsenic), most sources of contamination are either from agricultural activities (pesticides, insecticides, fertilizers, animal feedlots), municipal wastewater and sewage overflow, and multiple types of industrial sources. The federal government provides financial assistance to states for improvements to municipal wastewater treatment facilities. CWA regulations focus on water quality standards and wastewater effluent limitations in an effort to eliminate the discharge of pollutants into waters of the United States and maintain the quality of water for the protection of aquatic and terrestrial wildlife, as well as recreation in and on the water.

Aggregate producers are affected by CWA regulations in a number of ways: discharge of groundwater from dewatering systems, wastewater discharge from processing plants, discharge from dredge and fill operations, accidental spills of petroleum and/or hazardous materials, and stormwater runoff. Any industrial operation (including aggregates) that allows industrial wastewater to leave the property via a point source is required to obtain a National Pollutant Discharge Elimination System (NPDES) permit. A point source is defined as "any discernible, confined and discrete conveyance, including but not limited to any pipe, ditch, channel, tunnel, conduit, well, discrete fissure, container . . . from which pollutants are being or may be discharged" (48 Fed. Reg. 14153). The USEPA has delegated the authority to issue and enforce NPDES permits to many but not all of the states.

Unlike much heavy industry, the potential for an aggregates producer to generate severe pollution is fairly minor unless faced with a serious fuel spill. The industrial waste from aggregate operations includes mostly sediment from overburden and fine rock dust washed off the crushed stone during processing. These are very minor pollutants in the grand scheme of things. Any discharge from a point source will require an application for an NPDES permit from the state or the USEPA depending on the authority in your state. A permit is required before any discharge into waters of the United States and it is a good idea to start the process 9–12 months prior to the planned discharge. Public comment may be required before the issuance of an NPDES permit and the permit application will be routed to state and local environmental authorities throughout the process.

Most states require a permit to install (PTI) a wastewater treatment facility prior to the actual issuance of the NPDES permit. Most water treatment at aggregate facilities consists of one or a series of settling ponds to slow the velocity of the effluent and remove the particulate matter by gravity prior to discharge. In some states a settling pond is considered a treatment facility and a PTI will be required (Figure 6.7). Whether or not settling ponds are technically considered to be treatment facilities, they do require some engineering. The size, depth, and construction of the ponds depend on the sediment load, velocity and volume of water, and specific gravity of the sediment particles. An engineering drawing of the pond(s) signed by a professional engineer will probably be required.

The multiple types of NPDES permits include industrial wastewater permits, stormwater permits, individual NPDES with stormwater provisions, and industrial general stormwater permits. Which type of permit will be required is a function of how the stormwater is handled at an individual operation. Some states offer general stormwater permits that are industry specific

FIGURE 6.7 Settling pond ahead of dewatering pumps

and are usually much easier and quicker to obtain than individual permits. A final NPDES permit will include effluent limitations that must be met as well as monitoring requirements. Monitoring may be required on a monthly, quarterly, or semi-annual basis, but monthly is the norm. The permittee will be required to test the effluent for specific parameters and report the results to the USEPA. The normal testing requirements for aggregate producers include the discharge rate, total suspended solids, pH levels, and quantities of oil and grease.

More stringent effluent limitations may be imposed if discharging into designated state or national scenic rivers. Some states have different classifications for their individual streams based on water quality, aquatic life habitat, and so on, that can also affect effluent limitations. Most NPDES permits are issued for a period of 5 years and must be renewed at that time. As with all permits, fees are required upon application, annually, and upon renewal after 5 years. All sampling, analytical testing, and reporting is at the expense of the permittee. A stormwater pollution prevention plan is normally required to accompany any industrial stormwater permit application.

A permittee can expect periodic inspections by USEPA representatives. These generally do not a follow a regular timed schedule, but you can expect an immediate visit if your monitoring report exceeds the effluent limits or if the agency receives a complaint from the public. Like any treatment facility, settling ponds require periodic maintenance. Over time, the ponds will begin to fill with sediment, which reduces the effectiveness of sediment removal. Systematic regular inspection with sediment removal and disposal is a necessity, so add settling ponds to your regular maintenance schedule. During sediment removal, make sure to close off the discharge pipe to prevent any discharge of dirty water while excavating the ponds.

GROUNDWATER ISSUES

According to the USGS, there is more than a thousand times more water underground than in all of the world's rivers and lakes (including the Great Lakes) (Figure 6.8). When precipitation falls to Earth, a certain portion returns to the atmosphere through a process called evapotranspiration, the amount of which is dictated by local climatic conditions. Another

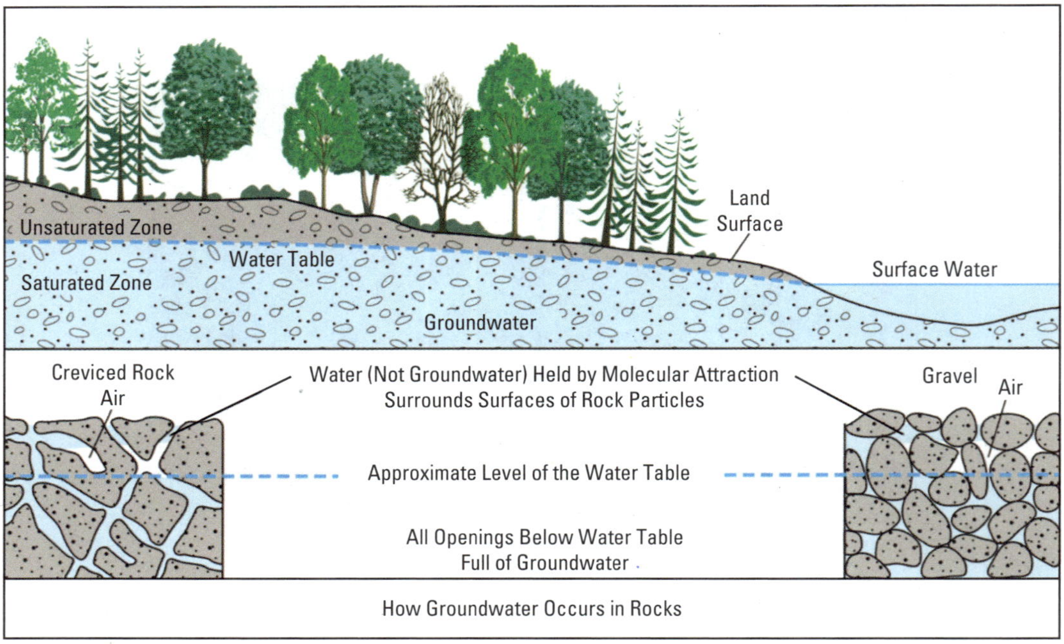

Courtesy of USGS

FIGURE 6.8 Saturated zone beneath the water table

percentage is taken in by plants and animals, while the remainder percolates into the ground. The unsaturated zone contains a mixture of water, soil, and air, which supports the vegetation. The area below the water table is the saturated zone where the pores, cracks, faults, and fissures in the underlying rock are filled with water. In wetter climates, the water table can lie a few feet (meters) beneath the ground surface, while in dry climates it can be several hundred feet (meters) below. The basic water cycle consists of precipitation, transpiration, evaporation, and saturation. Gravity is the force acting on the water that pulls it toward the center of the Earth. The density and porosity of the underlying formations controls the flow rate and direction of the underground water flow. Some of this underground water will eventually reach the surface once again as springs, and seeps or recharge to streams and lakes (Figure 6.9).

This is an oversimplified explanation of a very complicated subject. The science of groundwater hydrology and hydrogeology is very complex. Certain rock types are dense and nonporous and transmit only very small amounts of water except through cracks and fissures,

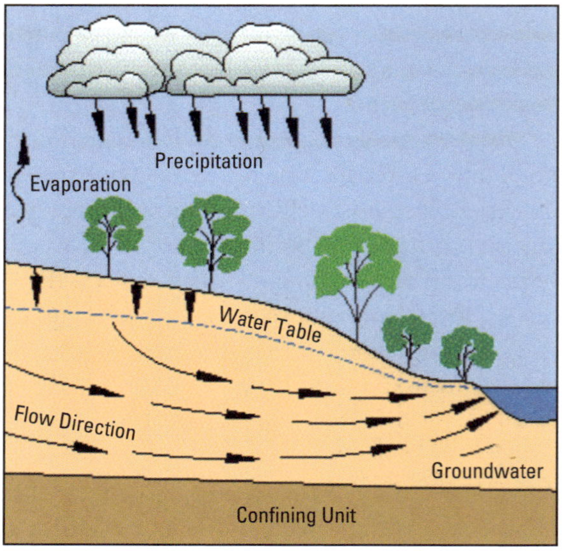

Courtesy of USGS

FIGURE 6.9 Groundwater flow recharging a stream

like granite. Other rock types like sandstone and unconsolidated sand and gravel can be very porous and permeable, able to move, store, and transmit large volumes of groundwater and provide recharge for surface streams and lakes. Limestone can store and transmit groundwater but is also easily dissolved by slightly acidic water and forms caves and sinkholes known as karst features. The underlying geology affects the flow of groundwater. Water forced downward by gravity through a porous formation may eventually flow horizontally if it encounters a dense nonporous layer like clay, shale, or granite. Groundwater can become confined between nonporous layers, causing significant pressure to build up due to the confinement. A well drilled into a confined formation can release pressurized water back to the surface as an artesian well with no need for pumping.

The water beneath the water table in the saturated zone is known as an aquifer and can be a huge storehouse for groundwater. When a well pump or sump is located beneath the water table and removes water from the aquifer, the water table is lowered at that point. The distance the water table declines as a result of pumping is known as drawdown. A well or sump that removes large volumes of water from the water table creates what is known as a cone of depression (Figure 6.10). The water table, which is generally flat lying, will exhibit a cone-shaped depression with the well at the center of the cone. Depending on the pumping rate, hours pumped, and the porosity, permeability, and transmissivity of the formation, the drawdown and the resulting cone of depression can either be small (e.g., domestic well) or extremely large (e.g., municipal well field, quarry dewatering). An extensive cone of depression can negatively affect neighboring wells if the resulting drawdown goes beyond of the depth of the neighbor's wells or a reduction of pressure within the aquifer results.

Groundwater laws vary considerable from state to state and generally become controversial when the dewatering of a new mine site is proposed. Presently, no federal regulations govern the use and withdrawal of groundwater other than the Safe Drinking Water Act (SDWA), which regulates water quality in public drinking water systems. More than 150,000 public

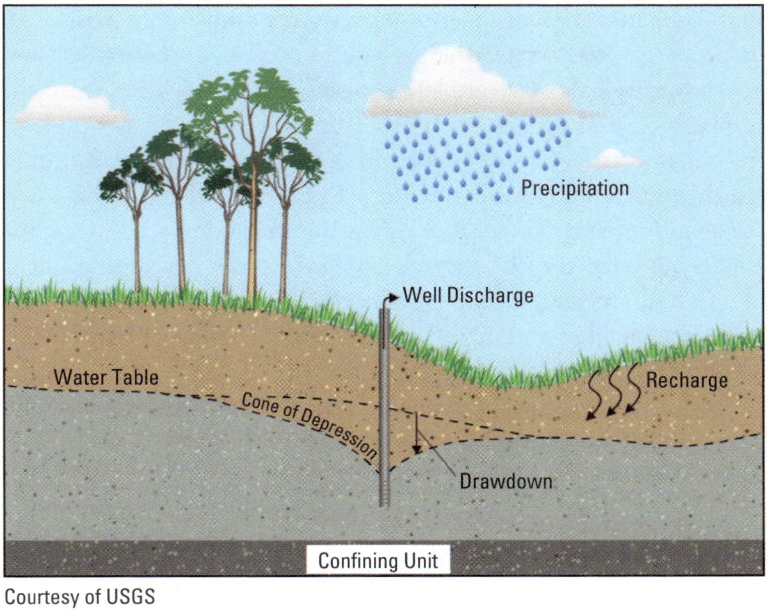

Courtesy of USGS

FIGURE 6.10 Cone of depression resulting from pumping a well

water systems provide water to a large segment of the population that are regulated under the SDWA. These regulations apply to both surface water and groundwater sources. Primary and secondary drinking water standards under the SDWA can be seen at USEPA (2022a). Individual domestic wells are not regulated under the SDWA.

Several different rules of law can apply to groundwater withdrawal and usage depending on the state involved:

- The doctrine of limited ownership apportions water to a property owner based on the acreage owned and the total amount of available water. Available volumes can change over time, such as during times of drought.
- The doctrine of capture allows a landowner to utilize however much water can be removed from beneath their property and put to beneficial use. This includes removal of water from beneath the neighbor's property if the cone of depression is large enough. The term *beneficial use* is sometimes defined but is often left up the courts to decide.
- The reasonable use doctrine allows unlimited extraction of groundwater by a landowner as long as it does not damage neighboring wells or the aquifer. This doctrine does not guarantee any specific volume of water but often gives weight to historical water usage.
- In some of the western states, water rights beneath vast acreages can be purchased, sold, and retained as property.

No matter what law applies, it can be changed by a judge's ruling in the future, setting a new precedent.

It is imperative that a mine owner who dewaters or has the potential to do so be well informed on the state regulations pertaining to groundwater usage. Some midwestern producers sell a portion of the water they remove from the aquifer to municipal water systems that then treat the water and resell it to consumers. As mentioned previously, the state of Ohio requires a producer to model the effects of their dewatering on the aquifer and the neighboring wells and restore a neighbor's water under certain circumstances.

It is prudent to be informed of the groundwater conditions that exist on, near, and surrounding your property. When test drilling prior to mining, converting some of the test borings to observation wells upon completion is always a good idea, especially any borings around the periphery of your property. An observation well can be as simple as some 2 inch (50.8 millimeter)–diameter PVC pipe fitted with a well screen and placed in an appropriate elevation within the test boring. This can be done in unconsolidated borings as well as in core borings prior to abandonment. The cost of this is minimal compared to the test drilling and can prove advantageous if faced with a dispute or a lawsuit over dewatering a neighbor's well. If someone 3 miles (4.8 kilometers) distant accuses you of dewatering their 75 foot (23 meter)–deep well and your on-site observation well indicates only a 20-foot (6-meter) decline in water levels, the neighbor's well screen is likely encrusted with mineral deposits, which is out of your control. Consulting hydrologists are available to assist with these issues if you have no one on staff qualified to do so. An outside opinion is generally more credible when dealing with complex hydrogeological issues.

AIR POLLUTION ISSUES

Prior to the establishment of the USEPA in 1970, the first federal legislation to address air pollution was enacted in 1955, administered by the U.S. Public Health Service and was entitled the Air Pollution Control Act. The Clean Air Act (CAA) has a long, evolved history beginning as the Clean Air Act of 1963 and amended in 1965, 1967, 1970, 1977, and again in 1990. The USEPA is the primary federal agency administrating and enforcing the CAA, although the individual states can elect responsibility for compliance with and regulation of the CAA in exchange for funding. This election is not mandatory and some states have chosen not to accept responsibility for enforcement of the act. Many states have additional air pollution regulations of their own.

The CAA established a list of hazardous air pollutants that includes mineral fiber emissions from manufacturing or processing glass, rock, or slag fibers; thus, it has the authority to monitor and enforce air pollution regulations within the aggregates industry. The CAA establishes air quality control regions and assigns the designation of either attainment or nonattainment areas. Attainment areas meet national standards for primary or secondary ambient air quality. Nonattainment areas do not meet the national standards, and some areas are unclassified based on the lack of available information (Figure 6.11). Under Title V, the 1990 CAA amendment authorized a national operating permit program and required industries to assess and measure pollutants released into the air and develop a plan to minimize and control them. The permit requirements within nonattainment areas are much more stringent. New Source Performance Standards (NSPS) established the maximum level of pollution that a new source

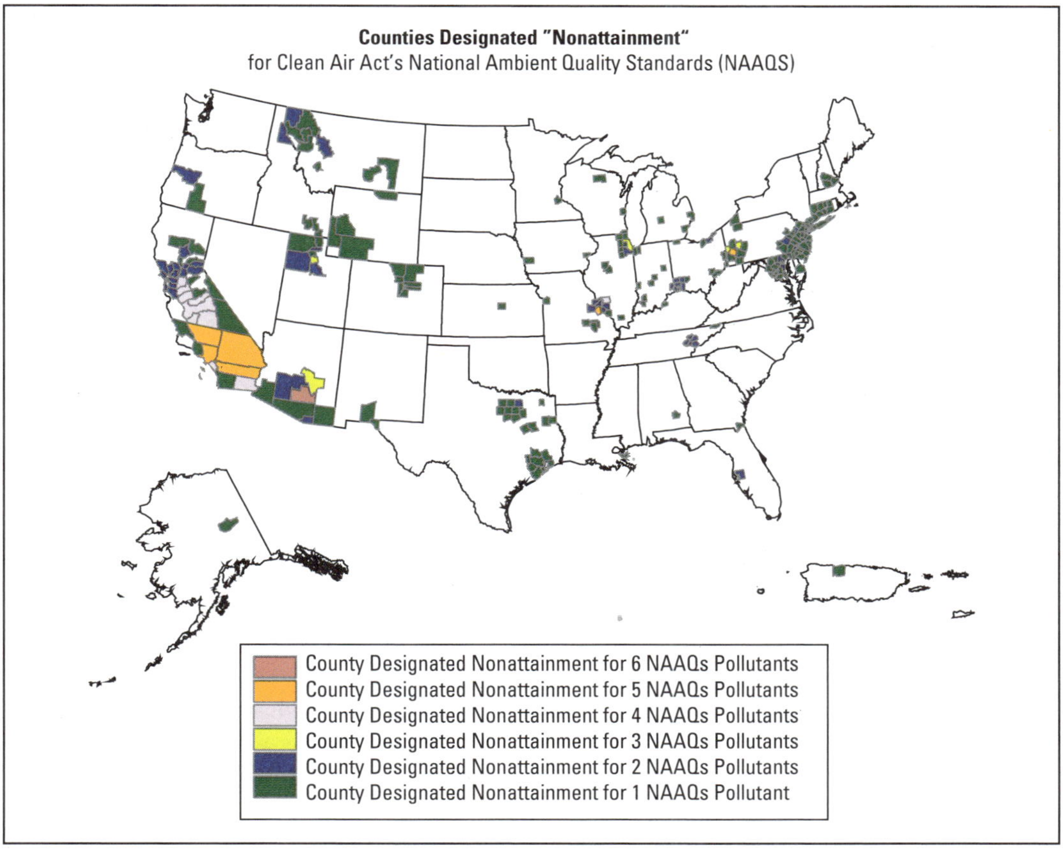

Courtesy of USEPA

FIGURE 6.11 Nonattainment areas

of air pollution may produce. These levels are generally significantly less than the levels allowed for pre-existing sources.

Air pollutants produced by the aggregates industry are normally limited to particulate matter (dust particles) emitted as fugitive dust. Sources of fugitive dust at an aggregates facility are from plant processing (crushers, screens, conveyor belts, etc.), aggregates stockpiles, roadways and parking areas, and mineral extraction operations (blasting, excavating, and hauling). PTIs and permits to operate (PTOs) are required prior to beginning any operations. Very large processing plants may be classified as a major source of air pollution if they have the potential to emit over 100 metric tons per year of particulate matter and will become subject to more stringent regulations under Title V of the CAA. Aggregate processing plants fall under Subpart 000 (nonmetallic minerals) of the NSPS, which defines the designation of affected facilities, standards for particulate matter emitted, reporting and record-keeping requirements, test methods, and compliance procedures. Each crusher, grinding mill, screening operation, bucket elevator, belt conveyor, bagging operation, storage bin, and enclosed load-out station installed after August 31, 1983, falls under the NSPS rules. This also affects any modifications, additions, or reconstruction of older equipment that happened after that date.

NSPS requires annual testing utilizing either stack testing or opacity by visible emissions for fugitive dust. Stack testing is used on discrete discharges like vent pipes and baghouse discharges. The most common method is opacity testing by visible emissions. A qualified visible emissions tester (smoke reader) has to be certified by the USEPA and recertified semi-annually. USEPA Method 22 and Method 9 are the most common opacity tests utilized (USEPA 2019, 2017). As a general rule, opacity readings from crushers must be less than 15%, conveyor belt transfers less than 7%, enclosed bins less than 7%, and wet screening 0%. Visible emissions testing is a qualitative method of interpreting opacity levels by observation methods. During the certification process, the smoke readers are required to identify various examples with known opacity levels. The USEPA will require notification of opacity testing 30 days prior to the test date. This gives them the opportunity to send an inspector to the site to review methodology, testing procedures, and so forth. Periodic random inspections will also apply. Employees can be certified as smoke readers, but it may be advantageous to use an outside testing source to establish credibility with the USEPA.

Once a PTO is issued, it is imperative to thoroughly read the permit immediately upon receipt. The permit will include specific terms and conditions that must be followed to maintain compliance. There is a 30-day period, from date of issuance, where you can appeal if you feel that any terms or conditions are not reasonable. Each PTO will contain the following sections: applicable emissions limitations, control requirements, additional terms and conditions, operational restrictions, monitoring record keeping, reporting requirements, and miscellaneous requirements specific to your individual permit.

Violations of the CAA can result in civil penalties of up to $25,000 per violation day. In the event of a violation of CAA air pollution control requirements, corrective action should be taken without delay. A credible, swift response to a violation will determine whether formal enforcement is pursued by the USEPA as well as the resulting amount of civil penalty imposed. Regular maintenance of air pollution prevention equipment (e.g., bag replacement in a baghouse, water trucks) is an absolute necessity and should be completed on a regular schedule rather than an accidental emission of fugitive dust. The possibility of an enforcement action is much greater if the USEPA receives a citizen's complaint or finds a violation during an unannounced inspection. Always be proactive with regular inspections and internal audits. If an inspector knocks on the door, make sure the employee responsible for environmental issues accompanies them on the inspection to answer any questions and request a close-out conference to determine the inspector's expectations of further action.

Air permitting is a very complex issue that requires a thorough understanding of the regulations applicable in your specific jurisdiction. It is definitely advantageous to have an environmental manager on staff that possesses intimate knowledge of the local regulations and issues. More than likely you will end up with several air permits; an NSPS permit for the processing plant plus separate permits for roadways and parking areas, stockpiling, and mineral extraction. Some states offer general permits geared to specific industries that usually can simplify and expedite the process when available. Other sources of air pollution that may require additional air permits are fuel storage tanks, used oil burners and furnaces, and stationary internal combustion engines (on generators, welders, pumps, etc.). Depending on USEPA workloads, CAA permitting can be a very long process (8–15 months). Any planned plant additions or modifications, such as new type crusher, additional crusher, larger screen, or larger conveyor belt, will require an approved PTI prior to installation and a PTO prior to operation. The need to plan well in advance is obvious.

FUEL AND OIL STORAGE

The Resource Conservation and Recovery Act (RCRA) went into effect in 1984, regulating the underground storage of motor fuels in an effort to prevent pollution of groundwater. This act was updated in 1988 and again in 2015. The 1988 update required all owners of underground storage tanks (USTs) to upgrade their tanks with spill prevention and leak detection within 10 years. The USEPA estimated that there were over 2,000,000 USTs in the United States at that time. Like most federal environmental regulation, the enforcement of the statutes varies depending on the state and local jurisdiction.

Most aggregates producers have a need to store diesel fuel to power the mining equipment and many store gasoline as well. Maintenance shops require large volumes of motor oils (of various viscosities), gear oils, hydraulic oils, and solvents. As with any other commodity, petroleum products are less expensive when purchased in bulk. A single semi-tanker of diesel fuel contains about 8,000 gallons (30,283 liters) of fuel. It is much more cost-effective to purchase a tanker load than to get multiple smaller deliveries. Most fuel storage tanks at aggregate operations are in the range of 10,000- to 15,000-gallon (37,854- to 56,781-liter) capacity to facilitate tanker load delivery. Bulk oil storage can range from racked 55-gallon (208-liter) drums to 500-gallon (1,892-liter) tanks or larger.

As a staff geologist for a large midwestern producer, I was assigned the task of supervising the removal and remediation of several USTs after the 1988 legislation was passed. We hired an outside environmental consulting firm to handle the paperwork and to supervise and document any subsequent remediation. In the 1960s and 1970s, tanks of any design with sufficient volume that could be purchased cheaply were buried underground and attached to aboveground pumps for fuel distribution. In our underground tank removal, we discovered more than one riveted steel railroad tanker car buried underground. The age of the tanks was unknown because they were purchased used, but the usable life of rail cars is 30–50 years if that is any indication of age. In this process we encountered several underground fuel spills that required remediation. It was a surprise to me, but none of the several old steel tanks we removed had actually failed. All of the contamination was a result of leaking plumbing between the tank and the pump.

In today's world, there are two alternatives for bulk storage of fuels; one is surface tanks and the other is underground tanks. Both have advantages and disadvantages, but both require regular inspection and maintenance. RCRA exempts buried tanks on farms, residential tanks <1,100 gallons (<4,164 liters) for storing fuel, and heating fuel tanks for noncommercial uses. USTs are defined as "one or any combination of tanks, including the underground pipes connected thereto, that are used to contain an accumulation of regulated substances the volume of which, including the volume of the underground pipes connected thereto, is ten percent or more beneath the surface of the ground" (USEPA 2021). Most of today's USTs are constructed of steel or aluminum that is wrapped in fiberglass, carbon fiber, or plastic for corrosion protection; a composite material with a metal liner; or a composite material with a thermoplastic liner. Cathodic protection is usually attached to the outside of the tank to protect against galvanic corrosion. Double-walled tanks (a tank within a tank) is the preferred system for underground storage. Double-walled tanks are equipped with leak detection between the two tanks that can detect leaks as small as 0.1 gallons/hour (0.38 liters/hour). Double-walled pipes are also available to install for the required plumbing to the surface.

The alternative to underground fuel storage is storing the fuel in aboveground tanks. Many of the almost 1,000,000 tanks removed between 1988 and 1998 were replaced with aboveground tanks. Aboveground tanks are required to either be of double-walled construction or be surrounded by a containment system to catch any spilled fuel. Containments can be constructed of concrete, aluminum, or steel and must be capable of holding the entire volume of the tank plus an additional 20% for rainfall accumulation. The pumps and fuel-dispensing apparatus should also be located inside the containment to minimize any fuel drips. Periodic maintenance of the containments is required to remove and properly dispose of any accumulated combination of rainfall and minor fuel spillage. Most states do not allow the use of aboveground fuel storage tanks at outlets selling retail fuel to the public. Most states have additional regulations for fire prevention and protection pertaining to aboveground fuel tanks under the enforcement of the state fire marshal's office.

Containments are also necessary surrounding the bulk storage of motor oils, hydraulic oils, and solvents. The same 120% volume requirement applies to the storage of bulk oil. In the past, many producers have had a tendency to store waste oil outside in old 55-gallon (208-liter) drums while waiting for the recycling service to pick it up. Not only is this very unsightly, it can ultimately result in environmental problems when drums accidently get knocked over and spill onto the ground, requiring additional remediation. This can also become an MSHA citation if the drums are not properly labeled. The best shop management practice would be a 500-gallon (1,893-liter) tank plainly marked "waste oil" inside a properly sized containment.

Like all other aspects of an aggregates operation, fuel storage must be managed and maintained on a regular basis. Because fuels and oils are considered hazardous materials, a thorough knowledge of the state and federal regulations is a necessity.

HAZARD COMMUNICATION

Originally promulgated by the Occupational Safety and Health Administration (OSHA) in 1985, hazard communication (HazCom) regulations were designed to inform employees of the nature and dangers of hazardous chemicals in the workplace. This also became known as the employee right-to-know law. The aggregates industry was not included under these regulations until 2003, when the MSHA followed OSHA's lead with its own set of HazCom standards. MSHA allows the producer to write their own program as long as it meets the basic requirements, is user friendly, and is easily accessible to all employees. A HazCom program must contain the following elements: a chemical inventory, warning labels on all containers, material safety data sheets (MSDSs) for all chemicals, a written program explaining the purpose and use, and employee training explaining how to use the program. The program also requires manufacturers of chemical products to publish MSDSs for their products, explaining the chemical composition, proper usage of the product, appropriate personal protective equipment (PPE) required, health hazards associated with the product, environmental cleanup protocols, and first aid treatment in case of an accident. Every chemical from diesel fuel to lubricants to office cleaning products is included in the program.

Many resources are available to assist operators in creating their own HazCom program such as the HazCom Tool Kit available on the MSHA website. The National Stone, Sand and Gravel Association (NSSGA) has HazCom guidance available as do many of the state aggregates associations. The requirements for the written HazCom program must include retention of MSDSs in a format available to all employees and independent contractors at all times, an inventory of all chemicals utilized on site, determination of all chemical health hazards, and labeling of all chemicals regarding their hazards. Employees and contractors must be trained on how to identify hazards and the necessary precautions to follow during use. The program should include language that each individual employee is obligated to bring to a supervisor's attention any chemical contact they feel could constitute a health hazard.

The most common form of hazard labeling is the National Fire Protection Association (NFPA) hazard warning labels. These labels have a color-coded system, with red indicating flammability, blue health hazards, yellow chemical reactivity levels, and white special hazards (Figure 6.12). A number rating

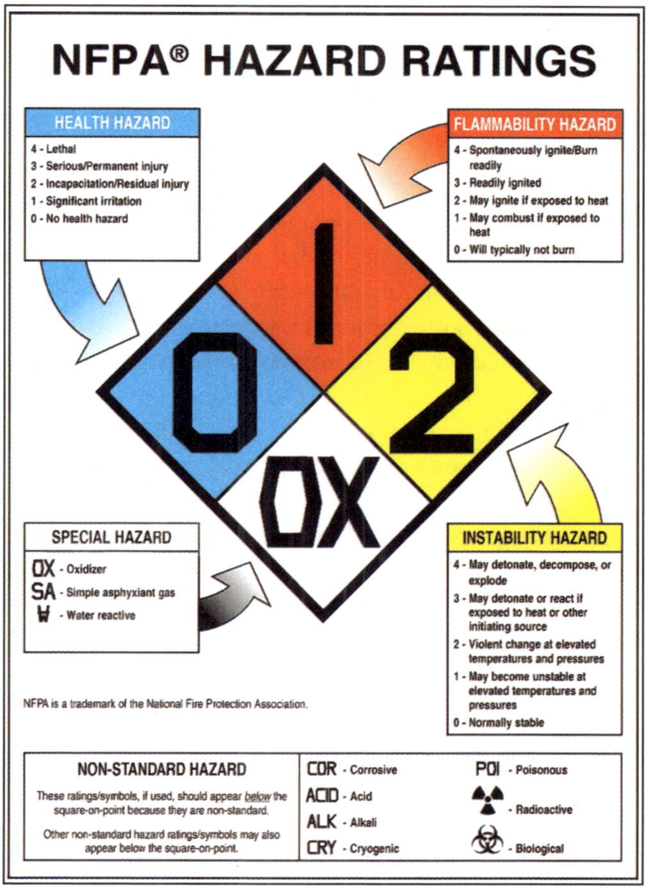

FIGURE 6.12 National Fire Protection Association hazard warning label

for each color from 0 to 4 indicates the severity of the hazard; zero indicates no hazard and four means extremely hazardous. The higher the number, the higher the hazard! The HazCom standard requires that chemical and material manufacturers, importers, and distributors properly label all shipments of hazardous chemicals or materials with the identity of the product, clearly noted hazard warnings, and the name and address of the manufacturer. Most chemical producers and suppliers comply with the labeling requirements for their products, but it may be on the outside of the case and not on the individual products within, and bulk deliveries like fuel and oil are difficult to label. Some aggregate producers automatically refuse delivery of unlabeled or mislabeled products. Other types of HazCom labels are available, but the NFPA is by far the most utilized. Labels are available from any supplier of safety equipment.

It is the aggregate producer's responsibility to see to it that all chemicals are properly labeled. If a chemical product is transferred from a labeled container to a portable container (e.g., from a 55-gallon [208-liter] drum to a 2-gallon [7.6-liter] container), the new container need not be labeled if only one person will handle it. If another employee will come in contact with the portable container, it must be labeled. Tanks receiving bulk shipments of fuel, oil, or dust suppression chemicals must have HazCom labels on the tank or at the dispenser to identify the chemical and the hazards.

The key to an effective HazCom program is the MSDS file of all chemicals on site. This file must be kept up to date at all times and be complete. It is a good idea to make one employee responsible for the MSDS files and they should update these files a minimum of once weekly. The MSDS files and written policy must be made available to all employees and contractors at all times and it is a good idea to keep a copy in the employee lunch area, locker room, or clock-in area as well as in the plant office. MSHA inspectors on their biannual inspections will generally perform a paperwork audit that will include your MSDS sheets. If an inspector finds a can of air freshener in the men's room not included in the MSDS file, that could generate a citation. MSHA only requires you to retain MSDS sheets as long as the chemical is utilized on site, so try to clear out unnecessary paperwork during your weekly updates.

Along with the MSDS file, an inventory of chemicals on site needs to be retained, available to the workforce, and updated regularly. MSHA requires that the inventory not only list the chemical but also the location where they are used (e.g., 10W30 motor oil, maintenance shop; dust suppression chemicals, roadways). It is simplest to keep the chemical inventory in the MSDS file to assure employee access. MSHA standards require periodic HazCom retraining of employees but do not specify the frequency. Your tailgate safety talks are a good forum to periodically go over the program, especially if new employees are on site. Management should also enforce the HazCom chemical usage and PPE requirements stated on the MSDSs. If the MSDS says rubber or nitrile gloves are required when using a product, it is the responsibility of management to supply the gloves and enforce their usage. MSDSs for your aggregates must be produced and available to your customers upon request as well. Examples of MSDSs for crushed limestone, sand and gravel, crushed granite, and so on, are available from NSSGA and/or state aggregates associations.

An employee right-to-know program should be viewed as participatory, with both the employees and management's welfare in mind. It's in the employee's best interest to protect their own health and in management's best interest to keep their most valuable resources working at full capacity.

RECYCLING AND WASTE DISPOSAL

Recycling is not limited to household recycling like separating cardboard, glass, yard waste, plastics, and papers into a separate trash can to be picked up weekly by the local trash service. Most states have regulations on the recycling of specific industrial wastes. The most common industrial wastes from aggregates operations that require recycling are used oil, lead-acid batteries (LABs), and tires. When I first started in the industry, it was not uncommon for aggregates producers to apply used oil to their internal roadways as a dust suppressant. They not only used the oil generated internally, but many times were paid by others to take used oil off their hands. Some became the dumping grounds for thousands of gallons (liters) of used oil annually, depending on how many miles of internal, unpaved roadways were on site. Thankfully this this practice is a thing of the past.

Regular preventive maintenance of off-highway construction equipment generates large volumes of used oil. Depending on the amount and size of this mobile equipment, these volumes can be substantial. Many equipment engines contain 5–15 gallons (18.9–56.8 liters) of oil in the crankcase, which is changed every 250–300 hours of operation. Hydraulic oils, in much larger quantities, need to be changed every 1,500 hours of operation. The larger the mining operation, the larger the equipment, and therefore the volume of waste oils generated. No longer is waste oil allowed to be deposited on the ground to prevent fugitive dust. Used oil is generally considered a hazardous waste because it contains the by-products of petroleum combustion and minute metal shavings from engine, cylinder, and pump wear.

Processing plants also generate used oil from crusher lubrication systems and gear boxes on conveyor drives, all of which require regular oil changes to operate at optimum performance levels. Most, if not all, maintenance shops utilize parts cleaners that generate waste solvents that require proper disposal and recycling. There are waste oil recyclers that will pick up your waste oil and then filter and recycle it for other uses for a nominal fee and sometimes at no cost. But this poses another problem for a producer: the need for bulk storage of waste oil until such time as it can be recycled. Some producers burn waste oil in specially designed waste oil furnaces to heat their maintenance shops in the winter, but usually produce more used oil than they can consume. The use of these furnaces is subject to air emissions regulation, which will vary by state. Used solvents are generally not permitted to be mixed with used oil and need to be recycled separately.

LABs are common items found in every piece of mobile equipment. These batteries are considered hazardous waste because on average each contains 8 pounds (3.6 kilograms) of sulfuric acid and 18 pounds (8.2 kilograms) of lead. Batteries should never be buried, disposed of in landfills, or incinerated. If LABs are stored on site, they should be stored indoors and not exposed to rainfall or freezing temperatures. Proper PPE should be utilized when handling these batteries. Licensed battery recyclers are usually available to pick up and recycle LAB for a minor charge. The easiest method of recycling these batteries is to return them to your battery supplier for recycling.

According to *Global Recycling* (n.d.), 4.19 million metric tons of used tires were generated in the United States in 2017. Most of the tires utilized in aggregates operations are the large, off-highway type for wheel loaders and trucks, which can be reused (recapped) multiple times, if managed properly. Eventually the tire carcass itself will be damaged or wear out and have to be disposed of. It was not uncommon in the 1970s to see aggregates producers bury tires on site or bury them in slopes as a means of completing reclamation. Regulations on tire recycling vary from location to location. Large stockpiles of tire carcasses can produce potential health hazards as the tires fill with rainwater and become breeding grounds for mosquitoes. Tires not only contain rubber but also nonbiodegradable polymers, steel cord, and other organic and inorganic chemicals. Some of these compounds are proven to kill beneficial soil bacteria. If a tire stockpile should catch fire, it is extremely difficult to extinguish, can burn for weeks, and significant environmental damage can result.

Tire recycling has come a long way in the last two decades. Tires are recycled into a number of products; some are burned for fuel, utilized to build artificial reefs or breakwaters, made into playground equipment, or used as an additive in asphalt, artificial turf in sports stadiums, in carpet manufacturing, and roofing materials. Almost all tires sold in the United States have an additional recycling fee attached to the purchase price depending on the state where they are sold. Tire suppliers will take used tire carcasses in trade for proper disposal upon the purchase of new tires.

ENVIRONMENTAL AUDITING

Environmental audits are not a regulatory or permit requirement, but they are a valuable tool to provide a producer with a level of confidence that their operations are in compliance with environmental regulations. Regulations are dynamic and change based on new legislation and precedent-setting legal decisions. Membership in organizations like the NSSGA or the various state aggregates associations that monitor environmental regulations is a way to stay informed of pending legislation changes.

A prudent buyer of industrial real estate would not culminate a purchase without an environmental audit to determine they are not also purchasing an environmental liability. It is also a good idea to do internal environmental audits on your own properties. This type of management tool is designed to identify any potential environmental deficiencies, liabilities, risks, and hazards, and to ensure compliance with environmental regulations. Audits can be completed to determine if environmental best practices are adhered to by the local managers and/or for strict environmental regulation compliance. Many companies will utilize their attorneys to contract for environmental audits so that audit results can be considered privileged information. If an environmental audit determines regulatory noncompliance and management does not act to remedy the deficiency immediately, they could face criminal charges if the deficiency is later uncovered by a regulatory agency that discovers the pre-existing knowledge.

A typical environmental audit will begin with a review of all environmental regulations that are applicable to the operation. From this review, a checklist should be developed to assist in the audit process. This will be followed by a review of the company's permits, permit conditions, compliance records, notices of violation, test reports, daily logs, observations, complaints, and maintenance records on environmental equipment, and any internal information pertaining to environmental issues. A review of the environmental agency's records covering similar information should also be undertaken. Any written company environmental policies should be intimately scrutinized at this time also. After a thorough review of all of the written records, key personnel should be interviewed: operations manager, environmental manager, plant manager, quarry manager, supervisors, maintenance personnel, and so on. A detailed site investigation team should be formed and led by an experienced environmental professional well versed in auditing. Other team members should include management and operations personnel knowledgeable with the specific operation.

Upon completion of the audit, a detailed report should be generated including maps, aerial photographs, diagrams, and specific photographs identifying deficiencies and/or compliance. The report should include an executive summary, a list of applicable regulations, results of the records search, determination of compliance with regulations, identified deficiencies, corrective actions necessary for regulatory compliance, methodologies to improve compliance, and recommended employee training.

The International Standards Organization (ISO) has developed two different standards on environmental management systems including information on environmental audits: ISO 14001 and ISO 19011. Some companies may wish to pursue certification under these ISO standards, which are generally quite stringent. The USEPA developed a series of notebooks summarizing key environmental indicators for specific industries. One such notebook, developed in 1995 for the stone, clay, glass, and concrete industries, contains helpful information pertaining to the pertinent regulations. This notebook can be found at USEPA (2022b). Numerous environmental consulting firms specialize in this kind of work including Phase I audits on new properties. An outside opinion may be the best option when completing an environmental audit, especially if you utilize a law firm as a go-between. Before hiring an outside consultant, make sure they have experience in auditing aggregates operations and obtain industry-specific references.

I have always appreciated the value of having an experienced environmental manager on staff. But as I write this chapter that realization changes from a value to a necessity. The myriad environmental regulations to understand, knowing the correct jurisdiction, making the right governmental contacts, and keeping up with new legislation and case law is a monumental task. Not to mention the permit paperwork, the testing requirements, the inspections, the audits, and the reporting requirements. Small producers with tight budgets need to utilize the expertise of the environmental committees of the state associations and the NSSGA and environmental consulting firms to assist with this daunting task.

CHAPTER 7

Quality Control of Aggregates Materials

Quality control (QC) plays a vital role in maintaining a satisfied customer base, attracting new customers and assuring quality construction projects in the local market area. In this regard, it serves a marketing function as well a QC function. The QC department should endeavor to act as a customer service providing information, test data, quality data, and product design assistance to customers, both internal and external. The entire perception of your company by your customer base is reflected in the quality of your finished products. Who would buy an automobile from a manufacturer that had a reputation for building poor-quality vehicles? Deliver inferior products and inferior service and you will be perceived as an inferior supplier. Quality products and service can easily result in higher selling prices, especially if the competition is lackadaisical about the quality of their materials.

The quality of a finished aggregate product delivered to a customer is not the sole responsibility of the QC technician or manager. The responsibility rests with all the employees in the company. Producing quality products needs to be the ultimate goal of each and every employee, from the CEO to the mechanic, the plant operator, the loader operator, and the delivery truck driver. Successful companies need to have high expectations of their employees, communicate these expectations regularly, and provide employee training geared toward the expected performance. Quality is a mindset that should be instilled in your employees from the day they are hired. The maintenance people need to know the ramifications of not immediately replacing screen cloth when they discover a hole, the loader operator must be trained how to properly remove material from a stockpile to minimize material segregation, and the QC technician must be well trained in sampling techniques and testing procedures.

It is important that operations personnel are kept informed of test results on a regular basis, especially any tests failing to meet specifications. Failing gradation test results can reveal deficiencies in the processing system. The screen cloth utilized to separate size grades is a wear item in the plant, has a limited life span, and must be replaced on a regular schedule. Maintenance personnel are required to inspect screen cloth regularly, but often a hole or split can develop when the plant is in operation and go unnoticed because of the continuous flow of material. A failing gradation test will quickly reveal such a deficiency and corrective measures can be taken immediately to prevent contamination of the inventory. Crushers need periodic setting adjustments to maintain the proper crushing ratios. Gradation test results reveal trends that can predict when a crusher adjustment is required. This is where the "control" portion of QC comes into play; identifying what the processing plant is producing so that adjustments can be made to control the process.

It is beneficial to have a computerized QC information system for data storage and reporting. Access to the information system should be made available to management, QC personnel, and operations personnel, and the system should be able to generate reports on specific products. A system to communicate test failures to the responsible plant personnel as soon as possible should be included. Several aggregates QC information systems are available for purchase on the market today. Internally maintained and developed systems are also common in the industry.

In many construction projects, the customer is not always the owner of the project. An aggregate producer sells material to a building contractor or a highway contractor who has been contracted to build a warehouse or an interstate highway for their customer. In many cases, the final owner is the one generating the specifications for the aggregates utilized on the project and not your immediate customer (the contractor). In other cases the customer may be an asphalt producer or a ready-mix concrete provider that blends and modifies your product to produce another product of their own. They will also resell your product to another final customer that ultimately has no direct connection to your company. Quality asphaltic and portland cement (PC) concrete products require quality aggregates as their primary ingredient.

Most contractors rely on published specifications from other sources; if the materials are delivered to a highway job, then the state department of transportation (DOT) (the final owner) will be the agent specifying the material requirements, not the contractor. Larger construction projects will publish bid documents with exacting material specifications generated by an architect or engineering firm or referring to a standard specification from ASTM International or another agency. Specifications on bid documents generated from an engineer or architect should be scrutinized because they may differ from the standard specifications you are used to seeing. The major agencies generating aggregate specifications include ASTM International, AASHTO, individual state DOTs (Ohio DOT, Kentucky DOT, Indiana DOT, Wyoming DOT, etc.), the Federal Highway Administration, and the Federal Aviation Administration (FAA).

There is a difference between QC and quality assurance (QA). QC is the process by which a producer samples and tests their material before, during, and after the processing and stockpiling to determine the material meets or exceeds project specifications. This should include periodic visits to customers' sites to make sure their material handling and stockpiling methods are not negatively affecting the quality and gradation of the delivered materials. QA, on the other hand, is the process of sampling and testing the material received by the owner of the project to assure them materials delivered equal or exceed the owner's expectations or specifications. State DOTs have their own laboratories and technicians, as do most asphalt and PC concrete producers, which they utilize to provide QA testing functions. QC personnel should communicate regularly and maintain good professional relationships with their QA counterparts in industry and government agencies.

The properties that quality aggregates need to possess to be incorporated into durable, long-lasting construction projects are individual gradation of both coarse and fine aggregates, combined aggregates gradation, maximum particle size, particle shape, bulk density, specific gravity (SG), absorption, surface texture, strength, soundness (freeze–thaw durability), pore structure, thermal expansion, deleterious materials, and organic impurities. The properties of aggregates can be divided into physical, mechanical, and chemical. The principal properties affecting aggregate quality include the following:

- Mechanical properties
  - Particle strength
  - Wear resistance
  - Resistance to degradation

- Physical properties
  - Particle shape and surface texture
  - Maximum particle size
  - Pore structure
  - Porosity
  - Permeability
  - Absorption
  - Specific gravity
  - Unit weight
  - Resistance to wetting and drying
  - Particle gradation
- Chemical properties
  - Resistance to attack by chemicals
  - Solubility
  - Surface charge
  - Chemical coatings
  - Chemical reactivity
  - Organic reactivity
  - Chloride content

QUALITY TESTING

Aggregates are utilized in many applications and for many different end uses. Aggregates are used as structural products, for erosion control, and in geotechnical and civil engineering applications, asphalt pavement, PC concrete, base products, railroad base products, and other industrial, agricultural, and chemical applications. Listing all the testing requirements for all uses is a monumental task and I refer you to the *Aggregates Handbook* (NSSGA 2013) for a more detailed analysis. The most common quality testing is discussed in the following section.

Aggregate soundness is the ability of an aggregate to stand up to the forces of weathering: freezing and thawing. This is a much larger issue in cold climates like Minnesota than warmer climates like Florida, although the in-between climatic areas that experience numerous cycles of freezing and thawing in a typical season are the most vulnerable to unsound aggregates. The sulfate soundness test (SST) was first used in Europe in the mid-1800s. This test is still widely specified, but its accuracy and repeatability are somewhat questionable. The theory behind the SST is that when an aggregate is saturated in a solution of either sodium or magnesium sulfate and then heated and dried, the sulfate salts dehydrate. When the sample is re-immersed and rehydrated, the salt recrystallizes and expands within the pores of the rock. In theory, this process is supposed to simulate the action of water freezing and thawing within the pore spaces of the aggregates. The test is explained in ASTM C88 with the added caveat that "the precision of this test method is poor and failing aggregates should not be rejected on this test alone without further additional testing."

In the SST, after the sample is washed, dried, and separated into specific size ranges, it is immersed in a solution of sodium or magnesium sulfate for 16–18 hours at a temperature of 70°F ± 2°F (21°C ± –16.6°C). The sample is then removed, drained for 15 ± 5 minutes, and placed in an oven at 230°F ± 9°F (110°C ± –13°C) until it reaches a constant weight. After five cycles of immersion and drying, the sample is washed, dried, and put through a certain sieve, depending on the specific size range tested, and a percentage of loss is then determined. A weighted average loss of the size fractions is calculated to determine the soundness loss percentage of the sample in question. Precise control and monitoring of the temperature specifications shown above is necessary to get consistent, correlatable test results.

In the 1990s, I was experiencing quality failures by the local DOT at a quarry operation when the results from our commercial testing laboratory and the state DOT lab did not correlate. Our test results from the commercial lab were consistently better than those produced by the state DOT lab on the same sample. In an attempt to determine the problem, our corporate QC manager and I toured eight commercial testing laboratories and all three DOT labs in a three-state area. The results of our laboratory tours were very disappointing. We determined that only one laboratory out the 11 toured (the commercial one we were utilizing) was following the specified temperature ranges in the ASTM C88 soundness testing procedure. This included the three state testing facilities. Faced with this dilemma, we eventually set up our own soundness testing tank with precise controls on the solution temperature range (70°F ± 2°F [21°C ± −16.6°C]) and the oven temperature range (230°F ± 9°F [110°C ± −13°C]). We also added chart recorders to keep accurate temperature records throughout the testing process. This was a pricey system at the time, but it was well worth the cost when we finally were able to repeat soundness tests with much improved accuracy, reliability, and repeatability. At the expense of sounding corny, soundness testing results should be taken with a grain of salt.

Actual freeze–thaw testing is another alternative to SSTs. AASHTO T 161 and ASTM C666 are freeze–thaw test methods widely in use today. Freeze–thaw testing is a very expensive test method that usually takes weeks or sometimes months to complete. A freeze–thaw chamber has to be capable of automatically and rapidly freezing and thawing samples of premade concrete beams. The modulus of elasticity is calculated using dynamic testing equipment to measure the transverse, longitudinal, and torsional frequencies of the concrete beam. The capital cost of the dynamic testing equipment, freezing chamber, heating and cooling equipment, and monitoring apparatus is substantial. Most DOT testing laboratories utilize freeze–thaw testing on structural concrete products only.

The concrete is cast into small beams that are 3–5 inches (76.2–127 millimeters) in width and depth and 11–16 inches (279–406 millimeters) in length and made with a specific concrete mix design including the aggregate blend to be tested. After the concrete is cured, the sample beam is subjected to 300 cycles of freezing and thawing, until the modulus of elasticity is reduced to 60% of the original, or the sample fails completely. Based on the modulus of elasticity and the number of freeze–thaw cycles, a durability factor (DF) is calculated. The change of length in the beam is also measured upon completion of the test and the maximum percentage of expansion is calculated. Some DOTs use a combination of DF and percent expansion as a specification, while others simply specify an acceptable DF range. This method also tests the mix design of the concrete and the quality of the concrete paste. Variations on the test are freezing and thawing in water, freezing and thawing in air and water, and freezing and thawing in a brine solution of water and salt.

SG and absorption is one of the best indicators of aggregate quality in my opinion. An aggregate with a very small percentage of water absorption is generally very sound and durable. When an object is submerged in water, it appears to weigh less than the same object in air. The following formulas define apparent SG, bulk SG, saturated surface dry (SSD) bulk SG, and water absorption. SSD is defined as a sample of aggregate that is immersed in water for 24 hours and then dried on an absorbent cloth until the visible film of surface water is removed:

$$\text{Apparent SG} = \frac{A}{A} - C$$

$$\text{Bulk (dry) SG} = \frac{A}{B} - C$$

$$\text{Bulk SSD SG} = \frac{B}{B} - C$$

$$\text{Absorption (water) (\%)} = \left(B - \frac{A}{A}\right) \times 100$$

where

A = weight of an oven-dried sample of aggregate in air
B = weight of an SSD sample of aggregate in air
C = weight of a saturated sample of aggregate in water

SG and absorption are extremely important factors in the mix design formulations for both asphaltic and PC concretes. AASHTO T 84, AASHTO T 85, ASTM C127, and ASTM C128 describe how these tests are to be performed.

The unit weight of an aggregate mixture is determined by filling a container of a known volume with layers of dried aggregate and compacting them with a steel rod. AASHTO T 19 and ASTM C29 outline the procedures for this test method. The weight of the material is then weighed and divided by the volume of the container to calculate dry unit weight. The percent voids in an aggregate mixture can then be calculated utilizing the following formula. The percentage of voids is an important factor in asphaltic concrete mix designs and is a function of gradation, particle shape and amount of compaction:

$$\%\text{ voids} = \left(\frac{(A \times W) - B}{A \times W}\right) \times 100$$

where

A = bulk SG
B = unit weight
W = 62.4 pounds per cubic foot (1,000 kilograms per cubic meter) (unit weight of water)

The particle shape of coarse aggregates can have a significant impact on the performance of PC concrete and hot mix asphalt mixtures. ASTM D4791 specifies the test method for determining flat, elongated, or flat and elongated particles. The ratios specified (length to width) vary depending on the final product and specifying agencies, but are generally either 3:1 or 5:1. A proportional caliper is used to measure these ratios. The particles defined as flat, elongated, or flat and elongated are separated and their percentage of the total sample (by weight) is expressed. Fine aggregates are tested using AASHTO T 304 where the material is compacted and the volume of voids calculated. Because of their irregular shapes, angular particles result in a higher void content.

Fractured particle count on coarse aggregate is specified by some DOTs that require a certain percentage of particles with one or two fractured faces. This is done by visual determination and separation following the ASTM D1252 test method. This test is normally required on crushed gravel because round particles resist compaction and fail to interlock in asphalt and concrete mixes. Cubical particles are preferred in all mix designs because of their ability to interlock. A large percentage of round particles in asphalt has a tendency to cause rutting of the pavement because the round gravel can act like ball bearings beneath heavily loaded tires. Depending on the crushing method utilized, round gravel can escape being crushed.

The two principal tests for determining aggregate degradation are very similar in nature. The Los Angeles (LA) abrasion test is probably the most commonly specified. A 5,000-gram sample of aggregate is washed, dried, and separated into individual size ranges by sieving. Four different size ranges for testing are defined in ASTM C131. Each aggregate sample size range is placed in an LA abrasion machine with 6–12 steel balls depending on the sample size range. The LA machine consists of a rotating drum with a shelf that lifts and drops the sample and the steel balls when rotated. The test is run for 500 revolutions of the drum. Upon removal from the machine the sample is dry sieved over a 12-mesh (1.68-millimeter) sieve. The minus-12-mesh material is weighed, the percentage of the total is calculated, and this is considered the percent of wear loss or degradation. A weighted average of the size ranges determines the final loss percentage. The Micro-Deval test is similar to the LA abrasion test but the aggregate is first soaked in a specified volume of water and the water and aggregate mixture is rotated in a drum at 100 revolutions per minute with 9.5-millimeter steel balls. The number of rotations and balls are outlined in ASTM D6928 for coarse aggregate and ASTM D7428 for fine aggregate. The percent of wear is determined in a similar fashion as in the LA test. This test originated in France in the mid-1950s, has good correlation to performance, tends to have repeatable results, and is becoming more widely accepted with DOTs and other testing agencies in the United States. Other less widely used tests designed to estimate aggregate strength are the mill abrasion test, the page impact test, the aggregates impact value, the aggregates crushing value, and the Schmidt rebound hammer.

Certain aggregate types provide better skid resistance than others. Highway safety is a high priority for all DOTs. High speed, heavy traffic volumes, and wet pavement contribute to many traffic accidents. The surface texture and mineralogy of coarse aggregates determines their friction-resistant properties. Some coarse aggregates that are very hard and durable tend to polish when subject to skidding tires on wet pavement. The fine aggregate in PC concrete tends to have a larger influence on skid resistance than in asphalt pavements. Fine-grained limestones tend to polish more than impure limestones containing a percentage of insoluble material and dolomites. The insoluble residue test (ASTM D3042) is used to identify aggregates susceptible to polishing. Other tests include AASHTO T 278 (the British portable pendulum tester) and AASHTO T 279 (the British wheel test), which directly measure the effects of polishing. Some state DOTs require specific aggregate types (e.g., dolomite, granite) to be utilized in surface asphalt mixes.

Certain materials have been determined to be deleterious (causing harm or damage) to aggregate mixtures. These include but are not limited to soft particles, clay lumps, coal or lignite, shaley materials, limonitic material, iron concretions, chert, metallic particles, and flat and elongated particles. Most quality specifications include limits on the amount of deleterious materials allowed in a sample of aggregate.

Quality requirements can vary from one product to another depending on the intended use. Quality requirements for asphalt aggregate are not necessarily the same as those for PC concrete aggregate or road base products. Quality specifications for aggregates used in superstructure PC concrete are generally the most restrictive, while base, subbase, and backfill products are the least restrictive. Chert is an amorphous form of silica (SiO_2) that tends to be highly absorptive and is defined differently in some state specifications. Typical quality specifications ranges are shown in Table 7.1.

TABLE 7.1 Typical aggregates quality specification ranges

Los Angeles abrasion (maximum % loss)	40.0–50.0
Sodium sulfate soundness (maximum % loss)	12.0–15.0
Micro-Deval (maximum % loss)	<20.0
Deleterious substances shall not exceed (% by weight):	
Soft particles	2.0–3.0
Coal and lignite	0.25–1.0
Clay	0.25
Elongated particles >5:1	10.0–15.0
Shale or shaley material	0.25–1.0
Limonite concretions	0.25–1.0
Alkalis	0.5–1.0
Metallic particles	0.5–1.0
Chert	0.5–5.0

Proper gradation is an important constituent of a durable, stable aggregate mixture, but it is not an inherent quality parameter of the stone itself. The gradation of an aggregate product is manufactured in a processing plant through the process of crushing, screening, and washing. Aggregates gradation is essentially the distribution of particle sizes in an aggregate mixture and is defined as the percent of the mixture retained on an individual sieve size; for example, percent retained on the 1-inch (25.4-millimeter) sieve, percent retained on the ⅜-inch (9.52-millimeter) sieve, or percent retained on the 4 mesh. This is most often expressed as the reciprocal of percent retained, which is the percent "passing" the individual sieve sizes and expressed as percent by weight. Gradation specifications are normally expressed as a percentage range passing a specific sieve (70%–100% passing 8 mesh). Coarse aggregates are considered to be any size that is greater than (plus) the 4 mesh and fine aggregates as any size that is smaller than (minus) the 4 mesh. U.S. standard sieves down to ¼ inch (6.35 millimeter) designate the individual opening on the sieve; the 1-inch (25.4-millimeter) sieve has 1-inch (25.4-millimeter) square openings and the ⅜-inch (9.52-millimeter) sieve has ⅜-inch (9.52-millimeter) square openings. Below ¼ inch (6.35 millimeter), the sieves are numbered from 4 through 400. These numbers refer to the number of openings per inch including the individual wire strands. A 4 mesh will have four (0.187-inch [4.75-millimeter]) openings per inch, while a 100 mesh will contain 100 (0.0059-inch [0.15-millimeter]) openings per inch (25.4 millimeters). See Table 2.4.

To determine the gradation of aggregate material, a sample must be dried and weighed and the material mechanically shaken through a stack of standard sieves (Figure 7.1). AASHTO T 27 and ASTM C136 define the methods to be utilized. A specified sieving time and sieve stacking arrangement are defined for individual products. Upon completion of the sieving process, the material retained on each sieve is weighed and the percent retained is calculated. This number is easily converted to percent passing each sieve size and a resulting gradation is determined.

Courtesy of Gilson Company, Inc.

FIGURE 7.1 Mechanical gradation testing device

The term *well graded* refers to the fact that the aggregates particles are well distributed with similar percentages of coarse-, intermediate-, and fine-sized particles. Uniformly graded aggregates are a blend of aggregates that are mostly the same size. When an aggregate blend has a majority of particles that are either coarse or fine with very little intermediate-sized materials present, it is known as gap graded. Tables 7.2–7.4 show gradation specifications for coarse and fine aggregates from AASHTO. The product is referred to by the size number in the first column, for example, #57 crushed stone or #8 gravel. Product numbers with double digits are a combination of two single-digit products; for example, #57 is a blend of #5 and #7.

It is often helpful to see a graphical representation of gradation on a gradation chart with sieve sizes indicated on the *x*-axis of the graph and percentage passing on the *y*-axis (Figure 7.2). These are generally presented on a logarithmic-scale graph. From the grain size distribution curve, a coefficient of uniformity can be calculated.

$$C_u = \frac{D_{60}}{D_{10}}$$

where

C_u = coefficient of uniformity
D_{60} = sieve opening size (mm) through which 60% of the aggregate passes
D_{10} = sieve opening size (mm) through which 10% of the aggregate passes

This can be a valuable tool for identifying gradation characteristics. A dense-graded aggregate will have a C_u value greater than 6, and open-graded material a C_u of less than 4.

When sieving very fine material, the percent passing the 200 mesh will require further testing because dry sieving is ineffective below the 200 mesh. AASHTO T 11 describes the wet method of sieving material minus the 200 mesh. The sand equivalent test (AASHTO T 176) is utilized to determine if the fines in a sand sample consist of silt or clay. In this test a flocculent solution is mixed with the sample of fines to loosen and remove any plastic particles coating the aggregate grains. After placing the sample into a graduated cylinder and shaking the solution, more flocculent is added through a tube and used to irrigate the sample and disperse the different sized material. The sample is then given a specified time to settle and separate, and, using the graduations on the cylinder, the ratio of sand to plastic fines (clay) becomes apparent. The sand equivalent is the ratio of the height of sand in the cylinder to the height of the plastic fines, reported as a percentage. The methylene blue test (AASHTO T 330) can be used to determine if the minus-200-mesh-sized particles are actually clay or in fact clay-sized dust particles. This test is sometimes also used to test pulverized materials for clay content.

Alkali reactivity can be a problem for some aggregates. Certain chemical constituents of aggregates can react chemically with the alkalis in cement. Alkali-silica reactivity (ASR) is a reaction of certain silica minerals in the aggregate with the alkali constituents in PC and alkali-carbonate reactivity (ACR) is a similar reaction with the alkali reacting with certain magnesium carbonates. ASR and ACR have similar effects on concrete structures and pavements. In ASR, a siliceous gel is formed inside the concrete by the chemical reaction, and if severe enough, pressure can build within the pores of the concrete and cause severe cracking over time. A petrographic examination (ASTM C95) of thin sections cut from the aggregate can identify potentially reactive aggregates for both ASR and ACR but needs to be done by an experienced petrographer. Other test methods to identify ASR are ASTM C227, ASTM

TABLE 7.2 Coarse aggregate gradations corresponding to AASHTO M 43

Size No.	Nominal Size* Square Openings		Amounts Finer Than Each Laboratory Sieve (Square Openings), Percent by Weight														
	in.	mm	4 in. 100 mm	3 ½ in. 90 mm	3 in. 75 mm	2 ½ in. 63 m	2 in. 50 mm	1 ½ in. 37.5 mm	1 in. 25 mm	¾ in. 19 mm	½ in. 12.5 mm	⅜ in. 9.5 mm	No. 4 4.75 mm	No. 8 2.36 mm	No. 16 1.18 mm	No. 50 300 µm	No. 100 150 µm
1	3 ½ to 1 ½	90 to 37.5	100	90 to 100		25 to 60		0 to 15		0 to 5							
2	2 ½ to 1 ½	63 to 37.5			100	90 to 100	35 to 70	0 to 15		0 to 5							
24	3 ½ to ¾	63 to 19			100	90 to 100		25 to 60		0 to 10	0 to 5						
3	2 to 1	50 to 25				100	90 to 100	35 to 70	0 to 15		0 to 5						
357	2 to No. 4	50 to 4.75				100	95 to 100		35 to 70		10 to 30		0 to 5				
4	1 ½ to ¾	37.5 to 19					100	90 to 100	20 to 55	0 to 15		0 to 5					
467	1 ½ to No. 4	37.5 to 4.75					100	95 to 100		35 to 70		10 to 30	0 to 5				
5	1 to ½	25 to 12.5						100	90 to 100	20 to 55	0 to 10	0 to 5					
56	1 to ⅜	25 to 9.5						100	90 to 100	40 to 75	15 to 35	10 to 15	0 to 5				
57	1 to No. 4	25 to 4.75						100	95 to 100		25 to 60		0 to 10	0 to 5			
6	¾ to ⅜	19 to 9.5							100	90 to 100	20 to 55	0 to 15	0 to 5				
67	¾ to No. 4	19 to 4.75							100	90 to 100		20 to 55	0 to 10	0 to 5			
68	¾ to No. 8	19 to 2.36							100	90 to 100		30 to 65	5 to 25	0 to 10	0 to 5		
7	½ to No 4	12.5 to 4.75								100	90 to 100	40 to 70	0 to 15	0 to 5			
78	½ to No 8	12.5 to 2.36								100	90 to 100	40 to 75	5 to 25	0 to 10	0 to 5		
8	⅜ to No 8	9.5 to 2.36									100	85 to 100	10 to 30	0 to 10	0 to 5		
89	⅜ to No 16	9.5 to 1.18									100	90 to 100	20 to 55	5 to 30	0 to 10	0 to 5	
9	No. 4 to 16	4.75 to 1.18										100	85 to 100	10 to 40	0 to 10	0 to 5	
10	No. 4 to 0†	4 75 to 0†										100	85 to 100				10 to 30

* Numbered sieves are those of the United States Standard Sieve Series.
† Screenings.
Where standard size of coarse aggregate designated by two or three digit numbers are specified, obtain the specified gradation by combining the appropriate single digit standard size aggregates by a suitable proportioning device that has a separate compartment for each coarse aggregate combined. Perform the blending as directed by the Laboratory.
Source: Ohio Department of Transportation 2019

TABLE 7.3 Fine aggregates for PC concrete

Sieve Size	Total Percent Passing
⅜ inch (9.5 mm)	100
No. 4 (4.75 mm)	95 to 100
No. 8 (2.36 mm)	70 to 100
No. 16 (1.18 mm)	38 to 80
No. 30 (600 μm)	18 to 60
No. 50 (300 μm)	5 to 30
No. 100 (150 μm)	0 to 10
No. 200 (75 μm)	0 to 5

Source: Ohio Department of Transportation 2019

TABLE 7.4 Fine aggregates for asphaltic concrete

Sieve Size	Total Percent Passing
⅜ inch (9.5 mm)	100
No. 4 (4.75 mm)	90 to 100
No. 8 (2.36 mm)	65 to 100
No. 16 (1.18 mm)	40 to 85
No. 30 (600 μm)	20 to 60
No. 50 (300 μm)	7 to 40
No. 100 (150 μm)	0 to 20
No. 200 (75 μm)	0 to 10

Source: Ohio Department of Transportation 2019

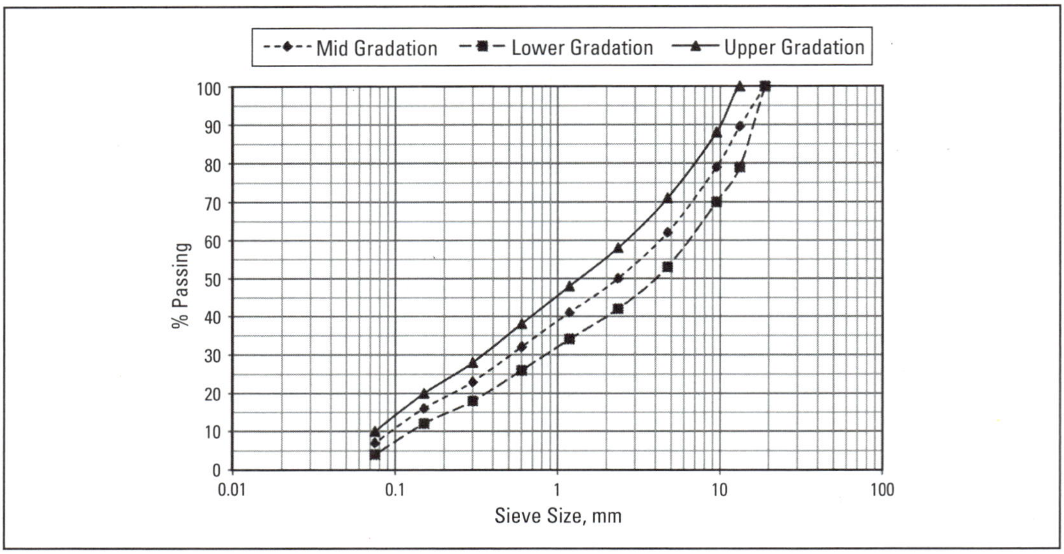

Source: Ohio Department of Transportation 2019

FIGURE 7.2 Typical gradation chart

C1260, ASTM C1567, ASTM C1293, and ASTM C289. Other tests utilized to determine ACR are ASTM C586 and ASTM C1105.

Electrochemical properties of aggregate materials come into play when aggregates are utilized for mechanically stabilized earth (MSE) applications. To reduce the volume of concrete and thus costs, MSE is extensively used in retaining wall applications along highways. Several methods are used to produce MSE: some systems use steel strips or grids and other systems utilize various polymer grids within layers of well-drained, compacted, granular backfill. This compacted mass of MSE resists the lateral forces exerted on a normal retaining wall, thus eliminating the need for massive, steel-reinforced concrete walls. Precast concrete panels form the outermost exposed surface of the MSE. Because of the use of galvanized steel grids and clips connecting the precast panels, the potential for galvanic corrosion exists. Besides needing to be well graded, the granular backfill must be tested for the following electrochemical parameters: pH, organic compounds, chloride salts, sulfate salts, and electrical resistivity. Requirements for pH levels also exist for various polymer grid applications. AASHTO specifications for MSE granular backfill are given in Table 7.5.

Aggregates used as compacted structural products (base materials) under highways, parking lots, sidewalks, structures, and so on, need to be made from 100% crushed material, either crushed stone, crushed gravel, or in some areas slag (a by-product of steel manufacture.) Structurally, the crushed particles in the base layer are compacted into a uniform layer thick enough to support the pavement and decrease the stress with depth caused by the loading. The crushed particles, specified gradations, and proper compaction provide interparticle friction and aggregate interlock for stability under heavy loading. For flexible pavements (asphaltic concrete) the base layer should be dense graded containing enough fines to provide compaction during construction. The base layer beneath rigid pavement (PC concrete) generally tends to be more open graded with less fines to promote drainage and prevent pumping under the slabs (Table 7.6).

TABLE 7.5 Mechanically stabilized earth (MSE) specifications

Granular Backfill for MSE Applications	
Corrosion Properties	
pH	Greater than 5 but less than 10
Organic content	<1%
Chlorides	<100 ppm
Sulfates	<200 ppm
Resistivity	>3,000 ohm-cm
Gradation	
Sieve Size	**Total Percent Passing**
4 inches	100
4 mesh	0 to 60
200 mesh	0 to 15

TABLE 7.6 Typical asphalt base gradation

Sieve Size	Total Percent Passing
2 inches (50 mm)	100
1 inch (25.0 mm)	75 to 100
½ inch (12.5 mm)	50 to 85
4 mesh (4.75 mm)	25 to 60
8 mesh (2.36 mm)	15 to 45
16 mesh (1.18 mm)	10 to 35
50 mesh (300 μm)	3 to 18
200 mesh (75 μm)	1 to 7

Source: Ohio Department of Transportation 2019

Proper compaction is the key to stable subbase and base layers underlying either flexible or rigid pavement. The standard Proctor test is a laboratory test for measuring moisture content and density of compacted materials. The standard Proctor test (AASHTO T 99) and the modified Proctor test (AASHTO T 180) are similar in nature but use different sized testing apparatus. The base aggregate with a known moisture content is placed in a mold and compacted in layers. The procedure is repeated with new material, incrementally adding more water to increase the moisture content by 1%–2%, and this is continued until there is a decrease or no increase in the compaction. This is normally repeated on five samples of the material. The actual moisture content of each sample is determined by drying. Dry density (in pounds per cubic foot [kilograms per cubic meter]) is defined as the wet density divided by 1 plus the water content expressed as percent as a decimal. A graph of the dry density on the *y*-axis and water content on the *x*-axis is generated and the maximum dry density is the peak density on the graph and the optimum moisture content is the moisture corresponding to the maximum dry density. These factors are then utilized in the construction design process. The standard and modified Proctor tests will give different results, so when reporting maximum densities the test method must also be specified.

During the construction process, a control strip is usually created and compacted in multiple passes until no apparent increase in the density is obtained. The most common method to determine in-place field density on a control strip is a nuclear density meter. Modern nuclear density meters can measure both moisture content and dry density for comparison with the targeted density specified. A typical highway base material specification can be seen in Table 7.7.

TABLE 7.7 Typical highway base material

Sieve Size	Total Percent Passing
2 inches (50 mm)	100
1 inch (25.0 mm)	70 to 100
¾ inch (19.0 mm)	50 to 90
4 mesh (4.75 mm)	30 to 60
30 mesh (600 µm)	9 to 33
200 mesh (75 µm)	0 to 15*

* Furnish steel slag that has 0% to 10% passing through the No. 200 (75 µm) sieve

Source: Ohio Department of Transportation 2019

SAMPLING METHODS

It is impossible to test 100% of the material produced through an aggregate plant. Gathering samples that are representative of the total stock of aggregates can itself be a challenge. The number of samples necessary to determine the actual quality parameters of an entire stockpile is always a question. Some state DOTs specify a minimum number of gradation samples per thousand tons of product production. The more samples taken from any one stockpile, the better the information. To eliminate any sampling bias, samples should be taken from random locations at random times during the production process. This can be very difficult to accomplish. ASTM D3665 attempts to establish methods to eliminate most sampling bias.

Unless there is a drastic change in the geology of a quarry or pit, the inherent quality of the stone should not change significantly. Processing and stockpiling functions can change some of the physical properties of the final product, for example, particle shape, moisture content, and gradation. Segregation of the individual particle sizes within a specific gradation can be problematic and must be controlled within the process. What makes a gradation sample representative of the entire individual product inventory is always a good question. Production personnel will want samples of the production taken directly from the conveyor belt, arguing that belt samples are indicative of what was produced prior to any outside influences of segregation or degradation. The customer wants the material sampled as delivered to their jobsite to ensure they are receiving what they specified (QA). Upper management wants assurance that the inventory in their stockpiles meets specifications for the work intended. Proper procedures must be developed with engineering controls, personnel training, and close communication between QC and production personnel to assure that the gradation produced (belt sample) closely matches the inventory gradation (stockpile sample).

Most aggregate processing plants stockpile the finished products using conveyor belts to move the material to storage piles or bins. Gravity and vibration play large roles in the segregation of the particle sizes within the products as they are transported and stockpiled. Well-graded products are more sensitive to segregation than uniformly graded ones and the greater the range in particle sizes, the greater the chance of segregation. A base product with particle size ranging from 100% minus 2 inches (50.8 millimeters) to 0%–15% minus 200 mesh has a much greater propensity to segregate than a #8 stone, which is 100% minus ½ inch (12.7 millimeters) to 0%–5% passing 16 mesh. As an aggregate mixture travels along a conveyor belt, vibration and impact with the conveyor idlers cause the fines to sink and accumulate on the bottom of the belt with the coarse material concentrated on the top (Figure 7.3).

When a conveyor belt discharges onto the ground, the coarser particles on the top of the belt are pushed to the front of the pile by gravitational forces and the fines, some of which, if wet, tend to stick to the belt, are dragged to the rear of the pile. The farther the conveyor head pulley is from the ground, the more time the force of gravity has to do its work and the worse the segregation problem becomes. To stockpile more material in a single pile, it is advantageous to have high conveyor belts, which exacerbates the segregation problem. Faster conveyor belt speeds can also exacerbate the problem by throwing the material further from the end of the belt. Mechanical mixing devices can be installed on the ends of the belts to help eliminate segregation. Automated radial stacking conveyors that can be programed to build small, low, radial stockpiles built in multiple layers and that are designed to eliminate segregation are available. These tele-stackers are manufactured in lengths up to 270 feet (82.3 meters) and can build piles with storage capacities up to 300,000 metric tons (Figure 7.4).

Some processing plants utilize bins to collect the aggregate products from the screens and then unload them into trucks that haul the product to stockpiles. Truck dumping can also cause segregation within stockpiles (Figure 7.5). Once again, the force of gravity separates the coarse from the fines when dumping from the trucks, and once again the longer the fall, the worse the problem. When stockpiling via truck, it is best to dump piles directly on the ground and close together to form a coalition of small piles rather than ramping upward and dumping the material over the edge. Training of wheel loader operators is necessary to overcome the segregation issues when loading customer trucks. Conveyor stockpiles should always be loaded from the side of the pile perpendicular to the conveyor discharge. Load-out should take place in multiple passes across the entire cross section of the pile to reblend any segregated materials while loading the trucks.

When segregated by gravitational forces, stockpiles are difficult to sample for accurate

Source: InDOT 2016

FIGURE 7.3 Segregation caused by conveying materials

Courtesy of THOR Global

FIGURE 7.4 Telestacker building nonsegregated stockpile

Source: InDOT 2016

FIGURE 7.5 Segregation caused by truck dumping

gradation analysis. Which portion of the pile is representative of the entire stock? Some state DOTs specify methods for sampling stockpiles. The Indiana DOT has a sampling method, ITM 207, which details using a front-end loader to remove approximately 10 metric tons of product from a stockpile. The material is then mixed by the loader operator by excavating the sample pile from each end and gently rolling over the loader bucket until the sample is thoroughly mixed into a tent-shaped pile. The final sample is composed of at least six shovels from different locations equally around the sample pile taken at the approximate mid-height point.

When sampling from bins, the best method involves using a loader bucket to catch a load from the bin discharge and gently rolling the bucket onto the ground for sampling. Never attempt to sample a bin from the top because this can be very dangerous. Samples can be obtained directly from conveyor belts in a number of ways. Automatic sweep sampling devices are available that will sweep off a section of running conveyor into a catch bin, but these are expensive and multiple devices would be needed on different belts to sample every product. Automatic belt discharge samplers are specialty chutes at the end of conveyor belts that can be rotated into the discharge stream and samples recovered from the chutes. An alternative method is to stop the individual conveyor belt and take a sample of the entire cross section of material from the belt, including sweeping up all of the fines from the belt with a brush. Make sure the conveyor has sufficient power to restart under load prior to attempting this method or you may be shoveling for several hours. This can also be dangerous depending on the height and accessibility of the conveyor. Never take this type of a belt sample unless the conveyor has been electrically locked out and is incapable of accidently restarting. This is a good method of getting samples, but it does interrupt the entire production process.

Sample sizes vary depending on the product and the test to be performed. It is generally good to obtain larger field samples than required for the specific test. Each test method will specify the amount of material needed for the test. Specially designed sample splitters are available for both coarse and fine aggregates that, when properly used, will evenly divide a sample. The sample can then be divided multiple times until the proper amount is obtained. ASTM C702 outlines the procedures for reducing sample sizes.

Contamination of aggregate products after they are produced and stockpiled can become an issue without proper management and QC. Contamination is the addition of deleterious materials or incorrectly sized material into a stockpile. This can be done in various ways and QC personnel must be alert to the possibilities. One of the most common sources of contamination are delivery trucks. Whether company-owned trucks, an outside contractor's trucks, or trucks owned by your customers, these trucks may be delivering other commodities or different products. It is not unusual to find buildup of different materials adhering to the inside of a truck bed, especially during freezing weather conditions. A contract trucker hauling topsoil for one customer in the morning and concrete stone to a different customer in the afternoon is a recipe for disaster. You can expect a call from your ready-mix customer the next day complaining that your #57 stone or concrete sand is contaminated with clay. Policies enforcing truck drivers cleaning their beds between loads must be stringently enforced and equipment made available for doing so, such as high-pressure water hoses and access platforms.

Deleterious materials can be also introduced on truck tires. Trucks hauling to a new construction site in wet weather are likely to bring wheels and tire treads covered in mud into the stockpile areas that can easily be flung into the aggregate piles. If conveyors cross on top of one another in the processing plant, a hole in the conveyor belt or an overloaded belt can spill the improperly sized material onto a lower belt, modifying the gradation. These crossover belts should be covered to eliminate this spillage possibility. Sticks can enter the process, especially in a wet sand and gravel operations, and are very deleterious in concrete. Constant vigilance by plant and QC personnel is necessary to prevent contaminants from entering the stockpiles at any stage during production.

Contamination can happen on the receiving end of the process as well. I distinctly remember a large airport runway job involving 70,000 short tons of #57 crushed stone from a quarry that I managed in the early 1990s. We were ahead of production with a good inventory in anticipation of the job and were adhering to very tight FAA specifications. Our testing and QC protocols were well documented and we had seen no evidence of any deleterious materials in the stockpiles. A portion of the inventory was already on site at the portable concrete plant located on the airport property. The second week of the job all hell broke loose when the contractor called saying that the stone was contaminated with clay balls. Our QC people and I diligently searched our records and stockpiles and found no trace of clay present. The next day we received a repeat phone call from an irate contractor threatening to throw us off the job and buy the material from our competitor. I grabbed my salesman and we proceeded to the airport to gain some more information. The clay ball problem was immediately evident. The contractor's wheel loader charging the plant from the stockpile of our stone was traversing a section of temporary roadway that was pumping large volumes of clay from the insufficiently constructed road. The tires of the wheel loader were dragging the clay into the stone stockpile. We immediately took pictures, stopped the loader operator, and summoned the job superintendent. The next 60,000 metric tons of product was consumed without a hitch. QC oftentimes spills over onto the customer's site.

Degradation of existing stockpiles can happen especially if stockpiling by truck. Large stockpiles formed by trucks and/or loaders running up ramps can cause changes in gradation by friction of tires and particle-on-particle wear. Running of equipment on stockpiles should be avoided if at all possible.

STATISTICAL QUALITY CONTROL METHODS

It is not my intent to teach a statistics course, nor am I qualified to do so, but a basic grasp of some fundamental concepts of statistics is necessary in QC. Test data for aggregate materials generally follow what is known as a normal distribution. The data when plotted versus frequency of occurrence will form a bell-shaped curve with 50% falling above and 50% below the mean value (Figure 7.6). Determining the mean of a number of test values is done by adding all the individual data points and dividing by the number (n) of data points (average). Once the mean is determined, a standard deviation (s) can be calculated. The standard deviation is a measure of the variability of the data. A minimum of nine samples is required to determine a valid standard deviation. I will not bore you with complicated mathematical formulas because a standard statistical calculator or Microsoft Excel spreadsheet is capable of calculating standard deviation from a series of data. What is important to remember is that the more data that are available, the more accurate the standard deviation will be.

Once you have the mean and standard deviation(s) of enough actual test data points from your production process, you gain the ability to predict the range where additional random samples will occur. Approximately 68% of the data will lie within the mean $\pm 1s$, 95% will lie within the mean $\pm 2s$, and 99.7% within the mean $\pm 3s$. If, for example, the mean of the

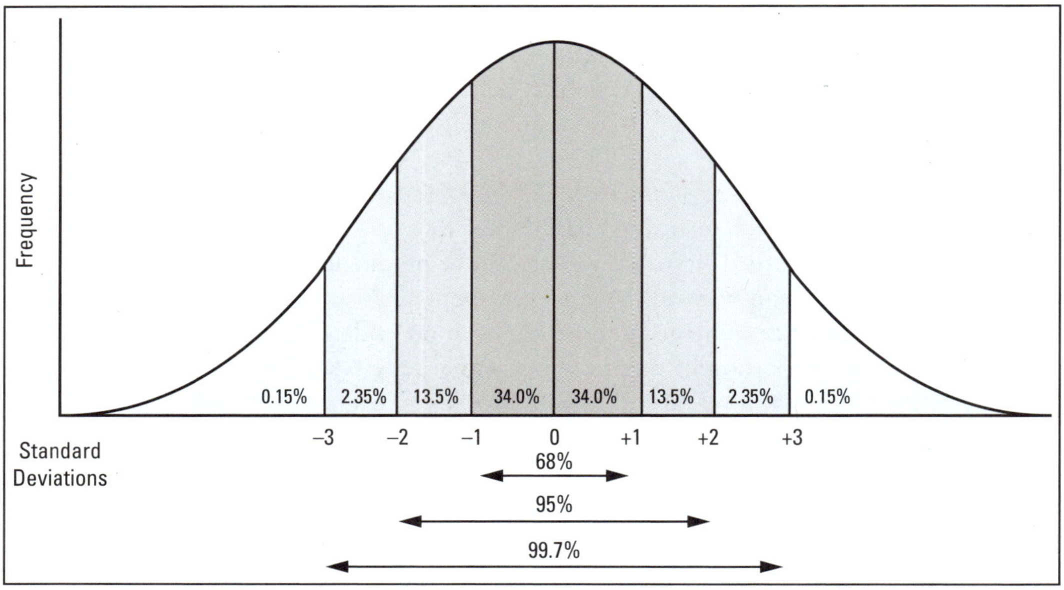

Source: University of Washington 2018

FIGURE 7.6 Bell-shaped curve

percentage passing a certain sieve size (in 125 tests) is 27.6% with a standard deviation of 2.3%, then you can predict with good confidence that 95% of time the test results will be within the range of 27.6% ± (2 × 2.3%), which is equivalent to between 23% and 32.2%. This is a powerful tool for using random testing to predict and control the quality of stockpile inventories. Again, I want to stress that the confidence level depends on the number of tests (*n*) from a stockpile. A total of five tests gives no confidence, but 105 tests provide a substantial level of confidence.

I was once asked to evaluate an offer where a manufacturer was selling off some excess inventory of lime at a very reasonable price. I was given a number of chemical test results for the inventory in question and the boss wanted to know whether to buy it or not. I did a statistical evaluation of the mean and standard deviations for the chemical test results. The evaluation of the mean plus or minus two standard deviations resulted in my recommendation to not purchase the inventory because there was too much deviation in the samples to have any assurance that the inventory would consistently meet specifications. No wonder the price was so reasonable! There are many areas where this evaluation can be applicable.

Other statistical calculations based on these parameters can be utilized to design a statistical QC program to instill confidence within the production department, management, customers, and project owners. Variance, coefficient of variance, and standard error of the mean are other useful statistical calculations derived from mean and standard deviation. The *Z*-value calculated from the mean, standard deviation, and population can be compared to a *Z* distribution probability table, which can then predict the percentage of time the sample will be expected to be within a specified range. Percent within limits is sometimes used to determine payment on certain contracts and utilizes a similar concept to the *Z*-value. Control charts can be a very beneficial in evaluating the fluctuation of test results. These are especially useful when beginning the process of making a new product. These charts graph the moving range, the upper and lower natural process limits, and the upper range limit and are used to evaluate the fluctuations in test results. A basic knowledge of statistics is a necessity for all QC technicians.

QUALITY CONTROL PLANS

I have always been a great proponent of planning and it is a good idea to establish a quality control plan (QCP) specifically designed for each operation. Every aggregate operation is unique. The deposits being mined differ based on specific geological parameters, and mining methods are based on the unique deposit, topography, property boundaries, depth to water, and so forth. Processing methods vary from plant to plant depending on engineering, capital availability, equipment availability, and plant upgrades. The customer base is another variable; certain plants may service primarily ready-mix customers, others may have an asphalt plant on site, and crushed stone operations will probably sell more base products than a sand and gravel plant. A unique QCP takes all these variables into account. A QCP should begin with a list of the principal products produced and what markets are being targeted.

At a minimum, a QCP should contain:

- Goals of the QCP
- Plant management structure
- Description of the deposit
- Mining method
- Processing method
- List of all products produced
- Specifications used for comparison (ASTM, AASHTO, state DOT)
- Test methods to be utilized
- Frequency of quality testing (soundness, LA abrasion, absorption, etc.) on a calendar or per ton basis
- Frequency of gradation testing for each product (times per day or tests per tons produced)
- Reporting methods
  - Frequency of reports
  - Personnel receiving reports
  - How, when, and to whom to report failing tests
- Person(s) responsible for testing

Crushed stone and sand and gravel are commodities that can only be distinguished by their quality. A well-designed QCP managed by knowledgeable technicians will assure your customers quality products and can provide you with the ability to distinguish your product from that of the competition. Quality products + outstanding service + professional personnel = financial success.

CHAPTER 8

Mine Planning and Overburden Removal

Mining is the process of extracting a mineral commodity of economic value from within the Earth by various means for some commercial use or purpose. Throughout this chapter, the terms *surface mining* and *quarrying* are utilized synonymously, as are the terms *quarry*, *pit*, and *surface mine*. Underground mining is discussed later in another chapter and in my mind it is a vastly different subject than surface mining. Many people do not think of a surface mine as being mining per se because they automatically associate the term *mine* with a dark, dangerous, dungeon-like underground cavern. This is a misnomer; both systems, whether surface or underground, extract valuable minerals from the Earth and therefore constitute mining. In Chapter 2 we covered the fact that the aggregates industry is the largest mining industry in the world by volume of commodity sold, although not necessarily by value.

Ironically, humans have been extracting minerals from the ground (mining) since the Stone Age (Paleolithic Era). Archeologists have dated, via radiocarbon methods, the oldest identified mining site in the world as 43,000 years old. This is the Ngwenya mine site in Swaziland, where Stone Age people mined the mineral hematite to produce a red dye called ochre. Other mining sites of a similar age have been identified by archeologists in the United Kingdom, France, and Poland, where flint was mined for the production of weapons and tools, such as arrowheads, spear points, knives, and scrapers. Fire was utilized by early humans to heat and cool rock, which caused it to eventually crack and break into more manageable pieces; a much slower process than today's modern blasting techniques. As long as humans have existed, they have been exploiting the Earth's resources. Perhaps mining is the world's oldest profession!

DEVELOPING A MINE PLAN

The mining of aggregates materials is divided into two distinct areas based on physical and geological properties and the different techniques employed in the extraction process. In Chapter 3, we discussed the differences between unconsolidated and consolidated aggregate deposits. Unconsolidated deposits of sand and gravel require less physical effort to extract from the ground because no blasting is required to reduce the particle size prior to excavation, but this does not always equate to ease of operation. Sand and gravel deposits can require as much or more effort to mine depending on deposit characteristics. Whether surface mining unconsolidated sand and gravel or consolidated rock, the basic process is the same with a few exceptions (Table 8.1).

Earlier chapters discussed exploration, land acquisition, and permitting issues, so I will not reiterate those points here. Detailed planning and engineering are imperative for a successful, profitable mining operation. Once a detailed geological model has been established from the exploration and initial testing data, a mine plan can be developed. The mine plan

TABLE 8.1 Comparison of mining processes

Processes	Unconsolidated	Consolidated
Exploration	X	X
Property acquisition	X	X
Permitting	X	X
Planning/engineering	X	X
Overburden removal/stripping	X	X
Dewatering	?	?
Drilling	—	X
Blasting	—	X
Excavation/loading	X	X
Hauling/transportation	X	X
Equipment maintenance	X	X
Reclamation	X	X

must consider not only extraction, but also local mining regulations, environmental controls and permitting, market fluctuations, neighbor relations, economics (cost), and equipment selection. Mine planning should begin by defining the deposit and selecting the appropriate equipment necessary for the greatest return on investment. In many cases in the aggregates industry, certain equipment may be available to the operator and the mine plan needs to be designed around the available equipment. Although not the ideal situation, economics sometimes dictate this approach.

The planning processes for sand and gravel mining versus crushed stone are very similar in nature; the main differences are in the areas of equipment selection, dewatering, and drilling and blasting. The following parameters need be analyzed during the mine planning process:

- Three-dimensional (3-D) geological model of the deposit
- Physical restraints: property lines, easements, utilities, and so on
- Location of processing plant, scales, entrance roads, and so forth
- State and local mining regulations such as setbacks from property lines and highways
- Environmental regulations such as water discharge, stormwater, and dust control
- Neighboring property owner's concerns from zoning hearing
- Annual and future production requirements
- Overburden removal and storage
- Dewatering
- Equipment selection
- Drainage
- Raw material transportation such as haul trucks, conveyors, and pipelines
- Mining direction, depth, and sequence
- Reclamation

All aggregate mining sites are unique and therefore so are the mine plans, but all mining plans have one requirement in common: flexibility. Aggregates sales can be variable and sometimes sporadic depending on the market location, customer mix, product mix, economic fluctuations, state and federal government budgets, and weather patterns. In many market areas the sales patterns are seasonal with moderate volumes in the spring, high sales in the summer and fall, and low volumes during the winter months. Severely cold winter climates require the operator to produce an entire year's production in a 7- to 9-month time frame. This can be accomplished by oversizing plant and mobile equipment capacities or running multiple shifts during the production season. Areas in the southern United States have the distinct advantage of mild winters, steady year-round sales volumes, and the ability to produce 12 months per year.

Because of the capital requirements necessary, mining operations are long-range investments. Long-term reserves are necessary to justify the investment in the plant and equipment. Most aggregate mining ventures start with 30–50 years of minable reserves. It is generally wise to have multiple phases built into a mining plan: a short-term plan for the next 2–5 years followed by a 10-year plan, a 20-year plan, and so on. This allows for the flexibility needed to deal with unexpected deposit changes and fluctuations in the market over time. If in year 4 a large multiyear highway project is won by the sales department, the stripping and mining plan can be accelerated for years 5–7 and the process reversed when demand goes back to normal in year 8. Normally a new start-up operation, especially if entering a new market, will start slowly and the sales ramp up after the first year or two.

A 3-D geological model must be constructed, whether using cross sections, fence diagrams, or more sophisticated computer models. Back in the 1980s when dealing with complicated glacial outwash deposits, we constructed 3-D models using dowel rods to indicate individual boring logs. A base map was mounted on plywood and the dowels attached at the appropriate boring locations. We marked the dowels, to scale, with paint; green for sand and gravel (indicating money), yellow for pure sand (caution), and red (stop) for the intervening clay layers. The water table was used as the common elevation datum on the dowels, so above- and below-water deposits and surface topography were evident. This was extremely time consuming but for complex deposits yielded valuable detail for the planning process and indicated where additional drilling detail was required. Today, 3-D mine planning software and computer-aided design (CAD) programs can quickly construct the necessary 3-D models. This software is common today but was virtually unavailable in the 1980s.

Along with the 3-D geological model, a detailed property map showing property lines, surface water features, utilities, neighboring properties, and easements will be needed as a starting point. The location of the processing facilities, stockpile areas, scale office, and entrance road must be determined. If there is a portion of the property that has been determined to be unminable, or contains excessive overburden or an inferior-quality deposit, that is the best target area for locating the processing infrastructure. There is no sense in covering up good reserves if avoidable. The size of these areas will depend on the plant design, the anticipated levels of inventory needed to meet sales demand, ability to connect with highways, length of entrance roads, and so on. This can entail substantial acreage when added together, especially if stockpiling with large radial stacking conveyors.

Once the footprint of the surface infrastructure is established, it's time to familiarize yourself with the state mining regulations and any local zoning conditions specific to the operation. This is where setback regulations are likely to be found. What setbacks are required from adjoining property lines, easements, buildings, public roadways, and so forth? Setbacks usually vary from 50 to 150 feet (15 and 45 meters). As the setback restrictions begin to add up, coupled with the area needed for surface infrastructure, the size of the minable reserve begins to diminish. Are you restricted from piling overburden in the setback areas? Are earthen site barriers permitted on the setback areas? Were there any specific restrictions or conditions associated with the zoning permit? These conditions can at times be more restrictive than state mining regulations. The height of highwalls is sometimes limited by mining regulations, which will affect the benching levels in a quarry. Some mining regulations have restrictions or special regulations pertaining to overburden capable of producing acidic runoff. Do the overburden piles need to be seeded when not actively in use?

Environmental restrictions need to be carefully considered during the mine planning process. What conditions apply to surface water sources on the property? Can a narrow stream be relocated around the operation? Are there any regulatory wetlands that will require mitigation? Where will wash water from the processing plant be discharged off site and what is the potential design of settling ponds? If dewatering is going to take place, where will the water be discharged and what pollution control devices will be required? Can the dewatering, process water, and stormwater be combined and routed through the same settling ponds prior to discharge? If the material transportation will be via trucks on unpaved roads, how will dust be controlled on the haul roads? Do we need to add a water truck to the equipment list?

Consider the concerns expressed by the neighboring property owners at the zoning hearing. Will site barriers be necessary to prevent the neighbors from seeing the unsightly mining operation? Are there already natural site barriers on the property that can be utilized to your advantage, such as natural screening, wooded areas, and favorable topography? Strategically placed earthen site barriers, tree lines, shrub lines, and hillsides can help deflect noise away from the neighbors and back toward the quarry. These can also be helpful in mitigating dust emissions, noise, and extraneous light from nighttime operations. Were there concerns expressed by the neighbors about dewatering? Do observation wells need to be installed surrounding the property to monitor the effects of pumping? Was blasting a concern? Pre-blasting surveys of adjoining buildings can be done by a qualified vibrations consultant, which can help alleviate neighbors' concerns and minimize future liability for the company.

The early question to answer when starting a mining plan is one concerning market. How many tons will be sold in the first year of operation? Will this sales volume be constant or will it increase in subsequent years? Once this question is answered and a net sales requirement is determined, it must be converted into a gross tonnage figure. How many tons have to be extracted from the ground to produce the required annual sales product mix and required inventory? Once this gross tonnage figure is calculated, the volume of stripping or overburden removal necessary to expose this amount of reserve will need to be determined. The initial amount of stripping in the early phases of mining will be much greater than in later phases because of the amount of slopes exposed when first opening an excavation. This initial area of overburden removal must be expanded significantly to include room for haul roads, ramps, safety ledges, berms, and benches necessary for equipment to safely access the next area to be stripped. Because of market fluctuations, it is always a good idea to have an additional year's worth of mineral reserves stripped in advance. An intelligent policy is to have a minimum of 2 years of reserves stripped and available for development at all times.

Dewatering must be considered early in the mine plan because of its multiple effects on the mining, the environment, neighboring properties, and reclamation. The volume of water that will need to be pumped to keep a quarry or pit dry will increase with time and the expansion of the pit. Estimates of water volumes required to dewater an operation can be made from pump tests on wells in the area. A qualified hydrogeologist can evaluate drawdown data from pump testing and make estimates to determine what size pumps will be necessary for initial dewatering. Some producers choose not to dewater their operations for various reasons; for example, regulatory requirements, neighbor relations, or costs. Specialized mining equipment is necessary to remove aggregates from beneath the water table and is further discussed in Chapter 9 on below-water mining.

It is possible to dewater a mining operation in cells, dewatering and mining a small area while developing another cell for dewatering and then leapfrogging from cell to cell. This is very labor-intensive and requires extensive planning and a lot of pumping equipment. Sand and gravel, being very porous and permeable, generally holds much larger volumes of water than solid rock. Certain rock types like limestone can contain channels, sink holes, and cave systems capable of generating very large volumes of groundwater. These karst features are not always apparent from the surface or from core drilling. Limestone deposits with obvious sinkholes on the surface may require additional hydrogeological investigation to predict the flow of subsurface groundwater.

Initial development work in a saturated stone deposit can be a frustrating experience requiring a large amount of tenacity. After the initial lift blasts, ramping downward will require multiple portable pumps and the establishment of numerous temporary sumps as the ramp slowly progresses. Developing a new ramp that would normally take 1 week's excavation may take 3–4 weeks under saturated conditions. Settling ponds will need to be utilized during this early ramp development because the water will contain appreciably more suspended solids than normal and the environmental protection officials and neighbors will be diligently watching during this initial developmental stage of the operation (Figure 8.1).

A comprehensive mine plan consists of numerous different individual operations that need to be coordinated in logical sequence. Overburden removal, dewatering, drilling, blasting, excavation, and transportation to the processing plant all need to be choreographed in an efficient manner to ensure profitability. Roadways and ramps need to be planned well in advance to minimize haulage distances and relocation of roadways. Ramp inclines will be dependent on the equipment utilized and should be kept to a minimum (10% or less) to ensure equipment longevity. Bench widths need to be planned to allow safe access based on equipment sizes and turning radius. Bench heights may be dependent on material quality or can be planned based on optimum drilling and blasting efficiency or ultimate pit limits.

FIGURE 8.1 Maneuvering a portable pump into position during ramp development

Mine plan maps should be updated on a regular basis to reflect stripping limits, active mining benches, and percentage of material coming from each individual bench. The mining sequence for the upcoming quarter, end of year, following year, and following 5 years should be outlined in the mine plan. Deviations from the mine plan should require approval from upper management and need justification as to the reasoning. Once a plan is agreed upon by engineering and production, it should be shared with all stakeholders to ensure compliance and ultimate long-term success. Periodic re-evaluation of mine plans should be scheduled on a regular basis (every 5 years) or whenever significant changes in quality, cost, or market occur.

Some deposits are much more complicated than others. A quarry with a 70 foot (21 meter)–thick dolomite ledge overlain by 9 feet (3 meters) of clay soil overburden is much simpler to plan and operate than a deposit with multiple different layers of stone separated by intervening deposits that require stripping. Deposits containing benches with different levels of quality or intervening layers of poor-quality material (e.g., shale between two limestone layers) that need to be removed prior to reaching the next bench require more detailed planning and frequency of review. Multiple benches of divergent quality will also require separate stockpiles at the processing plant and careful quality control to prevent stockpile cross contamination.

I managed a quarry in the Midwest at one point that contained some interesting geology and very challenging mine planning, managing, processing, and inventory issues. The state department of transportation (DOT) at the time allowed for several different quality classifications for aggregates, with Class A approved for all specification uses and Classes B through D allowed limited uses based on quality test results. The overlying sand and gravel at this site was mined and processed through a separate processing plant from the crushed stone products, which were produced from the underlying bedrock deposits. Between the outwash sand and gravel and the first limestone ledge was an organic shale layer that averaged 27 feet (8 meters) in thickness. This shale bench was shot and hauled away to a dump site. The first limestone bench was 72 feet (22 meters) thick and classified as Class A by the DOT.

Underlying the Class A ledge was 44 feet (13 meters) of dolomite, and because it was highly absorptive it would not meet specifications for DOT-approved Class A aggregates. This dolomite was only approved for use in base products or as a commercial aggregate product. Because the dolomite was underlain by an additional 135 feet (41 meters) of Class A material, it had to be produced to access the next two proposed ledges in the future. The logistical challenges of excavating and processing the sand and gravel, dewatering the quarry, stripping and wasting the shale layer, mining and processing two different quality levels of crushed stone benches, and managing multiple different stockpiles were enormous. Meticulous mine planning is an absolute necessity, especially in complex operations like the one detailed above (Figure 8.2).

OVERBURDEN REMOVAL

Very few mining operations are not affected by overburden. Overburden is any material overlying the ore zone that is not usable or salable in the mining process. Overburden consists of any weathered materials, soil, or subsequent surficial materials deposited after the original mining (ore-bearing) horizon (Figure 8.3). After defining the area to be stripped, the next question is what to do with the overburden removed during this process. As overburden is excavated and removed, whether it is topsoil, subsoil, or solid rock, the material swells from the original in-place volume. Most soils and clays will expand by approximately 20% when excavated; solid

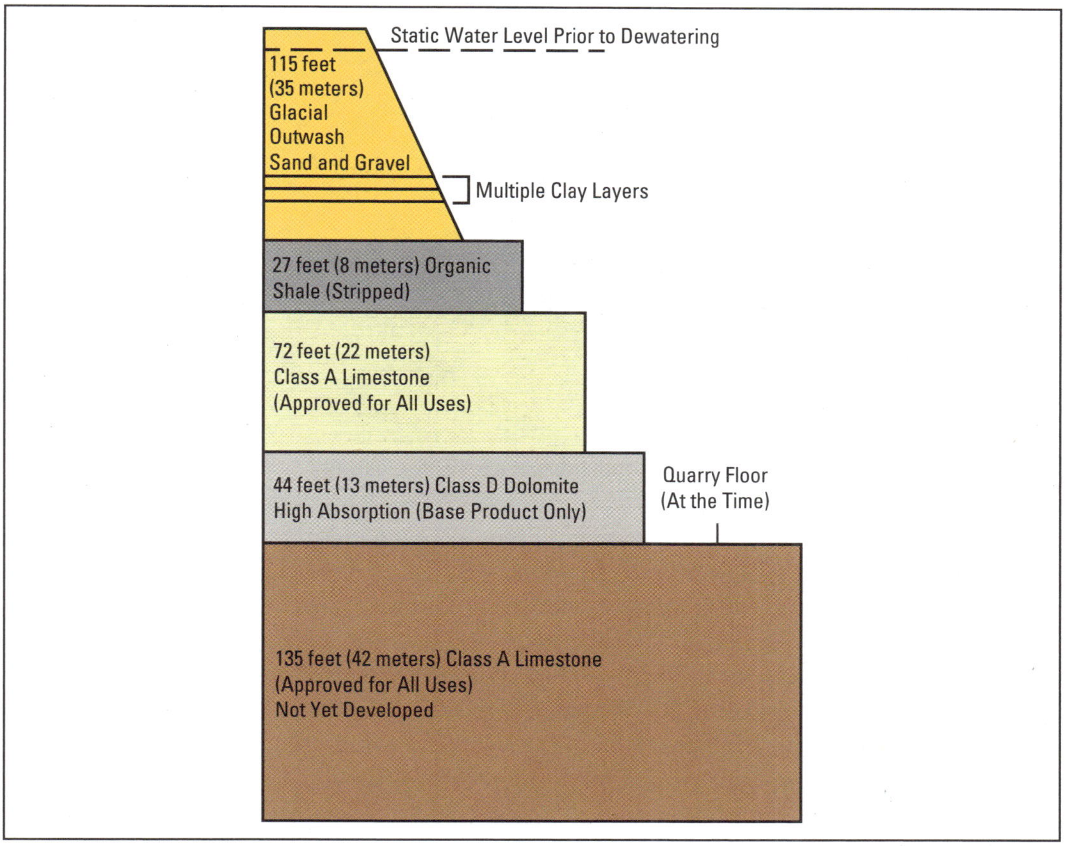

FIGURE 8.2 Complex mining section

rock overburden that needs to be blasted or ripped may increase in volume by as much as 30%. Topsoil should be stockpiled separately from the less fertile subsoils. Topsoil can be a source of additional revenue in many market areas but may need to be mechanically shredded prior to selling. A good source of initial overburden disposal is to utilize it in building earthen site barriers at strategic locations around the property. These site barriers will need to be seeded to prevent erosion and off-property sediment pollution. Topsoil should be utilized to top off the site barriers to encourage initial vegetative growth.

Many producers will hire outside earthmoving contractors to do their stripping. Overburden removal is generally a seasonal operation but requires a substantial capital investment in the necessary mobile equipment. Most contractors will quote earthmoving work on a price per cubic yard basis. Pre-stripping and post-stripping surveys are a necessity to determine the actual volume of overburden moved before final payment. These surveys can be done with conventional survey methods but newer drone-based technology can do this accurately and relatively inexpensively. Always be sure to specify in your contracts with the earthmoving companies that in-place cubic yards will be used to determine payment, so there is no question at the end of the job what volumes are the basis for payment. This puts the responsibility on the contractor to determine the swell factors involved. Using a contractor to strip during the winter off-season when construction projects are at an ebb can sometimes help decrease the contracted price per cubic yard. Many contractors will charge less per cubic yard during

FIGURE 8.3 Stripping shale overburden at a limestone quarry

the off-season in an effort to keep their employees working and cover the depreciation costs on their equipment.

I have had many heated disputes with excavating contractors over the years concerning the volumes of materials excavated at the conclusion of a contract. Our contracts always plainly stated that we were paying for in-place cubic yards and that the volume would be determined by a survey performed by the owner. We encouraged the contractors to do their own surveys for comparison. I vividly remember one dispute where we disagreed by approximately 30% on the volume of overburden removed on one contract. Naturally, the contractor believed he had moved 30% more yardage than did the owner. Most contractors will keep a daily load count multiplied by the truck capacity to calculate the volume they have moved. We had both completed pre- and post-stripping surveys of the property. The contractor sent his survey crew chief to see me with drawings and cross sections that he had used to complete his volume determinations. The dispute was quickly settled when I asked the crew chief how many survey points he had used to obtain his results; his answer was 160. When I produced our in-house survey, which was completed utilizing GPS with more than 1,000 survey points, he quickly acquiesced and agreed to our total survey volume.

Overburden removal can be a costly proposition depending on its composition, its thickness, and the distance it has to be hauled. Rehandling overburden multiplies the cost and should be avoided as much as possible. In the first few years of mine development, overburden may need to be stockpiled if it cannot be utilized in earthen site barriers or possibly sold for fill. Most state reclamation legislation defines the maximum final slope angles that must be created to receive final release of bond. Developing the perimeter of a pit or quarry early in the mining process provides a place for the disposal of overburden with minimal rehandling. As the next area is being stripped, the current overburden can be used to fill the perimeter area previously mined or utilized to slope highwalls for final reclamation (Figure 8.4). This type of mine planning reduces the amount of overburden rehandling and substantially reduces costs over the life of the mine. If overburden disposal is not carefully planned and incorporated into the reclamation process, a producer can be faced with a huge reclamation liability at the end of the mining life cycle. Contemporaneous mining and reclamation must always be incorporated into the mine plan. This will also reduce the amount of bond money held in escrow by the reclamation regulatory agency. Some state mining laws limit the amount of land that can be actively mined in any one year.

Overburden removal can be accomplished in several ways. Above-water earthen materials can be excavated and moved by push-pull scrapers (Figure 8.5) and bulldozers, but this type of equipment is extremely large and very difficult (costly) to move from one site to another. Most of today's excavation contractors utilize articulated haul trucks and excavators as seen in Figures 8.3 and 8.4 rather than incur the costs of moving the oversized scrapers. Earthen material below the water table can also be excavated using articulated haul trucks and excavators, but finding the interface between overburden and aggregate deposit can become somewhat difficult if underwater.

FIGURE 8.4 Slope reclamation using overburden at a sand and gravel operation

FIGURE 8.5 Caterpillar push-pull scraper

Inferior-quality rock overlying the mining horizon (shale, sandstone, tuff, etc.) may need to be drilled and blasted prior to removal. This obviously increases the cost of stripping. This type of overburden can be excavated by the use of articulated haul trucks and excavators or by more conventional mining methods using wheel loaders or shovels to load rigid-frame haul trucks. Depending on the composition of the rock to be stripped, bulldozers with rippers may eliminate the need for the drilling and blasting process. If the rock is fairly thin, laminated, friable, or heavily weathered, the use of dozers with rippers may be a viable option.

Mine planning in underground mines requires even more detailed planning than surface mining. Mining with the room-and-pillar method, which is most often used in underground aggregates operations, requires detailed analysis of the mechanical properties of the host rock to design pillar size, opening widths, room heights, and benching options. Mechanical properties of compressive strength, triaxial compressive strength, indirect tensile strength, modulus of rupture, and point load strength will need to be determined early in the design process. A qualified mining engineer will be needed in the design and planning process of underground mines. Regional structural characteristics of the host rock will also need to be considered (faulting, regional jointing, etc.). This is discussed in more detail in Chapter 11 on underground mining.

CHAPTER 9

Below-Water Mining

The type and size of equipment required for a specific operation varies considerably depending on the type of rock, production requirements, property size, thickness and depth of deposit, mine plan, dewatering, climate, and numerous other factors. We'll start by discussing the types of underwater excavating and mining equipment. Underwater deposits are saturated when excavated and need to be drained prior to transport and processing. It is generally necessary to pile up the run-of-mine deposits adjacent to the shore in windrows and allow a 12- to 24-hour drainage interval before loading into trucks or feeding on to a conveyor belt.

Types of underwater excavation equipment include:

- Hydraulic excavators
- Draglines
- Clamshell cranes
- Wireline scrapers
- Bucket ladder dredges
- Dipper dredges
- Grab dredges (clamshell dredges)
- Suction dredges

Mining beneath the water table can be a difficult proposition because it is impossible for the equipment operator to see what has been excavated and what has not. The bottoms of lakes excavated to recover sand and gravel deposits when examined tend to be very uneven, with ridges of mineral deposits left behind because of the inability to see what is being mined. Periodic lake-bottom surveys are extremely beneficial in analyzing recovery rates of underwater deposits. Sonar systems linked with GPS survey equipment can be utilized to efficiently map lake bottoms. A sonar sensor is submerged beneath a boat and readings taken at specific intervals while the GPS system marks the location of each reading. The boat is then steered across the lake in a grid system to obtain as many sonar depths as reasonably possible. To complete the survey, GPS readings are taken along the shoreline for an accurate map of the lake bottom.

During a due diligence project on a potential acquisition, I used this survey method in an attempt to quantify reserves at the site. The site had been mined for more than 20 years and consisted of four separate lakes with very little solid ground except for a minimal amount of shoreline. The owner of the company for sale, of course, claimed a substantial amount of reserves were still available at depth beneath the lake bottoms. His equipment had a limited digging depth capacity and he assumed the gravel extended considerably deeper than he had mined. Our reserve analysis consisted of drilling test holes using sonic drills around the perimeter and between the lakes. While performing the test drilling process, I had a survey crew complete lake-bottom surveys of all the existing lakes. Based on the test drilling, we compiled a total estimated reserve for the property. The next step was to calculate the volume of reserves previously removed by the owner based on the lake-bottom survey and subtracting that from the total reserve. The final analysis concluded that the reserves were insufficient to justify the purchase. Too bad for the seller, good for us!

MECHANICAL EXCAVATION

Tracked hydraulic excavators (large backhoes) have the ability to efficiently excavate unconsolidated deposits below the water table. This type of machinery is very common in the construction industry, readily available, and easily acquired. The larger class of excavators range from 175 up to 525 horsepower and they can weigh up to 200,000 pounds (91,000 kilograms). These machines have a digging depth capacity in the range of 25–30 feet (8–9 meters) depending on bucket size, boom length, and stick length (Figure 9.1). Because of the weight distribution of these machines, the larger the bucket capacity, the lower the digging depth. Too large of a loaded bucket will cause the machine to topple onto its nose into the water. Larger-capacity machines are limited to 5–6 cubic yard (3.8–4.6 cubic meter) buckets, which can excavate between 7.5 and 9 metric tons per pass, a percentage of which is water. Special excavator buckets with holes or slots designed to facilitate water drainage are used in this application.

Long-reach excavators are available to make deeper cuts, but the production becomes very limited because of the small bucket size; 1 cubic yard (0.8 cubic meter) or less. Hydraulic excavators are designed to dig trenches and special safety training for operators is necessary to make certain they do not undercut the machine while excavating unconsolidated material from a lake. Specifications and capacities, including range-of-motion charts, are available from equipment manufacturers to assist in sizing the correct machines. Hydraulic excavators can be mounted on pontoons and floated on the water (dipper dredges), but this extends the limitations of their depth capacity by the amount of freeboard of the pontoons.

Draglines are commonly used to mine wet sand and gravel deposits and come in a variety of models and sizes (Figure 9.2). A dragline is a type of mobile crane with an additional winch drum utilized for the drag function on the bucket. Draglines can be relatively small, portable, diesel-powered machines or very large and electrically powered. The motive force on some draglines is via crawler tracks (like a bulldozer) or the larger machines have cam-operated

FIGURE 9.1 New 190,000-pound (86,000-kilogram) excavator being delivered to a gravel site

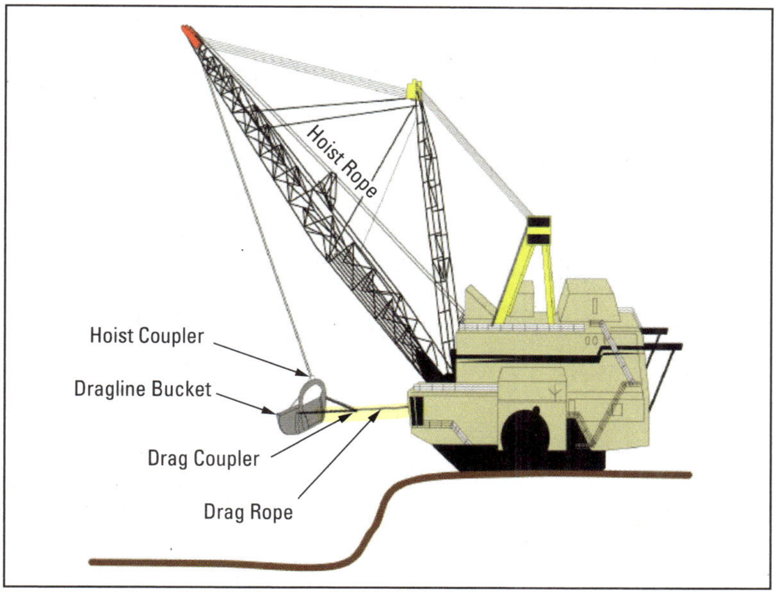

Adapted from Toy and Foster 1998

FIGURE 9.2 Dragline schematic

walking mechanisms that lift the machine off the ground and move it forward one step at a time. The dragline boom is a lattice structure with a large pulley known as the point sheave on the top end for the hoist rope (wire rope) to operate over. The angle on the dragline boom is usually set at about 30° off horizontal. Draglines are specifically designed to dig beneath the level on which they operate and are not very efficient at digging above the operating level.

Draglines designed specifically for mining operations are some of the largest pieces of machinery ever built and can cost several hundred million dollars. The largest draglines are commonly used in the coal mining industry to remove massive amounts of overburden from the top of coal seams. Big Muskie was one of the largest draglines ever built, weighing 13,800 metric tons and having a bucket capacity of 169 cubic yards (129 cubic meters). It was owned and operated by the Central Ohio Coal Company and was retired in the early 1990s. The largest production dragline available today is the Caterpillar Model 8750, which can be ordered with an operating weight ranging from 13.1 to 15.9 million pounds (5.8 to 7.1 million kilograms), a boom length from 360 to 435 feet (110 to 133 meters), and a bucket capacity from 100 to 152 cubic yards (76 to 116 cubic meters). The majority of draglines operating in sand and gravel production are much smaller, with bucket capacities in the range of 6–15 cubic yards (4.5–11 cubic meters) and are commonly diesel powered. However, several machines in Florida used for excavating blasted limestone from underwater are electrically powered, walking draglines in the 20–40 cubic yard (15–30 cubic meter) size range.

The specially designed dragline bucket is dropped to the bottom of the cut by the hoist rope and the drag rope (dragline) is then pulled toward the machine, filling the bucket. When full, the bucket is lifted by the hoist rope while the machine is rotated 90° to 120° (Figure 9.3). The drag rope is then released, allowing the bucket to tilt and empty its load onto the run-of-mine stockpile. The length of the dragline boom controls the digging depth as well as the stockpiling capacity of these machines. When excavating underwater, a good operator is capable of digging

FIGURE 9.3 Loaded dragline bucket breaking the water's surface

FIGURE 9.4 Large dragline in an underwater limestone quarry (note the bucket size in relation to the service truck)

to a depth of approximately one-third the length of the boom and half the boom length in a dry bank operation.

While normally associated with sand and gravel and coal mining, draglines can be used to excavate blasted limestone from beneath the water table as well. In Florida and Georgia where the groundwater is plentiful and very close to the surface, limestone is drilled and blasted underwater (a specialized form of blasting) and the blasted stone is excavated using large draglines (Figure 9.4). These limestones are generally poorly cemented and fragment easily into small pieces. The blasted stone is stockpiled into windrows by the draglines and after draining is loaded into haul trucks or onto conveyor belts for transport to the processing plant.

Another type of crane-based excavating equipment is a clamshell excavator (Figure 9.5). The term *clamshell* applies to the type of bucket used, which resembles the shell of a clam with two halves hinged at the top, closing together at the bottom. These machines are more commonly utilized in stockpiling operations or barge unloading, but under the right conditions can be used in the excavation process. Clamshells are best suited for use in free-flowing materials containing a minimal amount of clay.

FIGURE 9.5 Clamshell crane

Wireline scrapers are another type of excavator utilized in sand and gravel excavation, but they are somewhat less common than draglines (Figure 9.6). The crescent-shaped bottomless bucket known as a Sauerman scraper was originally patented in 1924 by Harry Roe. The Sauerman Brothers Company improved on the original design many times over the years. Several varieties of this bucket still exist today. A wireline known as the skyline is extended from the tower on a portable winch system to a heavy weight (known as a tailhold or deadman) usually located on the opposite side of the cut. The bucket is suspended from the skyline by a pulley system and when the bucket sinks to the bottom of the lake a dragline attached to the winch system pulls the bucket toward the winching apparatus. The bottomless, crescent-shaped bucket drags and scrapes the sand and gravel from the lake bottom and deposits the material in front of the machine. The empty bucket is then lifted and travels back toward the tailhold to repeat the process.

The tailhold on the opposite side of the lake is movable and the cut progresses in a fan shape with successive moves of the tailhold until the final depth is achieved. Many systems use a bulldozer as the tailhold or deadman to facilitate faster moves of the system. Sauerman buckets range from 3 to 16 cubic yards (2.25 to 12 cubic meter) in size. The cycle times on the bucket are dependent on the length and depth of the cut. As the cut gets deeper or longer, the cycle times decrease. The capital and operating costs of this type of system are moderate. A similar type of equipment utilized in the logging industry to drag felled trees up mountain sides, known as "yarders," has been converted to excavators utilizing Sauerman scraper buckets. Large cut lengths can be achieved by adding more cable, but again the longer the cut the longer the cycle time. Cable replacement costs on these units are very high because of the abrasive nature of the sand in the deposits. This system requires rehandling of the material by a wheel loader to remove the sand and gravel from in front of the machine. The area in front of the machine used for stockpiling is very limited and removal by a second piece of equipment is a constant necessity.

Much like draglines, dredges are extensively used to mine underwater deposits of sand and gravel, and at times underwater blasted limestone. Dredges were originally invented by the Dutch in the 1600s to reclaim their lowlands from the North Sea (Figure 9.7). Dredges float on top of the water supported by a number of pontoons. The density of water is 62.4 pounds

Courtesy of Chris Nelson

FIGURE 9.6 Wireline scraper excavation of sand and gravel

Photographer unknown. National Dredging Museum at Sliedrecht.

FIGURE 9.7 Model of a 1600s Dutch human-powered dredge

per cubic foot (1,000 kilograms per cubic meter); therefore, every cubic foot of air displacing the water inside of a pontoon has the ability to support 62.4 pounds (1,000 kilograms) of machinery. Determining the weight of the equipment that needs to be floated (including the weight of the pontoons) will determine the size, depth, and number of pontoons required. Additional pontoon capacity will be needed to provide sufficient freeboard (the required height of the deck above the water line) to prevent capsize. If 5 feet (1.5 meters) of freeboard is required, an additional pontoon capacity will be needed to raise the level by 5 more feet.

Several different types of dredges are in use in the aggregates mining industry. Dredges fall into the following categories:

- Bucket ladder dredges
- Clamshell or grab dredges
  - Digging well (box design pontoos)
  - Catamaran-style pontoons
  - Luffing
- Suction dredges
  - Hull-mounted pump
  - Ladder-mounted pump
  - Cutterhead
  - Bucket wheel

Bucket ladder dredges have been widely used in the mining industry from sand and gravel deposits to placer gold mining (Figure 9.8). These machines are generally capable of excavating very dense materials as well as soft soils depending on the bucket design. Bucket ladder dredges consist of a series of cast iron or welded buckets attached to a chain that is pulled, on rollers, along the length of an angled, adjustable structure known as the ladder. A large sprocket at the top of the ladder pulls the bucket chain upward. As the buckets are tipped over the drive sprocket, the contents fall into a chute that can be directed over the side into a barge or onto a conveyor belt for transportation to the shore. These dredges can be powered by either diesel, diesel electric, diesel hydraulic, or direct electrical power.

Production from a bucket ladder dredge is a function of the speed of the chain drive, the horsepower on the drive, the bucket size, number of buckets, the density of the material, and the fill factor of the bucket. The fill factor depends on the angle of the ladder as the buckets meet the mining face. Bucket ladders are very inefficient at shallow depths because the angle of attack of the ladder prevents maximum bucket fill. Most bucket ladder dredges are limited to an excavation depth in the range of 70–80 feet (21–24 meters) because as the size of the ladder and number of buckets increase, the power required to move the extra load increases substantially. Bucket ladder dredges require a large capital expenditure, are maintenance intensive, and consume a large amount of power. The mining direction is controlled by six winches anchored

FIGURE 9.8 Feed end of a bucket ladder on a retired gold dredge in Oregon

Courtesy of ROHR-IDRECO

FIGURE 9.9 Shore-mounted bucket ladder

to shore or to the lake bottom by wire rope, three on front and three on the back. The two swing winches on the front maneuver the dredge from left to right along the mining face and the center front winch pulls the dredge forward into the cut. The three rear winches provide stability, maintain the course, and help hold the dredge against the cut. Shore-mounted bucket ladders are sometimes utilized and are propelled parallel to the shore on crawlers (Figure 9.9). These shore-mounted bucket ladders are much more commonplace in Europe than in the United States.

The technology for grab dredges, also known as clamshell dredges, arrived in the United States from Germany in the 1980s. These machines are essentially a combination of an overhead crane and clamshell excavator mounted on a floating platform. These dredges are a sophisticated combination of components engineered to operate together to enable high production rates and deep digging depths. The components of clamshell dredges consist of the following:

- Flotation (pontoons)
- Main hoist
- Bucket
- Receiving hopper
- Grizzly system
- Hopper feed gates
- Coarse dewatering screen
- Fines dewatering screen
- Sand pump and cyclones
- Steering winches
- Hydraulic system
- Electrical control system
- Ground-fault circuit interrupter (GFCI) mine switch
- Waste pump (optional)
- Crusher (optional)

Grab dredges are electrically powered. The power can be supplied via power line or can be provided by a diesel generator on a pontoon floating beside the dredge. Utility-supplied power is the preferred option because it eliminates the environmental liability of transporting diesel fuel across the water to feed the generator and is generally a less expensive power option. A GFCI or mine switch mounted on the shoreline is a necessary safety precaution when running power lines across the open water and is also a Mine Safety and Health Administration requirement.

Mounted high above the pontoons is a gantry structure, on top of which is mounted a winch house containing the main hoisting winch, winch brake controls, and a separate winch for the bucket power cable. The winch house is mounted on rollers and can be driven along the length of the gantry very much like an overhead crane in a factory or warehouse. The digging depth capacity is determined by the size, wire rope capacity, and horsepower of the main winch. Suspended by large-diameter (1.5–2 inches [38–50.8 millimeters]) wire rope from the main hoist is the dredge bucket. Dredge buckets range in size from 3 to 20 cubic yard (2.2 to 15 cubic meter) capacity (4.5–30 metric tons). The dredge bucket is a principal component of the system and contains, inside a sealed compartment, a self-contained hydraulic system for the operation of the hydraulic cylinders that open and close the bucket. This hydraulic system within the bucket contains a reservoir of hydraulic oil, hydraulic oil filters, a hydraulic pump, a complex hydraulic valve system, and an electric pump motor. The bucket motor is powered by an electrical cable spooled by a winch located in the winch house.

The bucket design and horsepower will be specific to the deposit application. Free-flowing sand and gravel without clay layers or hardpan can easily be dug with a standard bucket. For deposits containing hard, dense material, massive clay layers, or layers of glacial till, a heavy-duty, aggressive bucket should be specified. A standard bucket will usually have bucket shells that have flat sides, straight lips, and square corners. The heavy-duty bucket will have additional horsepower on the hydraulic system, larger hydraulic cylinders, heavier steel plates, massive gussets, and aggressive, replaceable bucket teeth. The bucket shells (actual digging surfaces) are wear items that need regular inspection, regular rebuild, and periodic replacement. Bucket shells on the heavy-duty models have rounded corners, rounded bucket lips, and aggressive teeth to help penetrate very hard formations (Figure 9.10). These shells are difficult to manufacture because of the complex geometry of their

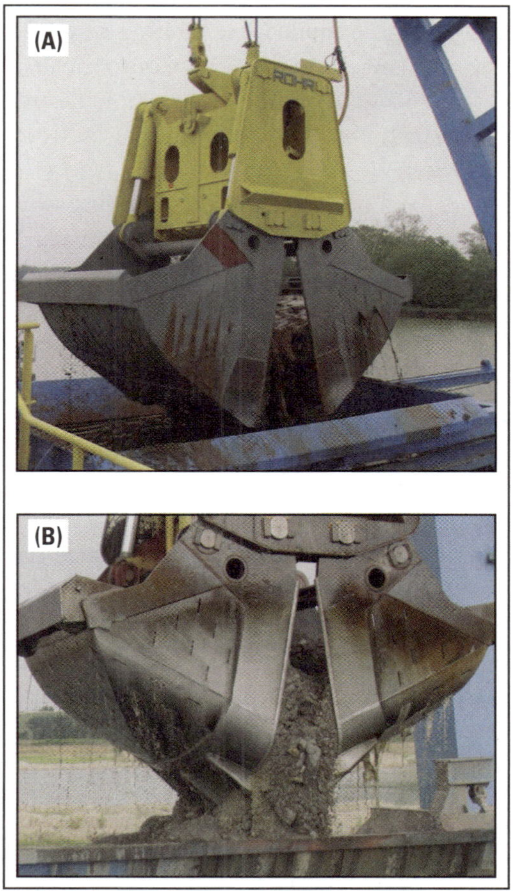

FIGURE 9.10 (A) Standard bucket, and (B) aggressive heavy-duty bucket

shape and are considerably more expensive than the standard models. All bucket shells have machined slots or holes designed to help drain the excess water from the bucket.

After excavating a bucket of material and hoisting it to the surface, the gantry crane moves the bucket over top of the receiving hopper on the dredge. Following a specified drainage interval to remove excess water, the bucket is opened and the load is dumped into the receiving hopper. The hopper usually has the capacity to hold two bucketloads of material. The dredge itself is a floating processing plant that screens the material into different size ranges to remove the water and then recombines them to enable conveyor transport. At the bottom of the receiving hopper is a series of finger gates across the discharge opening. These gates are operated by a separate hydraulic system that controls the feed rate of the mined material onto the coarse dewatering screen. The material being excavated needs to be dewatered to be conveyed to the shore via the floating conveyor belt system. If the material contains too much moisture, it will flow off the conveyor belts back into the lake.

The coarse dewatering screen usually is a two-deck horizontal screen with the top deck scalping the oversize material to a crusher (if the dredge is so equipped). The second (bottom) deck separates the gravel sizes from the finer sand sizes, with the gravel being sent onto the conveyor. The fine material (sand), together with all the water, falls through the bottom deck into a sump where it is pumped to a series of cyclones for further dewatering. The cyclones discharge onto a high-frequency dewatering screen for final water removal prior to being recombined with the gravel and discharged onto the floating conveyor for transport to the processing plant (Figure 9.11). If mining a deposit with excess fines, some dredges utilize an additional waste pump to pump the excess fines away from the dredge to minimize recycling them through the system.

Affixed to the top of the receiving hopper will be a grizzly scalping system that prevents unwanted oversized material and large clay lumps (which could clog or bridge) from entering the hopper (Figure 9.12). If these systems are set up to dump the oversized material back into the lake beside the dredge, the same large boulders may be recycled through the system numerous times until the dredge moves to another location (Figure 9.13). Some producers utilize a separate, powered, rock or clay barge to move the grizzly oversized material to shore or dump it in another part of the lake that has already been mined. Large clay lumps can be broken up by this reoccurring cycle. Rake grizzlies are cable or hydraulically operated and the teeth remove oversized material from between parallel grizzly bars that are spaced 8–12 inches (203–305 millimeters) apart using a raking motion. The bars on dump grizzlies generally are formed into a grid system with 10 to 12 inch (254 to 304 millimeters) square openings. When the oversized material builds up on the grizzly, hydraulic cylinders tilt one side of the hinged

FIGURE 9.11 Sand discharged from a dewatering screen

FIGURE 9.12 (A) Rake grizzly, and (B) dump grizzly

FIGURE 9.13 Recycled boulder in wobble feeder

system and dump the material into a chute that can either feed a rock barge or dump the material directly into the lake to be recycled back to the dredge.

The dewatered material containing 10%–15% moisture is then conveyed via floating conveyors across the lake to a surge pile or directly to the processing plant for further reduction, screening, and classification. Floating conveyor systems, like the dredges, depend on pontoons for their flotation. Floating conveyors come in various pontoon designs, conveyor widths, and lengths to fit the application. Long floating conveyors need to be anchored to the shore via wire ropes, especially during windy conditions. It is desirable to have winches on the pontoon nearest the conveyor tail section to easily maneuver the conveyor during dredge moves. The water-to-land transition conveyor section is designed so the tail end of the conveyor floats and the discharge end is attached to a stanchion on the shore where the conveyor pivots on both the vertical and horizontal planes. If utility-supplied electrical power is utilized, the floating conveyor also carries the power supply cable in a specially designed, isolated cable tray attached to the conveyor truss. The electrical supply can be of any voltage and is usually transformed to 480 V prior to the motor control center (MCC). I personally prefer conveyor pontoons that run parallel to the conveyor and serve as a walkway like those seen in Figure 9.14. This type of pontoon is less likely to flip over during violent wind events than those mounted on the head and tail sections only.

Various crusher types can be mounted on dredges for dealing with coarser deposits. Small-jaw crushers are the most common, but horizontal impact crushers are sometimes used. Generally, a small-jaw crusher will be mounted on a separate pontoon alongside the dredge fed by a chute and the crusher will discharge its crushed product back into the lake to be picked up by the dredge bucket on the next pass. Utilization of impact crushers and

FIGURE 9.14 Floating conveyors feeding a radial stacker on shore (note the power cables in cable trays)

vibratory feeders can put unexpected vibrations into the pontoons and eventually cause leakage along the welds. This type of equipment requires some extra engineering to dampen vibration and/or more robust pontoon construction.

Most grab dredges are equipped with the latest electrical systems with variable-frequency drives to save power on motor start-ups. Automation using programmable logic controllers (PLCs) is common on most of these machines. PLCs enable the operator to program the following parameters:

- Bucket descent speed
  - Above water
  - Below water
- Bucket settle time
- Bucket closing time
- Bucket hoist speed
  - Above water
  - Below water
- Bucket drain time
- Bucket opening time
- Over hopper time
- Feed gate open/close times

Some manufacturers offer optional, on-screen computer graphics showing bucket diagnostics, bucket position and tilt, and GPS mapping functions. The electrical systems on the entire dredge system, the floating conveyors, and the onshore stockpiling system need to be electrically interlocked. This interlocking protects all downstream components by shutting them down if one portion of the system fails for an unexpected reason.

Pontoon configurations on gantry-style grab dredges can vary considerably, with a box-style design around a center digging well being the most common. Many dredges have safety sensors in each of the four corners that instantaneously shut down all the machinery on board if for some reason the dredge lists too heavily to one side. Large catamaran-style dredges have two separately spaced pontoon sections with enough room in between to allow a barge to run through the center. Catamaran-style dredges have a very large footprint and are very stable in the water. At times, additional ballast must be added on top of or inside the pontoons to provide a stable, level work platform (Figure 9.15).

Dredging production rates are a function of the bucket size and the digging cycle time of the dredge. As the mining progresses deeper, the cycle time increases and the production is reduced because of the additional bucket hoisting and lowering times. When deep deposits are being mined and large production capacities required, a dredge system with two separate buckets and two hoisting systems can be utilized. As one dredge bucket is being raised and dumped, the second bucket is lowered and digging, doubling the production capacity of the system (Figure 9.16). Double dredges with 20-cubic-yard (15-cubic-meter) buckets are capable of large production volumes, even at digging depths in excess of 200 feet (61 meters).

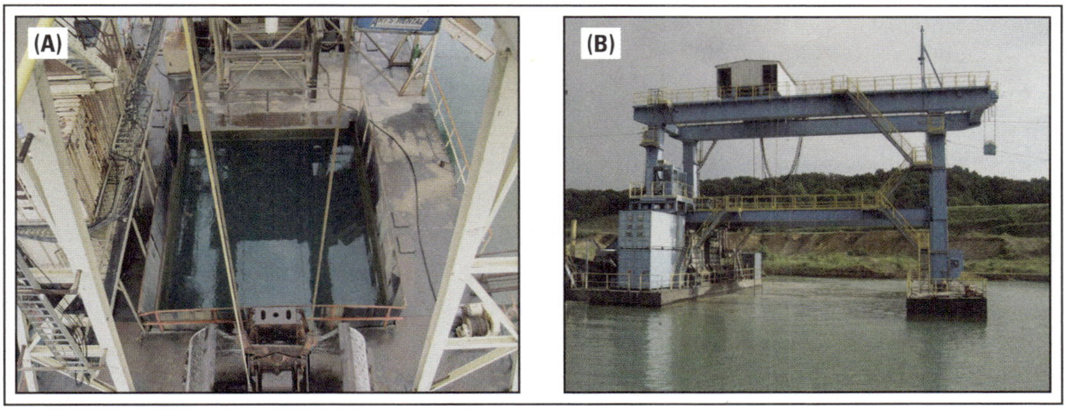

FIGURE 9.15 (A) Digging well, and (B) catamaran dredge

Courtesy of Chris Nelson

FIGURE 9.16 Double-gantry grab dredge

FIGURE 9.17 Luffing dredge in the dumping position

An alternate type of floating grab dredge is the luffing design, which has no gantry and thus no overhead crane (Figure 9.17). A luffing dredge has all the components of a gantry style without the winch house and overhead crane. Instead, the winch is deck-mounted and the cables are reeved over a luffing (tilting) jib. The jib is hinged to the deck and moved forward and backward by a long hydraulic cylinder. In the dump (forward) position the bucket is over the receiving hopper, and in the digging (rear) position the bucket is above the digging well. These dredges are usually smaller in size and capacity than the gantry style. Most luffing dredges have buckets in the 3 to 8 cubic yard (2 to 6 cubic meter) size range. Dewatering screens, cyclones, and floating conveyors are also necessary components on luffing dredges. The same automation is available on the luffing machines and like the gantry-style dredges they are steered by four winches attached to the shore by wire rope, one winch on each corner of the dredge.

Grab dredges are most productive in free-flowing material that will migrate toward the bottom of the cut, which is essentially funnel shaped. To get the best production, the dredge should be continually digging at the bottom of the deposit. As a bucket of material is removed from the bottom of the funnel, the slope becomes unstable and additional material flows down the slope to be excavated by the next bucket. When sand and gravel recovery begins to diminish, the dredge is moved 15–20 feet (4.6–6 meters) into the stabilized slope of the cone and the process is repeated, enlarging the size of the cone. This process of downslope flow also tends to mix the various layers within the deposit. Sand and gravel deposits containing interbedded clay layers can be excavated utilizing heavy-duty buckets. The more aggressive buckets are capable of digging through thick layers of glacial till, which is made up of very densely packed, hard silt and clay with some sand and gravel. Many deposits in the midwestern United States contain thick layers of glacial till interbedded with the sand and gravel. These massive chunks of till can be excavated with a heavy-duty bucket but tend to interfere with the production of salable product until cleaned away.

Grab dredges are capable of mining this type of deposit, but great care must be exercised in the process. After penetrating an underwater layer of till and proceeding downward into the next underlying sand and gravel layer, the funnel-shaped cone will again form with the sand and gravel flowing to the bucket. Problems can occur when the thick till layer forms a bridge.

As the cone beneath the till layer expands, the rigid till layer is no longer supported from beneath and massive pieces of till plus the overlying sand and gravel can break off and fall to the bottom of the cone. As the old saying goes, timing is everything. If the timing is unlucky and the dredge bucket happens to be on the bottom of the cone when a massive break in the bridge occurs, the bucket can end up being buried under tens of tons of material and can be severely damaged in the process (Figures 9.18 and 9.19).

Recovery of a buried bucket can take days or weeks to accomplish and some may never be recovered. Needless to say, when a dredge bucket has a value of between $250,000 and $500,000, a massive effort will be undertaken to recover the bucket. The faster the effort is undertaken, the better, because the material over top of the bucket will settle and compact over time and make recovery efforts even more difficult. Sometimes the recovery

FIGURE 9.18 Bucket shell damaged by a caving layer of glacial till

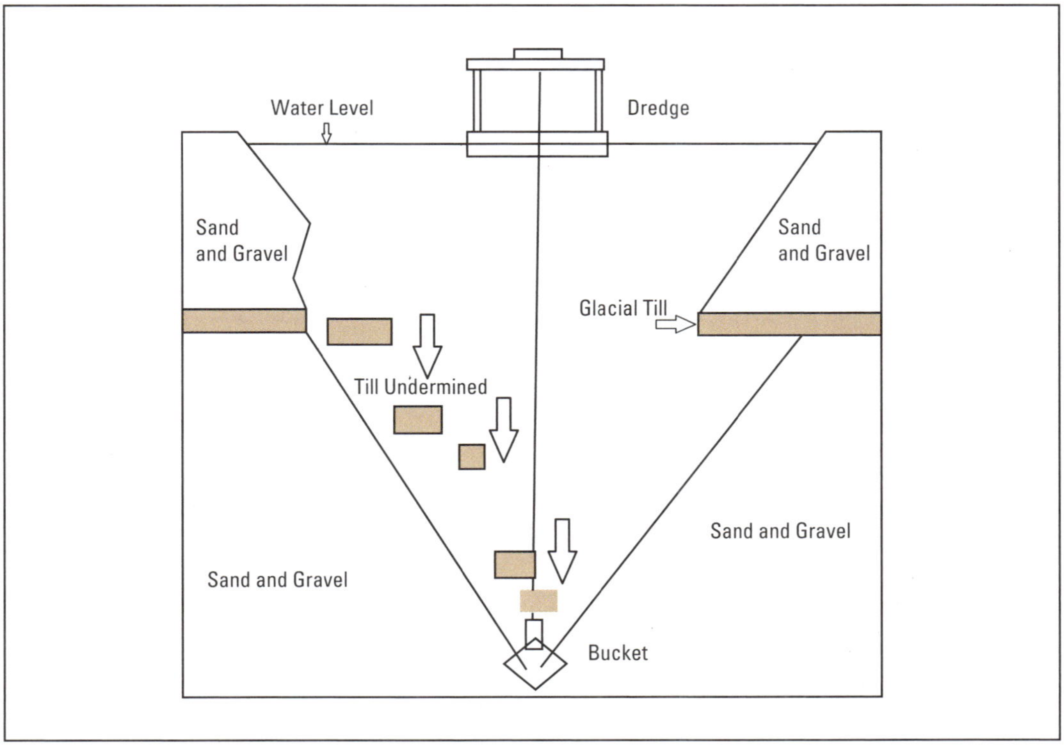

FIGURE 9.19 Cross section of an underwater cave-in

FIGURE 9.20 Capsized dredge with airbags attached

requires the aid of industrial divers and the use of massive airbags attached to the bucket (if accessible) or to the hoist cables. These airlift recovery bags are specially designed for underwater equipment recovery and are rated for their lifting capacity (10 metric tons, 20 metric tons, 50 metric tons). The bags are submerged and attached to the equipment and then inflated with compressed air and are capable of creating a large lifting force depending on the number and size of the bags utilized. Once again, 1 cubic foot (28 liters) of air is capable of lifting 62.4 pounds (1,000 kilograms).

If this type of effort fails, the recovery needs to be attempted by acquiring another bucket and carefully excavating around the buried bucket until enough of the material overlying the bucket is removed to facilitate recovery. Never lose access to the cables attached to the bucket or future recovery will be impossible. If the bucket cables need to be removed from the dredge for recovery, be sure to attach them securely to another pontoon. These buckets are not readily available on-the-shelf items, and few producers want to incur the cost of having a spare bucket on the ground just in case. A producer having multiple grab dredges may have a spare available. These bucket burials are somewhat common when mining layered deposits containing glacial till. I have spent many anxious hours on the decks of different dredges during bucket recovery efforts; another risk in the underwater mining business. The law of averages says that if you operate a grab dredge long enough, you too will experience a buried bucket.

Even more disastrous results can happen when attempting to recover buried buckets. A day I'll never forget started with an early morning phone call from one of my superintendents informing me that his clamshell gantry dredge had just capsized in the lake. After making sure everyone on the dredge was unhurt, I proceeded to drive to the mine site, which was about 30 miles away. The three employees on the dredge at the time were badly shaken and soaking wet, but thankfully unharmed. Luckily, at the time of the accident, the dredge was close to shore in only about 20–30 feet (6–9 meters) of water. The employees on the dredge were attempting to recover a buried bucket and had a maximum lifting force applied to the hoist cable when a second cave-in occurred, forcing the cable to twist radically to one side with enough force to eject the winch house from on top of the gantry and the sideward force of the cables on the gantry capsized the dredge (Figure 9.20).

The recovery effort took almost a month of attempting various methods to refloat the machine. We first attempted lifting the gantry with a 500-metric-ton crane on the shore with no success. Next, we added 100,000 pounds (45,359 kilograms) of lift force using air bags, again without success. The final solution was a combination of several factors. The pontoons below the water line had filled with water, creating additional weight on the bottom. We pumped out and sealed the submerged pontoons to lower the weight on the submerged portion of the dredge. We then filled the pontoons that were cantilevered above the lake level with water to increase the weight above and create a fulcrum. We attached (via a 2-inch [51-millimeter] cable) two wheel loaders and a bulldozer to the gantry pulling laterally across the top of the pontoons. This effort along with the 500-metric-ton crane and the 100,000 pounds (45,359 kilograms) of airlift bags finally flipped the dredge back up to a floating position.

Additional recovery effort was needed because the winch house had detached from the gantry and required separate recovery, as did the bucket that was still buried on the bottom. The cost of the recovery effort was staggering because most of the electrical wiring, many of the motors, and numerous other components needed repair or replacement. Luckily, the MCC remained above the water or the electrical rebuild costs would have been exponentially more. The main hoist winch and power cable winch had to be totally rebuilt before going back into production. This was a very trying time for the many people involved in the recovery effort and a huge relief when it was completed. Needless to say, on top of the recovery costs, the loss of production for several months was very costly also.

HYDRAULIC EXCAVATION

Suction (hydraulic) dredging is another method of removing unconsolidated material from beneath the water table. The same obstacles exist in hydraulic dredging as in clamshell dredging; the risk of cave-ins on top of the cutterhead or ladder, lack of the ability to visualize what is being excavated, and so on. As mentioned in the previous section on grab dredges, suction dredges are also most efficient when operating at the base of the deposit. The initial cut should be taken to the bottom of the deposit so the toe of the natural slope of the material can be removed, allowing the overlying material to flow toward the dredge. This increases efficiency by limiting the number of dredge moves.

The basic components of commonly used hydraulic suction dredge are:

- Flotation (pontoons)
- Power plant (diesel or electric)
- Dredge ladder
- Digging/mechanical agitation
  - Cutterhead
  - Bucketwheel cutterhead
  - No cutterhead
- Centrifugal pump
  - Hull-mounted
  - Ladder-mounted
  - Jet-assist pumps
- Discharge pipeline
- Mooring/propulsion system
  - Spuds (Figure 9.21)
  - Three wire
  - Five wire

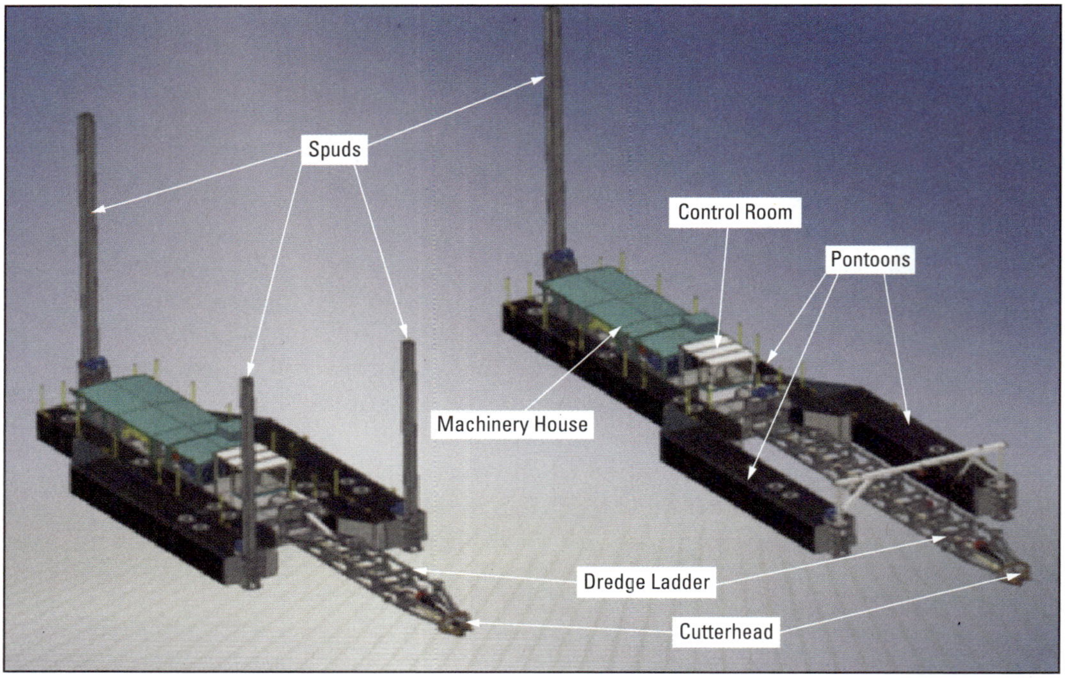

Modified from Dredge Supply Company

FIGURE 9.21 Hydraulic dredges with spuds

The flotation system on suction dredges is very similar to other dredges but is generally constructed in a U-shape to facilitate the dredge ladder, which is hinged to the pontoons at the rear of the ladder. The pontoons are usually rectangular and provide 3–4 feet (0.9–1.2 meters) of freeboard to the deck. The power plant is deck-mounted and can be either all electric, diesel electric, or all diesel powered. If utility-provided electrical power is used, specially designed waterproof power cable is floated on small plastic floats and a GFCI mine switch is required. The dredge ladder is normally a truss structure, the length of which is dependent on the depth of the deposit to be mined. The ladder is hinged to the pontoons on the rear and moved up and down by a deck-mounted cable winch attached to an A-frame mounted between the two front pontoons. The optimum angle of the ladder when mining is in the 40°–50° range from horizontal. Some dredge ladders on larger machines will have a downward turn at the bottom of the ladder to aid the angle of attack into the deposit. If the deposit becomes shallower than expected, the angle of attack can be negatively affected and the percent recovery can diminish. I have, on one occasion, removed a section of dredge ladder when an unmapped and unexpected layer of glacial till reduced the deposit depth from an average of 45 feet (14 meters) to about 20 feet (6 meters). Shortening the ladder actually increased the production rate fairly substantially.

Most deposits require mechanical agitation to increase flow to the pump and to loosen hard-packed layers within the sand and gravel. In deposits containing only sand and in some very free-flowing sand and gravels, suction from the pump alone can be capable of lifting the unconsolidated materials to the surface. I would not recommend the purchase of a dredge system without the ability to mechanically assist in feeding material to the pump.

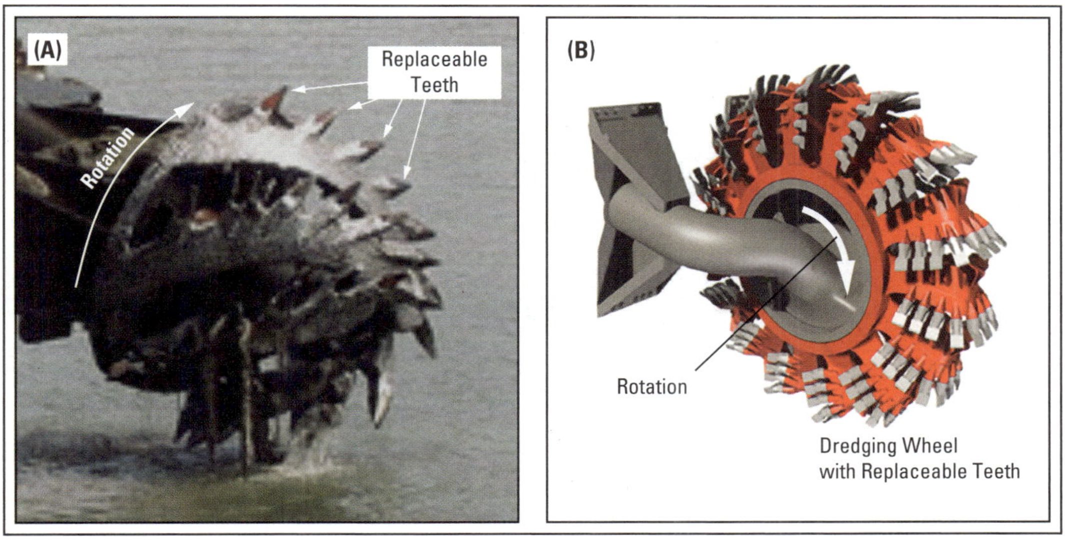

Source: (A) William Alden/Flickr; (B) Courtesy of Royal IHC

FIGURE 9.22 (A) Aggressive cutterhead, and (B) bucketwheel cutterhead

Standard cutterheads are usually hydraulically operated from a hydraulic system mounted on deck. The cutterhead is mounted on the front end of the dredge ladder, usually at a downward angle of 10°–15°. Standard cutterheads that rotate horizontally come in many styles: some are more heavy duty and aggressive than others (Figure 9.22). Most cutterheads are barred off or built like a basket with 6 to 10 inch (152 to 254 millimeter) openings to prevent oversize material from entering the pump and pipeline. The ability of the standard cutterhead to penetrate hard materials is a function of the weight of the ladder, the cutterhead horsepower, the cutterhead diameter, and the number and shape of the teeth. Thin layers of hard material can be excavated with this type of agitation. Because of the horizontal rotation, when encountering thick, hard layers of material like glacial till, standard cutterheads tend to skip across the top of the hard layer rather than penetrating.

Bucketwheel cutterheads work on a similar principle to bucketwheel dredges, but instead of dragging buckets along a ladder with a chain, the buckets are configured in a wheel shape. The buckets rotate vertically, agitate and excavate the deposit, and feed the excavated material to the pump suction intake known as the fish mouth. Bucketwheel cutterheads, if provided with sufficient horsepower and aggressive replaceable teeth, are capable of excavating very hard material including glacial till deposits lying between layers of sand and gravel.

Jet-assist (high-pressure) pumps can also be utilized to aid in the agitation of the deposit. Cutterheads with jet-assist nozzles that rotate are sufficient to agitate certain types of deposits. Jet-assist pumps can also be utilized ahead of the dredge mining face to facilitate the caving of above-water deposits into the lake and thus increasing the flow of material toward the dredge by adding additional feed to the system. There are many types of hybrid dredge cutterhead systems, such as drag chains and auger type, but the standard cutterhead is by far the most common in the mining industry.

The centrifugal pump is the heart of the suction dredge. The dredge pump not only lifts the material from the bottom of the lake, it also transports the sand and gravel in suspension to the processing plant on shore (Figure 9.23). A centrifugal pump converts mechanical energy

Courtesy of American Aggregates

FIGURE 9.23 Two dredges using jet pumps to cave a sand and gravel highwall in Michigan in the 1970s

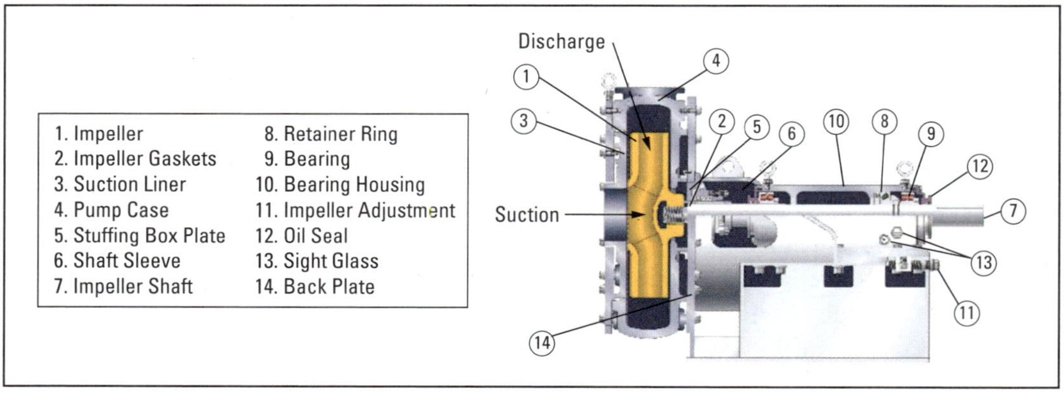

Original source unknown

FIGURE 9.24 Schematic of a centrifugal dredge pump

into kinetic energy according to the Bernoulli equation. Fluid enters the pump through the eye of the impeller, which is rotating at high speed in response to the motor turning the input shaft (mechanical energy). The fluid (water slurry) accelerates outward from the pump casing in a radial motion. The tip speed of the impeller controls the velocity of the liquid energy. A vacuum is created at the eye of the impeller that continuously sucks more slurry into the pump. The process continues until the vacuum at the inlet is broken, the flow at the eye is disrupted, or the discharge is restricted by some pressure greater than the pressure produced by the pump. The discharge pressure is a function of the speed and diameter of the impeller, and the volume of water pumped is a function of the horsepower applied to the drive shaft.

A centrifugal dredge pump is also known as a slurry pump because it must be capable of pumping not only water but also the sand and gravel in suspension (slurry) (Figure 9.24). Dredge pumps are the same basic design as centrifugal water pumps but the impeller and casing (shell) need to be constructed from heavy, abrasion-resistant materials that are easily replaced when worn out by the constant abrasion from sand and gravel particles. Impellers and shells are usually high-chromium steel and some casings are lined with ceramic or rubberized materials. The eye of the volute, which is the distance between the tip of the impeller and the pump casing, is larger in slurry pumps than in standard water pumps to allow large gravel and cobbles to pass. The impellers in slurry pumps generally have fewer vanes than water pumps.

The intake of a centrifugal pump is usually slightly larger than the discharge; an 18-inch (457-millimeter) pump discharge will normally have a 20-inch (508-millimeter) intake (Figure 9.25). The additional diameter helps facilitate the necessary suction (vacuum). Suction dredges are normally classified by the diameter of the pump discharge, for example, 18-inch (457-millimeter) dredge or a 24-inch (610-millimeter) dredge. A centrifugal pump generates much more discharge pressure than negative suction (vacuum) pressure. To get the maximum vacuum, the pump should be located as close to the water as possible inside the dredge hull (Figure 9.26). Most newer dredges have the pump mounted on the dredge ladder beneath the water level when in operation. This adds additional suction head to the pump via the pressure that the overlying water exerts on the system and also eliminates the need to prime the pump. Ladder-mounted (submerged) pumps are capable of almost doubling the percentage of solids in the slurry pumped by the dredge. Mounting the pump on the ladder requires a substantially heavier ladder structure to support the additional weight of the pump and the drive line.

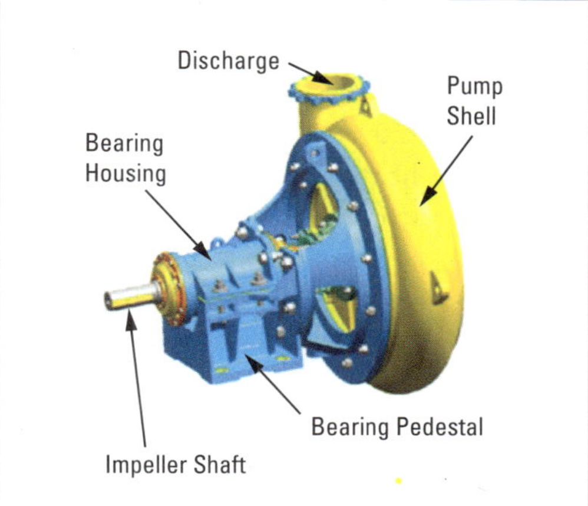

Original source unknown

FIGURE 9.25 Centrifugal dredge pump

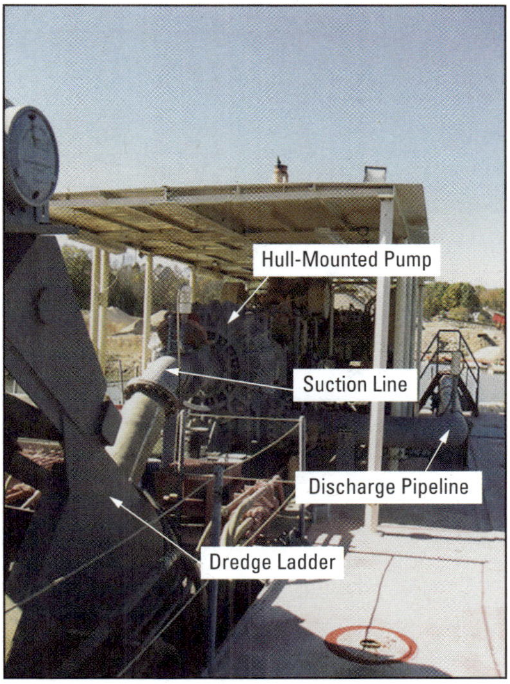

FIGURE 9.26 Hull-mounted dredge pump

The pressure measured on the discharge side of a pump is actually the resistance to flow, which happens immediately as the liquid flowing through the pump casing encounters friction. The slurry flowing through the pump and onward through the attached pipeline is affected by the friction between the slurry and the walls of the pump and pipe. As this pressure increases because of pipe friction with distance from the pump, eventually it will exceed the pump's capability and the flow will cease. Depending on the pump, wear on the impeller, and specific gravity of the slurry, the distance the slurry can be moved is limited. As a general rule,

1,500 feet (457 meters) is about the practical limit for slurry transport. This can be increased by adding a booster pump on the shore or floating a booster pump on the lake using pontoons. The function of a booster pump is to add additional kinetic energy to the system.

Slurry pumps require a considerable amount of power and maintenance to operate at peak efficiency. When adding a booster pump to a dredge system, the power demand (and cost) will increase in the range of 40%. Although booster pumps are widely used in the industry, it has been my experience that it is much more cost-effective to pump the slurry to the shore, dewater it, and utilize conveyors to move the material the remaining distance to the processing plant. This requires more initial planning to maintain dry ground for the conveyor system. In this situation, locating the processing plant in the center of the deposit and mining (dredging) around the processing plant is a good solution. This also quickly opens a perimeter where overburden from the next mining area can be disposed of in the process of slope reclamation (Figure 9.27).

The discharge pipeline is an important component of the transport system of the slurry to the shore (Figure 9.28). The pipeline needs to be supported on the water using pontoons evenly spaced to keep the pipeline afloat. Large-diameter steel pipe is very heavy, especially when filled with a slurry containing 15%–30% solids. Various types of pipe couplings are

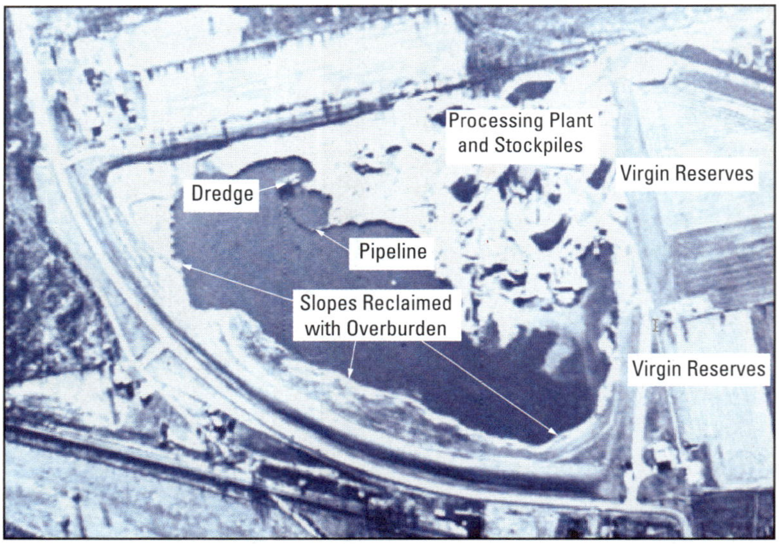

FIGURE 9.27 Dredging the perimeter with reclaimed slopes in foreground

FIGURE 9.28 Floating discharge pipeline feeding processing plant

available for heavy-walled steel pipe or flanges can be welded to the ends for bolting the pipe together. Periodic ball joints in the pipeline can facilitate turns in the line if necessary. To lengthen its longevity, the pipe should routinely be rotated by 45° about every 3 months. The larger, heavier particles in the slurry tend to sink and abrade the bottom of the pipe more than the sides and top. Regular rotation of the dredge pipe can increase its usable life by up to 20%. Some operations have started to use heavy-wall high-density polyethylene fused pipe in lieu of steel, but highly abrasive sands can sometimes limit its life span.

Dredges move in various ways but are generally not self-propelled with motors except some very small contractor dredges. The most common propulsion/mooring systems are either spuds or cable winch systems. Spuds are long, heavy steel structures pointed on the bottoms and mounted perpendicular to the flotation. The spuds can be raised and lowered vertically into the lake bottom (see Figure 9.21). Spuds can only be used in the shallower deposits because the length of the spuds is limited and they add considerable structure and weight to the flotation. Spud systems utilize anywhere from one to four spuds, but the system with two front and one rear spud is most common. The rear spud will usually be mounted in a hydraulic or cable-operated carriage that is capable of moving the spud horizontally and propelling the dredge 6–8 feet (1.8–2.4 meters) forward at a time.

Steering winches are very common, especially in deeper deposits where spuds cannot reach the bottom of the lake. Three-winch systems with two winches on the front corners and one in the center of the rear are probably the most common arrangement. The winch cables are attached to heavy weights on shore and used to steer the dredge into the deposit. Worn-out dragline buckets are often used to anchor the winches to the shoreline. In hard deposits where additional forward pressure is required, a five-winch system can be used, with two winches on the front pontoons and three winches mounted on the back of the dredge and reeved through a Christmas tree structure to elevate the cables. The additional two swing winches on the rear of the dredge are used to assist in pulling forward and holding the machine into the mining face. There are also hybrid systems that utilize both spuds and winches.

Production from a suction dredge is a function of the percentage of solids in the slurry being pumped and the speed the material flows within the pipeline. Monitoring equipment should be installed on the pipeline to monitor flow rate and slurry density. Doppler meters installed on the outside of the pipeline can, with fair accuracy, monitor the rate of flow through the pipe. These meters can also provide the operator with a warning of an impending plugging of the pipeline. Finding and removing a plug from an 18- to 20-inch (457- to 508-millimeter) floating pipeline is not an experience you will want to repeat once you have done it. A flow rate of 15–18 feet per second (4.6–5.5 meters per second) is optimal for production; a flow rate below 5 feet per second (1.5 meters per second) is dangerously close to a pipe plugging situation. Slurry density regulation (SDR) systems monitor the suction and discharge pressures simultaneously and can open a valve gate that adds water to the system in front of the pump. These systems regulate the density of the slurry, alleviate pipeline plugging, even out slurry flow, and increase production. I recommend that an SDR system be specified on all suction dredges.

No matter what type of underwater mining equipment is selected, the difficulties of mining underwater are exacerbated by not having the ability to see the deposit and know exactly what is happening at any given moment. Unconsolidated materials will always have a tendency to cave at unexpected times. Recovery of 100% of the deposit is problematic without the ability to see what is happening below the water. I have experimented with remote-controlled submarines equipped with underwater cameras with little success because it can take days for the water to clear sufficiently to achieve reasonable visibility. I always wanted to try underwater side-scanning sonar to see if it could predict cave-ins by mapping the underwater slopes, but the cost was always prohibitive.

In the early 1990s, I was promoted to operations manager of two aggregate divisions containing 18 separate plants. The second Monday on the job, I was called to a plant site when the superintendent reported that his suction dredge had just sunk in 45 feet (14 meters) of water. During the 6:00 AM shift change, the night dredge operator had left the dredge pump idling, pumping only fresh water (to prevent the day shift from having to prime the pump), and took the service boat to shore to turn over the job to the day-shift operator. Before the day-shift operator could reach the dredge, it was on the bottom of the lake. They reported that the dredge sunk in less than 3 minutes. Divers were sent into the lake, where they discovered the dredge had settled upright on the bottom, but there was no visible evidence that any of the pontoons had failed. The dredge was never recovered after multiple efforts were attempted and the actual cause of the disaster was never determined. This dredge had a hull-mounted pump and, apart from sabotage, the only two viable explanations were that the pipeline split near the pump and the running pump filled the hull with water, or that a massive failure occurred on the bottom of one of the pontoons that was invisible to the divers. This is just another example of the risks of underwater mining.

Suction dredges are not immune to cave-ins. At one operation we had a suction dredge with a 210 foot (64 meter)–long ladder. A cave-in occurred on night shift that buried much of the ladder. This dredge had a hull-mounted pump that was able to continue pumping throughout the process. Although the ladder was halfway covered with sand and gravel, we were able to continue producing. It took more than 2 weeks of constant pumping to free the ladder from its grave on the bottom of the lake. When the ladder was finally freed and hoisted to the surface, we were very lucky to have suffered only minimal structural damage, which was readily repaired.

CHAPTER 10

Above-Water Mining

Excavation of surface-mined materials can be accomplished using some of the same equipment discussed in Chapter 8 for underwater deposits. Hydraulic excavators and draglines are capable of digging either dry or dewatered deposits of sand and gravel, as well as blasted rock (shot rock) (Figure 10.1). Draglines can also be used to remove (strip) thick overburden overlying minable deposits (Figure 10.2). I once operated a limestone quarry where the unconsolidated overburden was about 70 feet (21 meters) thick overlying 50 feet (15 meters) of stone. The property was mined in fairly narrow parallel cuts with a large dragline doing the stripping. After each limestone cut was blasted and removed using wheel loaders and trucks, the dragline cast the overburden from the next cut into the previously mined cut. This required a dragline with a long boom to reach from one cut to the next. A 10 foot (3 meter)–thick ledge of stone was left between each cut to temporarily retain the overburden cast by the dragline within the completed cut. Upon completion of a cut and stabilization of the fill, a bulldozer shaped and leveled the cast overburden for reclamation purposes.

By far, the most common method of excavating above-water deposits in the aggregates industry is utilizing wheel loaders, also known as front-end loaders. Wheel loaders are extremely versatile and come in numerous sizes and specifications depending on the application. Pit loaders operating at the mining face have bucket capacities that range from 6 to 20 cubic yards (9 to 30 metric tons) and operating horsepower ranging from 375 to 1,500 horsepower. Operating weights on these machines range from 100,000 to over 500,000 pounds (45,359 to 226,796 kilograms). The two primary specifications used in determining the required machine

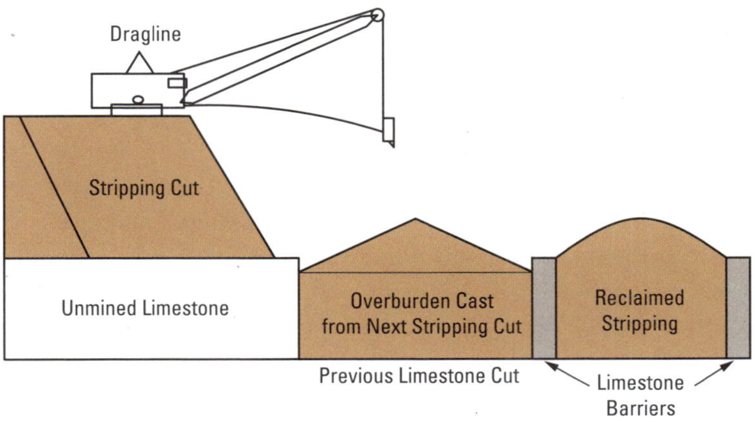

FIGURE 10.1 Stripping thick overburden in a limestone quarry with a dragline

FIGURE 10.2 Draglines working in tandem to excavate 115 feet (35 meters) of dewatered sand and gravel

model are the bucket size and the breakout force. The breakout force of the large wheel loader category depends on the machine weight, the bucket type, and the hydraulic system capacity, and ranges from 75,000 to 271,000 pounds (34,000 to 123,000 kilograms) of force.

On one specific occasion, the owner of the company asked me to figure out why the production at a certain small limestone quarry was not meeting expectations. This particular quarry was operating from a 70 foot (23 meter)–tall highwall. After spending some time evaluating the processing plant and finding no specific problems, I concentrated on the quarry operation. My immediate reaction was that the pit loader was too small for the application. I spent most of the following day doing a time study on what was a load-and-carry operation from the mining face to a dump hopper feeding the portable primary crusher. After separately timing the digging cycle, the haul cycle to the crusher, the dumping cycle, and the return cycle to the face, it was obvious that my first impression was correct. The bucket size of the existing wheel loader was not a problem based on the required production, but the machine was seriously struggling to dig the shot rock. The blast design was appropriate and the muck pile was well fragmented with minimal oversize. The breakout force of the wheel loader was insufficient to efficiently excavate the shot rock. The loader had a breakout force rating of about 36,000 pounds (16,329 kilograms) of force. We upsized to the next larger model of wheel loader with a breakout force rating in the range of 60,000 pounds (27,215 kilograms) of force and reduced the cycle time by over 40%, increasing the production to a record level.

Wheel loaders can be operated in dry bank sand and gravel operations as well as hard-rock quarries, but need to be specified for the particular application. Sand and gravel unless very densely packed or cemented will generally require less breakout force than a quarry operation. Dry bank sand and gravel should be excavated much like the dredge operations discussed previously by digging out the toe of the slope and allowing the deposit to migrate or cave toward the loader. These sand and gravel slopes can be somewhat unstable, especially when very tall, and care must be taken by the loader operator to avoid major cave-ins. Wheel loaders require stable footing on a hard surface and soft patches on the floor of a sand and gravel pit need to be avoided or stabilized to maintain a safe workplace.

FIGURE 10.3 Wheel loader loading shot rock into a rigid-frame haul truck

Wheel loader specifications also need to be matched to the system of transport from the pit to the processing plant. Heavy equipment manufacturers match wheel loader sizes to specific haul truck sizes and specifications for the best efficiency pairing. Matching wheel loaders with the haulage fleet will depend on the required production capacity, the haul distance from pit to plant, the capacity of the dump hopper at the plant, and the number of operating shifts.

Off-highway, rigid-frame haul trucks come in a variety of sizes and capacities (Figure 10.3). The most common haul trucks in aggregate operations start at a nominal 40-metric-ton capacity utilizing 500 horsepower with a gross weight of 150,000 pounds (68,000 kilograms), and range up to 100-metric-ton capacity utilizing 1,000 horsepower and having a gross weight of 360,000 pounds (136,000 kilograms). Much larger haul trucks with capacities of over 400 metric tons are utilized in large open pit mines around the world, many of which operate with diesel electric power. The size of the fleet required will depend on the production capacity needed, the distance from the quarry face to the processing plant, maintenance of the haul roads, and the grade of the ramps that the trucks must traverse. To maintain an efficient operation, the trucks need to be spaced out so the wheel loader or shovel is never without a truck to load and the trucks are not waiting to be loaded or waiting at the dump hopper to feed the plant.

Articulated trucks are sometimes utilized in a similar capacity as rigid-frame haul trucks, but these vehicles are primarily designed to operate in soft ground conditions (Figure 10.4). They generally are all-wheel drive and the larger models have three axles. Hauling capacities of articulated trucks are typically limited to 25 to 35 metric tons, although larger models with a capacities in the range of 50 metric tons have been produced. They are very good as stripping trucks when moving unconsolidated overburden in wet conditions because of the additional traction provided by the all-wheel drive. These vehicles are not recommended for use at a quarry or pit that has good hard-surface roads. The additional maintenance required on the extra drive axles and the lower haulage capacity limit the ability of these trucks to operate cost-effectively in a quarry or pit operation unless very poor ground conditions require the added traction they can provide.

Although haul trucks are very versatile and can be used to do many things, I have always been a proponent of eliminating haul trucks whenever possible. Off-highway haul trucks

FIGURE 10.4 Articulated haul trucks

require a large capital expenditure, are not particularly fuel-efficient, and are very time consuming and expensive to maintain. These trucks require regular lubrication, regular oil changes, brake maintenance, safety inspections, component overhauls, and lots of tires. Unpaved haul roads also require a lot of maintenance, including grading, berming, and dust suppression, all of which require additional equipment and personnel. Although the equipment manufacturers are experimenting with autonomous haul truck fleets at large surface mines, each haul truck and maintenance vehicle still requires individual maintenance personnel and employees at this point. Operating personnel receive hourly wages, insurance benefits, retirement benefits, worker's compensation, Social Security, vacations, sick leave, and training, which can be a significant cost depending on the number of personnel required.

Field conveyor systems are a very efficient alternative to haul trucks (Figure 10.5). Capital costs for conveyor systems are moderate, power cost is low, and conveyors require only a routine maintenance effort. If good-quality idlers (CEMA C or better), high-quality tapered roller bearings, heavy-duty pulleys, and quality conveyor belt are specified, these components can operate for many years. Routine maintenance including daily inspection, greasing bearings, regular oil changes on gear boxes, proper belt scrapper adjustment and maintenance, and regular cleanup will keep a conveyor system running almost indefinitely. The production from a conveyor system is a function of the width of the conveyor belt and the speed it is run. A slower belt speed will increase the longevity of all the conveyor components but will require a wider belt to get the same production capacity. A belt speed in the range of 300–350 feet per minute (1.5–1.8 meters per second) is a good starting point and belt speed can be increased with only a minimal cost if production requirements later dictate the need.

Conveyors are capable of handling material with a top size of about 8–10 inches (20–25 centimeters). In quarry blasting, the shot rock generated is usually produced with a top size in the range of 20–24 inches (51–61 centimeters), which is too large to efficiently convey. Obviously, the larger rock particles need to be reduced for the conveyor to transport the material to the processing plant. This will require a portable primary crusher in the pit or quarry to reduce the top size and deposit the material onto the conveyor system. Primary crusher types are discussed in further detail in Chapter 13 on crushing equipment. If utilizing a wheel loader

FIGURE 10.5 Field conveyor being installed at a limestone quarry

to excavate the raw material and feed a portable primary crusher in lieu of trucks, the mining face needs to be kept within about 300 feet (91 meters) of the crusher or the cycle times become too great to maintain production with one wheel loader. This is known as a load-and-carry operation.

Thus the need for the primary crusher to be portable. When the production slows because of increased haul distance, the crusher must be moved closer to the mining face and additional conveyors added to or removed from the system. Production could be supplemented with an additional wheel loader, but the additional operating costs are much larger than the costs associated with a crusher move. Portable crushers come in many forms and sizes; some are skid-mounted, some wheel-mounted, some are on powered crawlers like bulldozers, and the largest models have walking mechanisms similar to big draglines. More information on portable equipment can be found in Chapter 15. Load-and-carry operations generally take more planning than putting a main haul road across the quarry floor and letting the haul trucks fly.

Larger portable primary crushers that move via their own propulsion systems sometimes have a series of powered (on wheels or crawlers) discharge conveyors that can be advanced and retracted in a scissors fashion to move the crusher toward the face and then move it away while blasting (Figure 10.6). This is the ideal portable primary situation but comes with a large price tag. Wheel-mounted or crawler-mounted primary crushers with rigid field conveyor systems are much more common and require less capital expenditure. The most common types of portable primary crushers are either jaw crushers or horizontal shaft impact crushers.

Single-bench quarries are the most conducive to the load-and-carry field conveyor system. Quarries with multiple benches are more expensive to operate with this method. Separate field conveyor systems on each level are required and connecting separate conveyor systems can be difficult depending on the bench heights and distances. The capital expenditure required for multiple conveyor systems must be weighed against the additional operating cost of utilizing a fleet of haul trucks. The versatility of haul trucks in a multibench operation may outweigh the normal savings of the field conveyor option. A hybrid operation of truck haulage to a centrally located primary crusher and then a conveyor to the processing plant is another viable option. Whether using a portable primary or a fixed primary location, it is usually more cost-effective

FIGURE 10.6 Portable primary crushers: (A) crawler-mounted primary jaw crusher, and (B) very large portable primary crusher on a walking mechanism

to locate the primary crusher on the quarry floor and convey the crushed material out of the quarry than to haul it with trucks up steeply sloping ramps to the surface.

Wheel loaders, although probably the most common, are not the only method of excavating above-water deposits. Large hydraulic excavators are sometimes utilized, but they are much more efficient when digging below the level on which they sit. The excavator digs by curling the bucket and then raising the stick and boom, and to dump the load the bucket is curled in the opposite direction (uncurled). This means that if digging a muck pile of shot rock with an excavator, a temporary road must be built to the top of the pile and the pile leveled prior to beginning the excavation process. If this type of equipment is already available, it is certainly a viable option. Hydraulic excavators are versatile machines and can be used in many other applications.

Hydraulic front shovels are very similar to hydraulic excavators, but they are equipped with a shovel front instead of a boom and stick. These machines work from the bottom of the muck pile much like a wheel loader except they are mounted on crawler tracks instead of wheels. Because of the heavy-duty nature of these machines, the life expectancy is probably three times that of a wheel loader. The bucket opens on the bottom and the entire machine rotates horizontally on a center pin. The front shovel excavates a bucket load of material and raises the bucket while swinging 90° to 180°, and is a capable of dumping the load into a

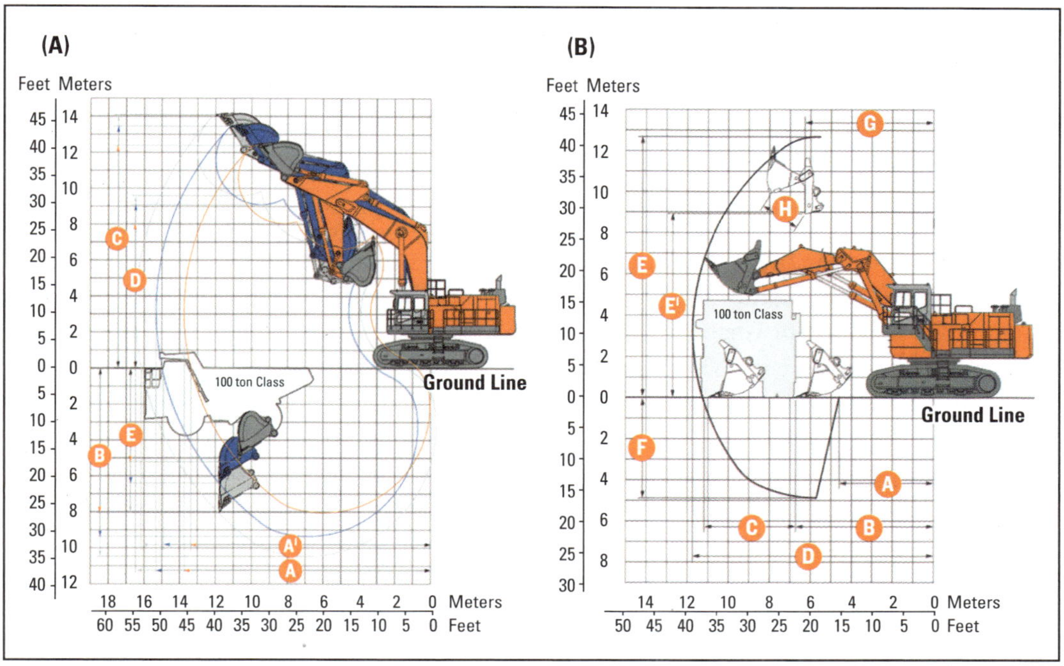

Courtesy of Hitachi Construction Machinery Company

FIGURE 10.7 (A) Excavator, and (B) front shovel specifications

haul truck or the hopper of a primary crusher. Hydraulic excavators and front shovels come in a variety of makes, sizes, and models, but most mining machines are the large, heavy-duty variety. Machines common to aggregate operations range between 500 and 1,500 horsepower and weigh between 188,000 and 570,000 pounds (85,000 and 258,547 kilograms). The larger variety used in open pit mining applications are even more massive and can exceed 3,900 horsepower and weigh in excess of 1.8 million pounds (816,000 kilograms). Cable-operated front shovels are also available but generally have longer cycle times than the hydraulic machines (Figure 10.7).

In certain above-water operations, the material can be transported via a pipeline much like a suction dredge. Some sand mining operations use high-pressure water hoses to wash the mined material into a series of sumps. Centrifugal pumps then pump the sand in a slurry to the processing plant. I have seen this mining method used in silica sand operations mining unconsolidated or very friable sandstone deposits.

DRILLING AND BLASTING

Just over 60% of aggregates sold in the United States are mined from deposits of solid rock versus 40% being unconsolidated sand and gravel. To mine the solid rock, some method is needed to break the formations into usable size materials. Neolithic miners used fire to heat the rock until it cracked into smaller pieces. In today's mining world, explosives are used to fragment the rock.

The earliest recorded use of explosives for mining purposes was in Hungary in 1627, although explosives were utilized in warfare for several centuries prior to this. For explosives to perform their task, they need to be inserted inside the rock by some method. In past centuries this was accomplished by excavating a large hole by mechanical methods and packing it full with kegs of black powder and covering it before detonating. This was called a glory hole or coyote hole and was extremely dangerous because there was no way to control the direction of force generated by the explosion. The fragmentation results were also very unpredictable.

Today's quarry blasting is much more scientific, much safer, and for the most part very well controlled. Instead of cramming massive volumes of black powder into large caverns, tightly controlled holes in specific measured patterns are drilled vertically into the quarry floor. Drills have evolved from handheld star drills being bashed by sledgehammers to very sophisticated truck- or crawler-mounted drilling rigs capable of high-production drilling. These drills are unlike those described in Chapter 3 on geology. Instead of being designed to sample underground deposits, these machines are specifically designed to produce holes to a specific depth quickly and accurately. They are known as rotary percussion drills because they operate percussion hammers that rotate while hammering and pulverize the rock into small chips and dust that is generally blown to the surface via compressed air. Some hammers operate hydraulically, but most are pneumatic. These drills can be used to aid in the description of the underground geology by analyzing the chips being blown to the surface.

Like all mining equipment, blasthole drills need to be matched to the application. Larger drills are capable of drilling holes that are 5–12 inches (12.7–30.5 centimeters) in diameter to considerable depths. Smaller models drill 2 to 4 inch (5 to 10 centimeters) diameter holes and are more limited in depth capacity. Rotary percussion drills can be equipped with either top hammers or down-the-hole hammers depending on the application. A top hammer hammers on the entire drill string, generating the percussive effort through the drill string into the bit on the bottom of the hole. Down-the-hole hammers are located inside the drill hole directly on top of the drill bit. Down-the-hole hammers tend to drill a much more accurate, straighter hole than their top-mounted cousins. Small-diameter holes cannot utilize down-the-hole hammers because the hole diameters are too small to accommodate them.

Specifying drilling equipment can only be done after the mine plan is completed and the bench thickness and production requirements are fully defined. A blasting plan defining the size of the hole, drilling pattern, and spacing of the holes along with explosive type and quantity needs to be generated by an explosive engineer. I will not go into a lot of detail about types and sizes of drilling equipment because I am not very qualified in that arena. At one point I managed an aggregate division that owned one blasthole drill to service four quarries. It was a scheduling nightmare and we ended up having to supplement our own drilling with the use of contractors, especially when a breakdown occurred. Since that time I have always utilized drilling contractors that had multiple drills available to service my drilling needs. Needless to say, numerous different drill manufacturers produce various drill models for every blasthole drilling need.

No one knows who first invented black powder, but it has been attributed to originating in either China or India in about the ninth century. By the 10th century it had become weaponized by the Chinese into what was known as the fire lance and the first cannons evolved in the 11th and 12th centuries. Black powder's first known use in the mining industry was in 1627 in Hungary. It caught on very quickly and was being used in the tin mines in Cornwall, England, by 1689. Black powder is a mixture of approximately 75% saltpeter (potassium nitrate), 15% charcoal, and 10% sulfur. By the year 1675, it was being manufactured in the United States near Boston for use in flintlock firearms for hunting. The safety fuse was invented in Cornwall in 1831 and quickly replaced the dangerous practice of igniting a trail of black powder laid out on the ground. Black powder exhibits relatively slow development of gas pressure but produces a hot flame and is classified as a low explosive.

Although the formulation was refined over the years, the use of black powder in mining was prevalent until 1866 when Alfred Nobel mixed nitroglycerine (discovered in 1846) with an absorbent (diatomaceous earth) to make a solid explosive known as dynamite. About this same time, Nobel invented the first efficient blasting cap, which consisted of a tin capsule filled with fulminate of mercury that could be ignited using a safety fuse. Nine years later Nobel dissolved collodion cotton in nitroglycerine, which resulted in a gelatinous mass more powerful than dynamite, which was latter refined into gelatin dynamite. The first successful electric blasting cap was invented in 1876 by H. Julius Smith, who later patented the first delay action blasting cap in 1895. The introduction of these high explosives revolutionized the mining, heavy civil construction, and blasting industries.

The relatively high freezing point of nitroglycerine (52°F [11°C]) caused many problems initially when dynamite had a tendency to freeze and fail to ignite. Attempts to thaw the frozen dynamite quite often proved to be fatal. Dynamite was subsequently improved by the addition of nitrated glycerin and ethylene glycol to overcome the freezing issues. Further refinements to dynamite included the addition of ammonium nitrate and the reduction of the percentage of nitroglycerin, which resulted in a product that was less hazardous to manufacture. In 1935 the DuPont Company introduced the concept of blasting agents that are insensitive to friction, blasting caps, shock, or being shot with a bullet. Blasting agents are a much safer alternative to high explosives and are easily and safely transported, but they do need to be detonated by a booster of more sensitive material. High detonation pressure (HDP) cast primers (boosters) were developed to detonate columns of blasting agents.

Blasting agents are made up of ammonium nitrate (NH_4NO_3), which when mixed with some type of fuel will produce large volumes of gas when detonated. The use of ammonium nitrate and fuel oil (ANFO) as a blasting agent has gained widespread use in quarry blasting applications. Ammonium nitrate is produced in two different grades; one for use as fertilizer, and the second used for explosive purposes. Ammonium nitrate, known in the industry as prills, when mixed in the correct proportion with No. 2 fuel oil becomes an excellent, low-cost explosive agent that provides a moderately high velocity of detonation and excels in gas production when initiated. ANFO prills are available in 50-pound (23-kilogram) bags that can be emptied directly into a drill hole over top of a cast HDP booster (high explosive) containing a blasting cap and will produce a very satisfactory explosion. On large blasting jobs, specially designed bulk trucks carrying ammonium nitrate and separate fuel oil tanks systematically mix the products to the specified ANFO concentration and load the blasting agent directly into the drill holes via augers and hoses. ANFO is sensitive to moisture and is not conducive for use in drill holes that have a tendency to fill up with water.

Water gel explosives are a gelatinous substance containing 10%–30% water with the same safety features as blasting agents but utilize a variety of sensitizers like TNT and therefore are considered a Class B explosive rather than a blasting agent. Water gels like Tovex developed by the DuPont Company come in packaged products as well as a bulk variety that can be pumped into the blastholes and have become very popular in quarry blasting in the last 10 years. They are not affected by moisture like ANFO and can be used in conjunction with other explosives and blasting agents. Packaged products like Tovex can be loaded into the bottom of wet holes

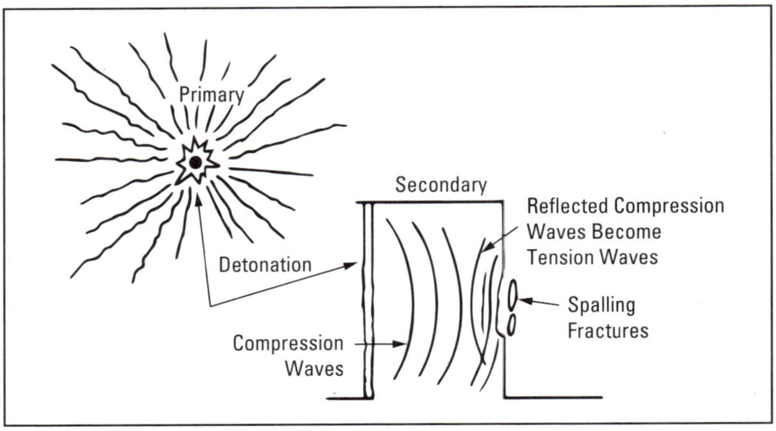

FIGURE 10.8 Blasting-induced stress waves

until above the water level and ANFO loaded on top where the hole is dry. Like blasting agents, they require a booster to detonate. Specially designed bulk trucks are able to pump water gels directly into blastholes on larger blasting jobs. Various combinations of water gels mixed with ANFO have also been successfully used.

An explosion is essentially a violent expansion of gaseous energy in which the energy is transmitted outward as a stress or shock wave. When explosives are detonated within a confined space such as a drill hole, the instantaneous expansion of gases produces enough energy to shatter the surrounding rock. The size and shape of the stress wave is dependent on the rock characteristics, the type of explosive, and the relation of the explosive's detonation velocity to the rock's propagation velocity (Figure 10.8). The stress wave moves at a speed ranging from 10,000 to 19,000 feet per second (3,048–5,791 meters per second) radially from the drill hole. The fractures propagate at speeds of only about 3,000–8,000 feet per second (914–2,438 meters per second) depending on rock density and fracture planes. The rock moves away from the face at a much slower pace than the fragmentation process in the range of 50–100 feet per second (15–30 meters per second). This contributes to the vertical movement of the rock.

One of my earliest experiences with blasting was in a limestone quarry in southern Indiana. The blaster in charge was detonating a series of blastholes using fuse cord to initiate the blast. The timing between the detonations of the rows of holes was determined by how many wraps (between his elbow and hand) of fuse he utilized. It is always an experience to see your first quarry blast, and this one was whopper, but it achieved its purpose to fragment several thousand tons of limestone. Modern blasting has come a long way in the last 30 years.

When loading blastholes, the drill hole itself acts much like a rifle or cannon barrel and unless the top is somehow plugged the gaseous energy would disperse out the top of the hole with little effect on the surrounding rock. When this happens, the phenomenon is known as rifling. For the explosion to cause fragmentation of the rock, it must be totally contained within the hole. The top several feet of the drill hole is plugged with "stemming," which consists of crushed stone in the ⅜- to ½-inch (9.5- to 12.7-millimeter) size range. Crushed material must be utilized because the angular particles lock together to form a relatively tight plug. If round particles like gravel were used, they would shoot out the top of the hole much like a shotgun, losing the explosive energy. Stemming consisting of crushed stone from the same geological formation that is being mined eliminates the potential for any contamination.

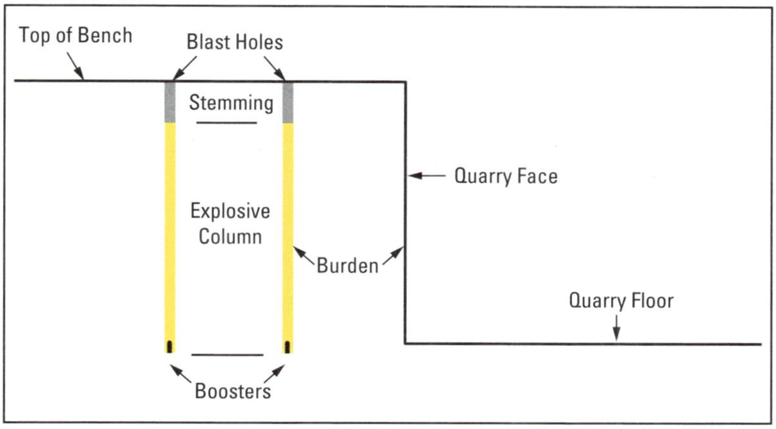

FIGURE 10.9 Cross section of a blast illustrating the concept of burden

Safety fuse is seldom utilized today to initiate the detonation of blastholes. Initiation systems have become very sophisticated to increase safety, maximize fragmentation, and lower the vibrations produced by blasting. Various initiation systems have been used over the years from the earliest trail of black powder, to safety fuse, electric detonators, detonating cord, gas tubing, detonation timing (blasting) machines, to millisecond-delay electric detonators. The newest, most efficient, and most accurate systems utilize blasting caps that contain programmable computer chips that control the timing of the individual blasts to within ±0.5 milliseconds.

Drilling and blasting is probably the most critical element in the cost of production in a quarry operation. Efficient quarry blasting requires multiple drill holes arrayed in rows parallel to the quarry face. The holes are evenly spaced and can be in a square, rectangular, or staggered pattern. The distance between holes in a row is called the spacing. The distance perpendicular to the nearest free face at the time of detonation (and separating the rows) is known as the burden and is not necessarily the same as the spacing (Figure 10.9). The pressure from a blast will take the path of least resistance, which will be toward the open quarry face because it is unconfined. The burden is the least distance to the unconfined highwall that each blast hole must fragment (and transport) the rock via the pressure wave exerted by the explosive. A blast design will be defined by the spacing and the burden such as an 8-foot (2.4-meter) burden by a 10-foot (3-meter) spacing, or 10 × 10 foot (3 × 3 meter) or 14 × 15 foot (4.3 × 4.6 meter) with the burden being the first dimension. The burden dimension can change based on the millisecond-delay sequence and the initiation point, for example, a square pattern shot timed in echelon (Figure 10.10).

To reduce the effects of off-site vibration, increase fragmentation, and control the throw distance and direction of the blast, not all of the blastholes are detonated at the same time (Figure 10.11). The timing of the initiation of each blasthole is controlled by delay timers within the individual blasting caps that initiate the boosters. This is accomplished by using millisecond-delay caps or the newer caps that contain programmable computer chips to control the delay. Instead of shooting 30 blastholes at the same time, creating one massive explosion, numerous small, separate, individual explosions, each separated by milliseconds, are initiated. This minimizes the effects of the vibrations (air and seismic waves) caused by the blast on any neighboring structures. Instead of hearing one giant bang, a rumble of numerous small explosions is experienced. Changing the timing between holes and between rows of holes can affect

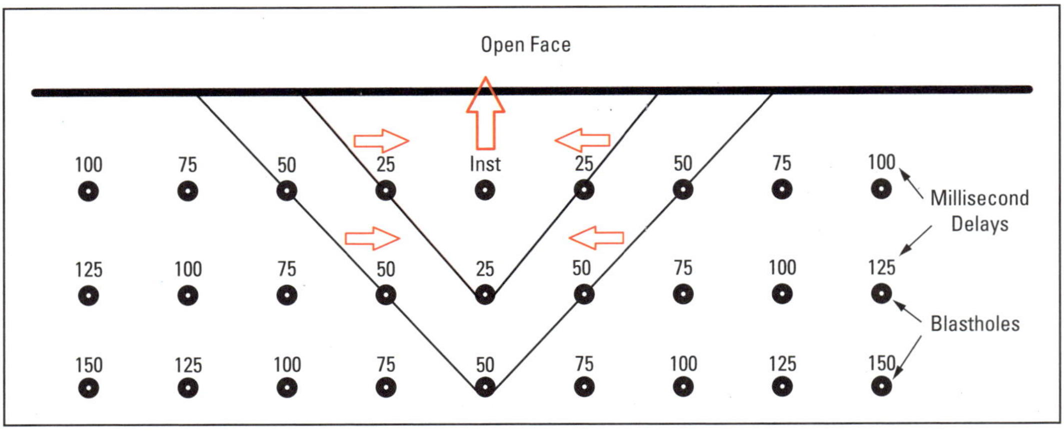

FIGURE 10.10 Square pattern blast delayed in echelon (25-millisecond delays)

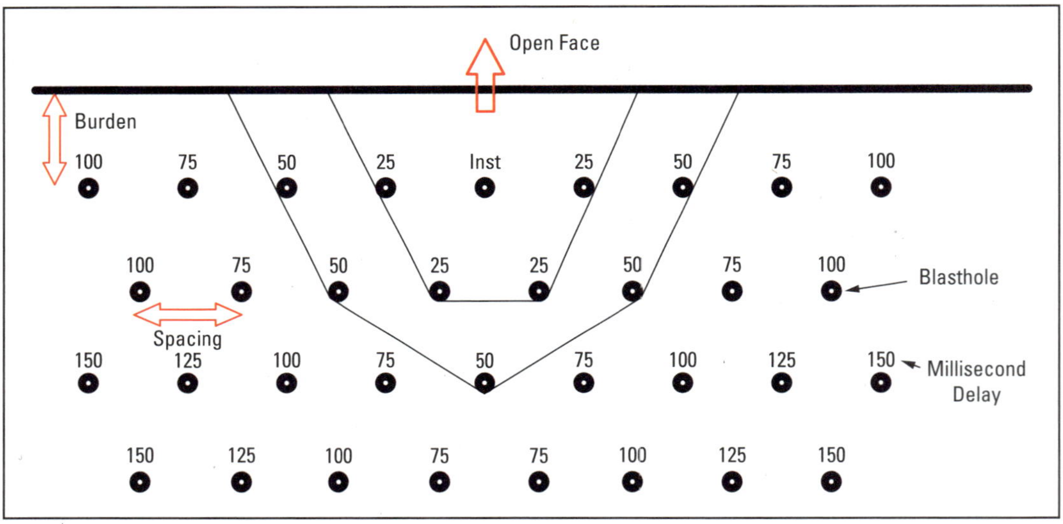

FIGURE 10.11 Staggered pattern blast delayed in echelon (25-millisecond delays)

fragmentation by allowing one explosion to impact the next explosion sooner or later in the initiation sequence depending on the timing specified. Where the initial explosion (instantaneous cap timing) is begun will also affect the direction of the blast and where the shot will lay out on the quarry floor.

A competent blasting engineer will be needed to design all the blasting. Most if not all states now require blasters to be thoroughly trained, tested, and certified. Blasting at a new operation will ultimately be somewhat of a process of trial and error. A blasting engineer can design the initial blasts based on previous experience, but every quarry and every geological formation is unique when it comes to blasting. The fragmentation of the muck pile will vary depending on the regional jointing of the bedrock, rock density, thickness of rock layers, and any imperfections in the rock such as mud seams, cracks, joints, small caves, or voids (Figure 10.12). Every quarry I ever operated had a preferential blasting direction in which the fragmentation was superior. It is always a good idea to have the blasthole driller keep a log of

FIGURE 10.12 Well-executed blast with good fragmentation

each drill hole noting any anomalies encountered in the rock such as voids, mud seams, and soft spots, and the log should be provided to the blaster prior to loading the shot.

Blasthole diameter is dependent on face height; shorter benches require closely spaced patterns and closely spaced patterns require smaller hole diameters. Larger-diameter holes are capable of holding more explosives and therefore are capable of breaking more rock. More explosives utilized in each hole increases the cost per hole, but this is offset by the larger spacing between holes allowed by the use of larger diameters and additional explosives. Powder factor (PF), usually expressed as pounds of explosive per metric ton of rock produced, is a generalized method of estimating blasting efficiency and relative costs. A PF of 1.0 is better than a PF of 0.75, and a PF in excess of 1.2 is desirable. Because different explosive types are not equal in the energy they produce, PF is only useful in comparing blast designs utilizing the same products (e.g., ANFO to ANFO). The velocity of detonation (VOD) of various explosives ranges from 8,000 up to 19,000 feet per second (2,400 to 5,791 meters per second). The VOD of ANFO varies with the drill-hole size and ranges from 8,000 to 15,000 feet per second (2,400 to 4,572 meters per second). Typical quarry blasthole diameters are from 3 to 8 inches (73 to 203 millimeters) but are dependent on the site-specific shot design.

Although face height is a function of the geology and the mine plan, at a minimum the face height should be two times the burden. When initially designing a blasting pattern, several rules of thumb can be applied to begin the design process and later be refined as field experience dictates. The burden is the most critical variable in the design of an individual blast. The following formulas should be considered when designing blasting and refined as experience dictates:

Borehole diameter

$$D = \frac{H}{10}$$

where D = borehole diameter, inches; and H = bench height, feet.

Burden

$$B = (25 \text{ to } 35) \times De$$

where B = burden, feet; and De = diameter of explosive column, inches.

Spacing between holes

$$S = (1 \text{ to } 1.8) \times B$$

where S = spacing, feet; and B = burden, feet.

Depth of stemming

$$T = (0.7 \text{ to } 1.3) \times B$$

where T = stemming, feet; and B = burden, feet.

Subdrilling

$$J = (0.2 \text{ to } 0.5) \times B$$

where J = subdrilling, feet; and B = burden, feet.

Geologic conditions within the formation to be mined have a large effect on blast design. Massive formations containing few laminations usually require closer burdens and spacings and relatively high explosives charges. Rock that is thinly laminated will generally break with good fragmentation results using wider burdens and spacings and lower explosives ratios. In sedimentary formations that contain both massive and laminated areas, this combination can be problematic. Massive formations of caprock overlying more laminated deposits are extremely difficult to break because the required stemming in the top of drill holes where the massive layers exist have little or no explosives. Short satellite holes penetrating the massive seams can be interspersed between the regular drill-hole pattern and loaded with a small charge to aid in breaking massive caprock formations. Massive layers in the bottom of the face may need to be loaded with higher-density explosives than the upper, more laminated layers. Hard, dense formations can sometimes provide better fragmentation then softer rock. The harder rock tends to be more brittle, while the softer rock has a tendency to absorb some of the energy generated by the explosives. In geological areas where the stratification is steeply dipping, blastholes angled parallel with the stratification will generally produce better fragmentation results.

Blasted material from the previous shot must always be completely removed prior to initiating the next blast. Previously blasted materials (muck) lying against the highwall will affect the burden calculation by adding additional confinement (weight) that the explosive is required to move. The highwalls (quarry faces) left behind after blasting are often irregular and somewhat jagged because of variations in the hardness or imperfections in the rock. Back break can also occur when cracking behind the blastholes because confinement of explosives opens joints or cracks in the highwall. If irregular quarry faces are commonplace, it is a good idea to profile the quarry face prior to determining the shot layout to equalize the burden. This is done using a laser profiling system. If a section of the quarry face has less burden than expected because of an irregular highwall, modifications in the layout need to be addressed to prevent excessive pressure due to the lesser burden from causing flyrock and face blowouts.

Numerous blast designs exist based on individual quarry conditions and changing geology. Multiple explosive decks with stemming in between can be placed within blastholes to avoid soft areas, limit blowouts, adjust the PF, and limit vibrations. Lift or raise shots to deepen a quarry or start a ramp have a different types of delay patterns designed to lift the shot vertically instead of throwing it horizontally. Defects and unexpected geological anomalies in the quarry face can cause a well-designed blast to go awry. This is why drill logs should be kept by the drillers. Flyrock incidents can be highly dangerous and are an unacceptable hazard that can result from geological variations, drill-hole misalignment, insufficient burden, or inadequate stemming.

I'll never forget the day when a neighboring property owner walked into the plant office and roughly dropped an 8- to 9-inch (20- to 23-centimeter) piece of newly fractured rock onto my desk. He then irritably explained that he had found the rock embedded in his side yard within 3 feet (1 meter) of his children's swing set. Being a father of young children at the time, I understood his indignation. After consulting with the blasting contractor and completing a thorough investigation, it was determined that a previously unidentified mud seam in the limestone highwall (a zone of weakness) allowed excess energy to escape (path of least resistance) and caused the flyrock incident. The 8-inch (20-centimeter) piece of limestone landed over 2,000 feet (610 meters) behind the shot and how high it flew to get there is anyone's guess. Luckily for all concerned, no one was injured in this incident.

Safety is the main concern when doing production quarry blasting. A detailed safety procedure must be developed and strictly adhered to for every blast initiation. Employees need to be assigned specific tasks when preparing for a blast. Every access to the quarry site must be blocked and monitored to prevent truck or pedestrian access to the area. Rigorous communication between the blaster and the on-site quarry supervisor is an absolute necessity. A detailed visual survey of the quarry and surrounding area must be diligently performed and an audible warning signal should be sounded 3 minutes, 2 minutes, and 1 minute prior to initiation. It is a good idea to have a written procedure that is reviewed with the employees on a regular basis. Detailed shot reports for every blast need to be generated by the blaster in charge. The shot reports should include the time and date of the shot, drilling pattern, delay period design, type and weight of explosives, weather conditions, pounds of explosives per delay, and any other pertinent information and should be signed by a certified blaster. These records should be kept on file permanently and are a requirement of the Mine Safety and Health Administration (MSHA).

One of the first quarries I ever managed was in southern Indiana not far from an Air National Guard base. This unit of the guard flew the A-10 Warthog jets, which were designed as tank killers. The area where the quarry was located was very hilly and mostly wooded and the Warthogs would appear out of nowhere, flying at near-tree-top levels mirroring the topography. You would not hear the planes coming until they were directly on top of you and then they would roar overhead at full speed. I had nightmares of setting off a blast just as three Warthogs in formation happened to fly over the quarry. My nightmare never materialized but thereafter I always looked to the skies before giving the heads up to the blaster for "fire in the hole."

Explosive storage is tightly regulated by the Federal Bureau of Alcohol, Tobacco, Firearms, and Explosives, which is now a part of the Department of Homeland Security. In most states the state fire marshal's office also has regulatory authority over explosive storage as does the MSHA. Detailed inventories must be kept on a daily basis and tight security maintained around explosive magazines. Blasting agents may be stored in silos, but high explosives, boosters, and blasting caps need to be stored in specially designed magazines, separated by berms and placed specific distances apart. The U.S. Department of Transportation has classified three different levels of explosives defined by their sensitivities: Explosive Classes A, B, and C and an additional classification for blasting agents. Each class has different safety, storage, and transportation requirements.

Underwater drilling and blasting poses a unique set of circumstances. The holes must be lined with cardboard tubes to keep the holes open and the explosive charges dry. Because of this, a specialized type of drill called a Kelly bar drill is utilized to insert the tubes. Nonelectric detonation is generally used with two primers, one on top and one on the bottom of the explosives column, initiated simultaneously. A toe of material is generally left behind from the previous shot to prevent excessive forward movement of the blast. Blasting into water intensifies the vibration frequency and must be considered in the blast design to ensure vibration limits are not exceeded.

For most of my career I avoided in-house blasting and relied on outside blasting contractors. I always felt that they were the experts and a dangerous business should be left to experts. This also insulated us one more step away from liability in case of an accident. With this in mind, I never allowed storage of explosives, other than blasting agents, on any of our sites to avoid the liabilities, regulations, and extra regulatory scrutiny associated with explosive magazines. Most blasting contractors charge their clients based on the weight of explosives, number of boosters, and blasting caps they detonate. Some blasting contractors will charge a specified cost per ton to blast. The exception to this rule would be in the case of underground blasting. By the nature of underground blasting, the shots are much smaller and much more frequent and blasting contractors are usually uninterested in this type of work.

MONITORING QUARRY BLASTING

Blasting can affect the neighboring properties in a number of ways. When a blast is detonated, two different phenomena occur. The first is air overpressure, also known as airblast, which is a sound pressure wave in the air emanating from the blast. The second is ground or seismic vibration. Airblast travels at the speed of sound (1,125 feet per second [343 meters per second]), depending on the air temperature), which is much faster than ground vibration (seismic waves) and is felt almost instantaneously depending on your distance from the explosion. The seismic wave propagates more slowly and is felt several seconds later. The human body is very sensitive, noticing and reacting to blast-produced vibrations at levels much lower than any damage thresholds. The effects of vibration characteristics are a product of the duration of the blast, peak level, vibration frequency, and frequency of occurrence of the blast. People describe their annoyance at an unexpected blasting event as startle, fright, fear of damage, or abrupt awakening from sleep. This fear of damage is a threat to the American dream of home ownership and thus elicits a very emotional response from the affected homeowners.

The U.S. Bureau of Mines (USBM) (now a part of the National Institute for Occupational Safety and Health) defines airblast as "an impulsive sound generated by an explosive blast and resulting rock fragmentation movement" (Siskind et al. 1980a). This is the phenomenon that produces the sound and, if large enough, is capable of rattling or breaking widows in nearby structures. In 1980, USBM did extensive research and measurement on actual structures near coal mine surface blasting to determine structural responses, damage, and human responses to blasting-induced soundwaves. Air overpressure levels exceeding 120 decibels will produce some annoyance from rattling windows. The results of these studies concluded that air overpressure of less than 134 decibels, although noticeable, is "low enough to preclude damage to residential structures and any possible human injury over the long term" (Siskind et al. 1980a). Levels of airblast are a function of the charge weight (pounds of explosives per delay), distance from the blast, delay intervals, quarry face orientation, confinement of explosives, and weather.

Wind direction and speed can significantly affect airblast levels. Strong winds blowing from the source of the blast toward the recipient can increase the overpressure by more than 20 decibels. Temperature inversions in the atmosphere can also have a significant effect on airblast events. A temperature inversion holds cloud cover close to the ground, creating what is known as a low ceiling. An airblast propagating from a quarry blast ignition can actually bounce off the low ceiling and ricochet to areas several thousand feet away from the normal, expected path. This low ceiling event can easily startle neighbors that would normally not feel the effects of the routine blasting. If possible, avoid blasting on overcast days with low ceilings.

During the same time period, USBM studied the response of damage produced by ground vibrations from blasting on 76 structures during 219 surface coal mine production blasts. USBM defined threshold damage as "the occurrence of cosmetic damage; that is, the most superficial interior cracking of the type that develops in all homes independent of blasting" (Siskind et al. 1980b). Older homes with plaster interior walls are somewhat more susceptible to blast-produced cracking then the gypsum wallboard used in the construction of newer homes. Peak particle velocity (PPV) is used to measure seismic waves and is the measurement of the maximum speed of a particle (point) in the ground measured at a specific frequency of movement. The type of subsurface material (bedrock, clay, sand) through which the seismic wave moves affects the PPV significantly. Ground vibration is measured in three different dimensions (transverse, radial, and vertical) by an instrument known as a seismograph. USBM concluded that "safe levels of vibration from blasting range from 0.5 to 2.0 inches per second PPV for residential type structures. The damage threshold values are functions of the frequencies of the vibration transmitted into the residences and the types of construction" (Siskind et al. 1980b). As a result of this research, USBM developed what is known as the *Z*-curve, which is a graph of PPV versus frequency (Figure 10.13). Any plot of PPV (in inches per second) versus frequency (in hertz) that falls below the curve should not result in damage to residential structures.

All responsible quarry operators and blasting contractors routinely monitor any blasting occurring on their property using seismographic instruments. Whenever a blast is initiated, one or more seismographs is set up at the quarry property line to monitor the effects of the blasting on the neighboring residences. If a specific neighbor has complained, a seismograph should be set up in that neighbor's yard at the time of the blast to monitor any potential harmful effects. Today's digital seismographs will print out a plot of all three PPV versus frequency graphs relative to the *Z*-curve, a reading of the maximum airblast decibel level, and a reading of the resultant PPV. The results of these readings can reassure the neighboring property owners and should be kept as a permanent record along with the shot reports for liability purposes. Permanent monitoring stations can also be set up around the periphery of a producer's property. Seismograph monitoring records should be attached to the blaster's shot report and kept on file permanently.

The human psychological response and annoyance factors created by blasting-induced vibrations has the potential to become a social and serious public relations issue if not properly

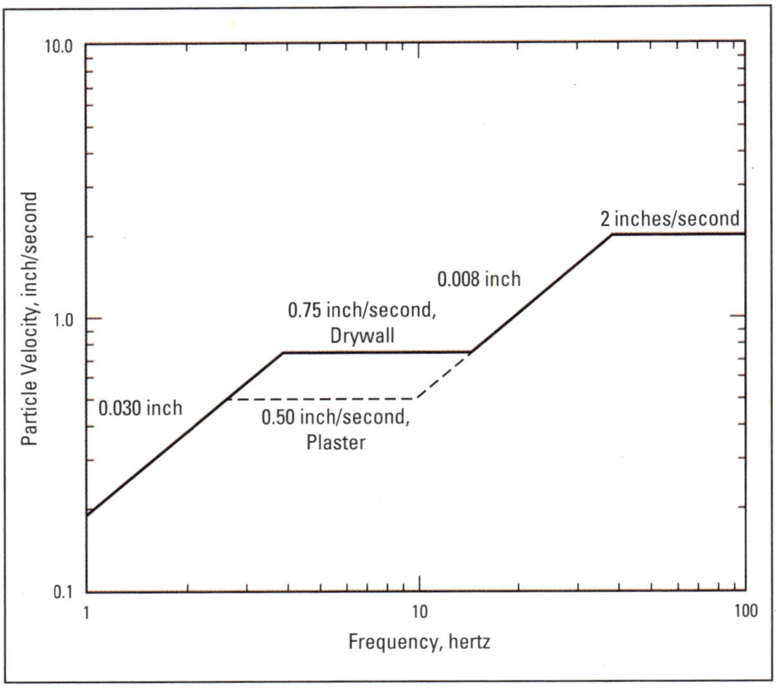

Courtesy of U.S. Bureau of Mines

FIGURE 10.13 The *Z*-curve

managed. Blasting should be kept to a minimum; it is better to set off larger shots less frequently than multiple small blasts because the neighbors are affected much less often. Every effort should be made to design blasts with the neighboring community in mind. A superintendent who reported to me for a number of years had a blasting theory that worked well for him: (1) never blast during the lunch or dinner hour, and (2) never blast when the school bus is due to pass by. The charge weight (pounds of explosive per individual delay) should be kept to a minimum, delaying blasts during adverse weather conditions (strong winds, temperature inversions) should be considered, and profiling highwalls to maintain the proper burden to eliminate blowouts from excessive gas pressure should also be considered. The startle factor can be minimized by calling close neighbors and informing them of impending blasting events. This is all part of what I referred to as a social license to operate in earlier chapters.

When attempting to obtain rezoning of mineral properties to develop a greenfield site, we would routinely offer to provide the neighbors with preblast surveys of their properties at our expense. A preblast survey consists of bringing a knowledgeable third-party professional in to thoroughly tour and inspect the property to establish and document (with photographs and videos) the existing structural condition of all buildings on the neighbors' properties. If, in the future, a neighbor decides to make a claim that they have experienced damage from blasting, both parties have documentary evidence of the pre-existing structural conditions prior to any blasting. Not all neighbors will accept the offer of a preblast survey, but these surveys seem to alleviate some of the fear of blasting. Blasting is a necessary part of the surface mining of consolidated mineral deposits but it does not need to be an annoyance to your neighbors.

My superintendent managing a quarry in rural Indiana asked me to accompany him to see a neighbor located about 1.5 miles (2.4 kilometers) from the quarry that had made a blasting complaint. We arrived to be greeted by a large (6-foot 2-inch [1.85-meter]) middle-aged woman (resembling the actress Bea Arthur) who had a 38-caliber revolver strapped to her waist. She lived in a remote log cabin that had been modified and enlarged several times over the years and explained how her late husband had personally constructed and later modified the structure. She proceeded to show us a number of cracks in the plaster and separations of the logs from the chinking and accused our blasting process of doing the damage. We set up a seismograph in her front yard for the subsequent five blasting events, and every time she greeted us wearing her revolver. We were able to hear the muffled explosion from the quarry blasts, but the seismograph was never triggered and the overpressure and PPV did not even register. During one of these monitoring events I clandestinely inspected her house. The original cabin structure was about 15 × 15 feet (4.5 × 4.5 meters) and all four walls had 6 to 8 foot (1.8 to 2.4 meter)–wide doorways cut into them to access additional rooms that were added later. The structural support provided by the original four walls was significantly reduced by the removal of the support in the areas where the doorways were cut, causing the roof to sag. We ultimately disavowed any responsibility for the structural problems and were threatened with lawsuits. Our insurance company, against my protestations, ultimately settled out of court without admitting any guilt.

Quarry blasting can never be accomplished without affecting the neighbors. If properly designed and executed, blasting will not cause physical damage to the neighbor's property or structures. The psychological effects of routinely experiencing airblast and ground vibration are an issue that will require management through monitoring (seismographs), documentation, education, and public relations efforts.

CHAPTER 11

Underground Mining

According to the U.S. Bureau of Mines (a division of the National Institute for Occupational Safety and Health [NIOSH]), in 2015 there were 4,303 operations actively surface mining crushed stone in the United States and 116 underground crushed stone mining operations (NIOSH, n.d.). Underground mining thus constituted only 2.6% of the total active mines. No data were available on what percentage of production volume this represented. Although only a small percentage of aggregate is produced from below the surface, the number of underground operations has been increasing on an annual basis. An aggregate producer may decide on subsurface mining for a number of reasons. The cost of underground mining is in the range of 25%–50% per metric ton more costly than surface mining.

When faced with certain obstacles, a producer must decide on the best option to sustain the level of business necessary to survive in the marketplace. Some of the reasons for going underground include excessive overburden removal costs, the cost of surface real estate, local market conditions, high cost of transportation, geological constraints, depletion of surface reserves, and lack of the ability to obtain zoning approval for a surface quarry. Environmental and neighbor issues can also influence these decisions. There is also something to be said for the old adage "out of sight, out of mind." If the cost of removing excessive overburden is appreciably more than the added cost of underground mining, the decision is fairly straightforward as long as sufficient capital is available for the development. If no surface stone is available within the market owing to geological restraints and the transportation cost into the market exceeds the additional underground mining cost, a favorable economic model can be built.

Underground mining is an entirely different world than surface mining with different maintenance challenges, additional safety issues, and new logistical obstacles, all requiring a new mindset on the part of management. Employees used to working in daylight may not be comfortable in the dark confines of a mine. Different types of equipment are required, accompanied by new learning curves for operation and maintenance. Many new safety procedures, inspection regimens, safety checks, training, and regulatory requirements will add to the learning curve. The number of Mine Safety and Health Administration (MSHA) inspections automatically doubles and thus so does the potential for violations and additional fines. If inspectors are so inclined, they can always find a loose piece of rock on a mine rib and write a citation. Unlike surface mines that can be essentially ignored during off-peak times, underground mines require continuous attention. This may seem daunting at first, but I have been through it and it is manageable with some diligent effort and an open mind.

There are many different mining methods for exploiting underground mineral resources, but by far the most common in the aggregate industry is the room-and-pillar method (Figure 11.1). This method is used primarily in relatively thick, horizontally oriented deposits.

FIGURE 11.1 Room-and-pillar limestone mine

Therefore, I will not go into detail on some of the other underground mining methods like sublevel stoping, shrinkage stoping, cut-and-fill stoping, top slicing, sublevel caving, block caving, or longwall mining. Room-and-pillar mining is a form of what mining engineers consider open stoping; a stope being a cavity left behind when the ore, in this case aggregate, has been removed. The open stopes are supported by leaving regularly spaced pillars of ore in the mining section as primary support for the roof. Depending on the depth underground, the engineering properties of the stone, and the size of the stopes, anywhere from 25% to 40% of the deposit must be left behind to support the vertical load. Secondary support such as steel sets, roof bolts, split sets, wire mesh, and shotcrete are sometimes necessary to aid in the support and prevent spalling of the pillars.

Like all businesses, underground mining has its own jargon to describe various aspects of a working mine. In an underground mine the ceiling is known as the "back." The pillars are called the "ribs," and the mining face is the "breast." The term "drifting" refers to advancing a mine face or heading. In relatively thick, horizontal deposits, multiple benches may be utilized to develop the mine. The first bench is normally developed on the top of the deposit; drilling and blasting horizontally into the formation and this level is known as the "breast heading" or "breast stope." This is done to enable easy access to scale loose rock and, when necessary, to bolt the back. The next level(s) are developed in a similar fashion to a surface mine by drilling and blasting vertical blastholes, and this level becomes the "bench" and the point where the floor of the upper heading meets the vertical bench is called the "bluff" or "bluffline" (Figure 11.2). Chasing specific veins in certain metallic mining applications, where irregular pillars are left behind for support in non-ore-producing zones, is known as "gophering."

MINE DESIGN PARAMETERS

To design safe room-and-pillar mines, undisturbed samples of the stone (rock core) need to be analyzed to determine their mechanical engineering properties. These properties will be

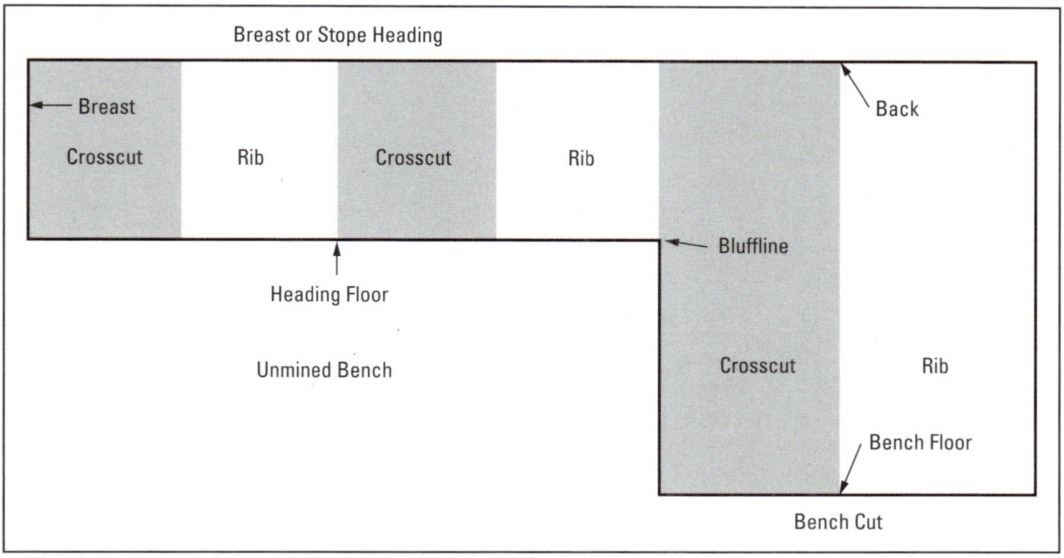

FIGURE 11.2 Cross section of an underground heading and bench

utilized by a mining engineer to design the heading sizes, pillar sizes, height of the openings, and any necessary supplemental support. The testing should include the following: uniaxial compressive strength, triaxial compressive strength, Brazilian indirect tensile strength, point load index strength, bulk density, and specific gravity. Further testing of the cores for aggregate quality and perhaps chemistry will probably also be necessary. Past experience has dictated that 40 foot (12 meter)–wide pillars with 40- to 50-foot (12- to 15-meter) openings provide fairly stable room-and-pillar mines.

Employee safety, and therefore mine stability, is the primary consideration in the design of underground mines. Maximum recovery (extraction ratio) is achieved in room-and-pillar mining utilizing square pillars. Rectangular pillars (rib pillars) are generally only specified under certain geological conditions involving zones of weakness in the bedrock like bedding planes, joints, slips, excessive weathering, or faults or when dealing with unequal horizontal stresses. A thorough geological evaluation of the local area is essential in identifying any geological anomalies or planes of weakness. Mapping the joints in the locally exposed bedrock should be accomplished in a systematic manner and recorded along with the number of observations. Joints should be mapped horizontally with an associated compass bearing as well as the vertical inclination of the jointing. If possible, the average spacing between joints should be measured as well. Predominant regional joint directions, which are always accompanied by a secondary joint direction at 90°, should be evident. Other less prominent joint systems may also be identified. A rose diagram indicating the jointing directions is a handy tool for graphically displaying the data. Identified fault zones should be avoided whenever possible. Geological discontinuities (weaknesses) can cause a number of problems in underground mines, which is why a thorough geological analysis prior to mining is a necessity (Figure 11.3). Low-angle jointing can cause wedge-shaped failures in the mine roof that are difficult to predict and hard to prevent using roof bolts. Undulating or wavy bedding planes in sedimentary deposits are capable of causing similar types of roof failures. High-angle joints or slips (>50° or 60°) can cause weakness and instability in pillars (Figure 11.4).

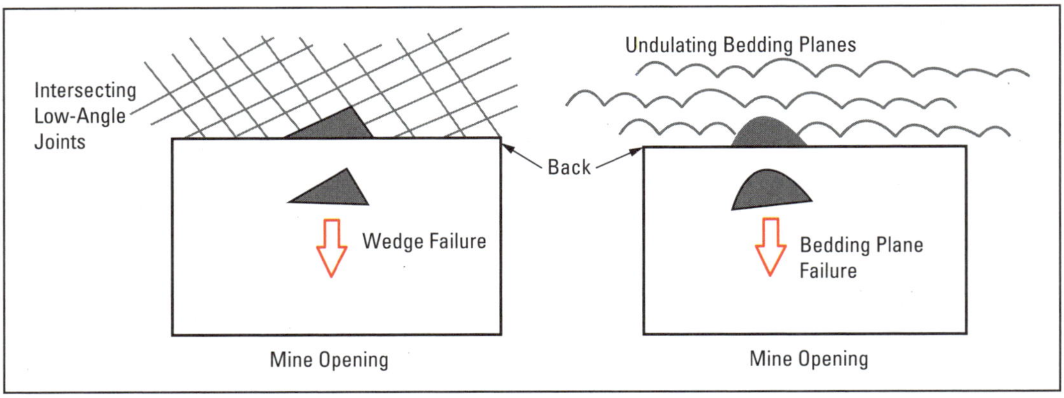

FIGURE 11.3 Geological discontinuities capable of causing roof (back) failures

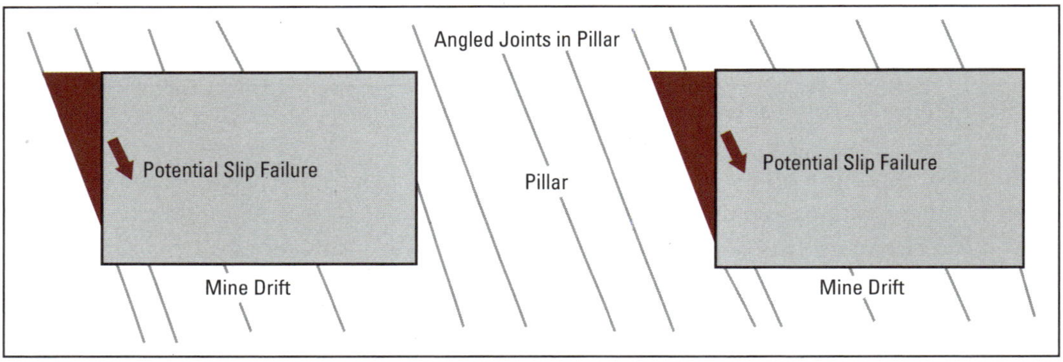

FIGURE 11.4 Potential slippage along joint planes

Mine pillars have a tendency to spall over time as the vertical stress is relieved and can result in an hourglass shape with the pillars thinner in the center and wider on top and bottom. Pillars should initially be oversized to prevent this thinning (spalling) over time. When benches are a part of the long-term plan, additional pillar width should also be seriously considered. When blasting underground benches, the line of blastholes adjacent to the ribs should be set inward toward the heading to eliminate burning the ribs, which contributes to the spalling. A pillar (rib) that is wider on the bottom than on the top is much more stable over time and will have less of a tendency to "hourglass" (Figure 11.5).

Certain areas of the country are subject to regional stresses that can produce either extension or compression stresses within the local bedrock. These inherent regional stress anomalies can cause excessive horizontal stress, which must be identified prior to initial mine design if possible. Several stress provinces have been identified in the United States and should be considered when designing mines located in these areas (Figure 11.6). Excessive horizontal stress (2–10 times greater than the vertical stress) will negatively affect the integrity of the mine roof via shear failure. Headings driven perpendicular to the direction of the stress field are subject to the greatest stress concentration and therefore the greatest potential for failure. Rib pillars oriented parallel to the horizontal stress field can minimize the potential for roof failures by limiting the number of headings running perpendicular to the stress field direction. Staggering

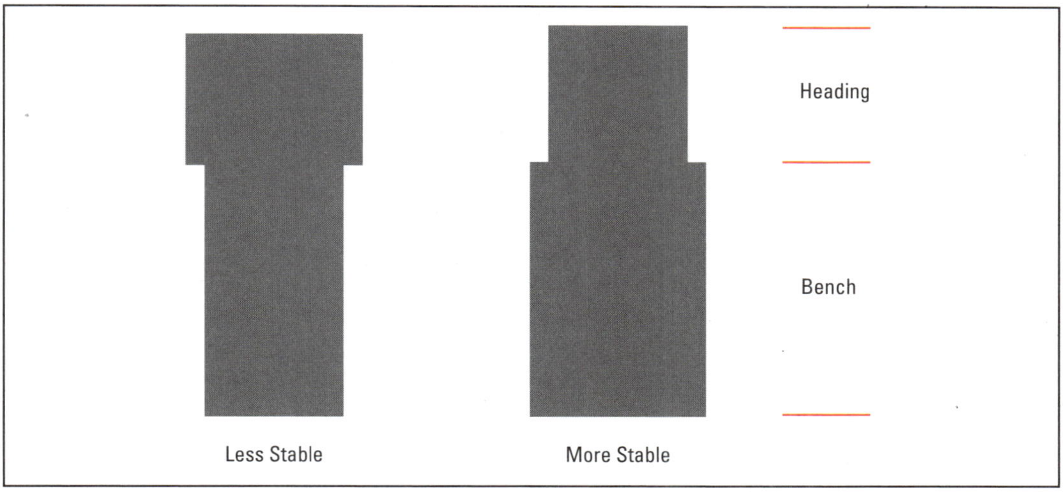

FIGURE 11.5 Pillar cross sections

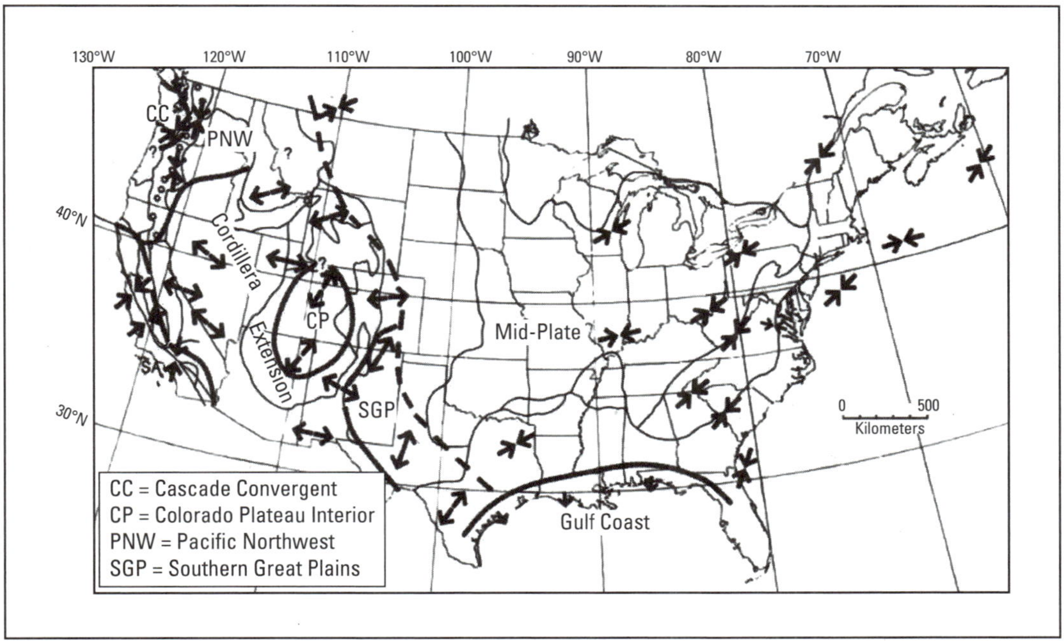

Source: Zoback and Zoback 1989

FIGURE 11.6 Stress provinces in the continental United States

the rib pillars may also help overcome the problems associated with horizontal stresses but makes efficient underground haulage much more difficult.

The recovery or extraction ratio reflects the volume of material mined versus the amount of rock left behind to support the mine openings. Although high recovery rates are desirable to utilize the highest percentage of the reserve, the higher the recovery ratio, the less material is left behind as roof support. The extraction ratio can be calculated as follows:

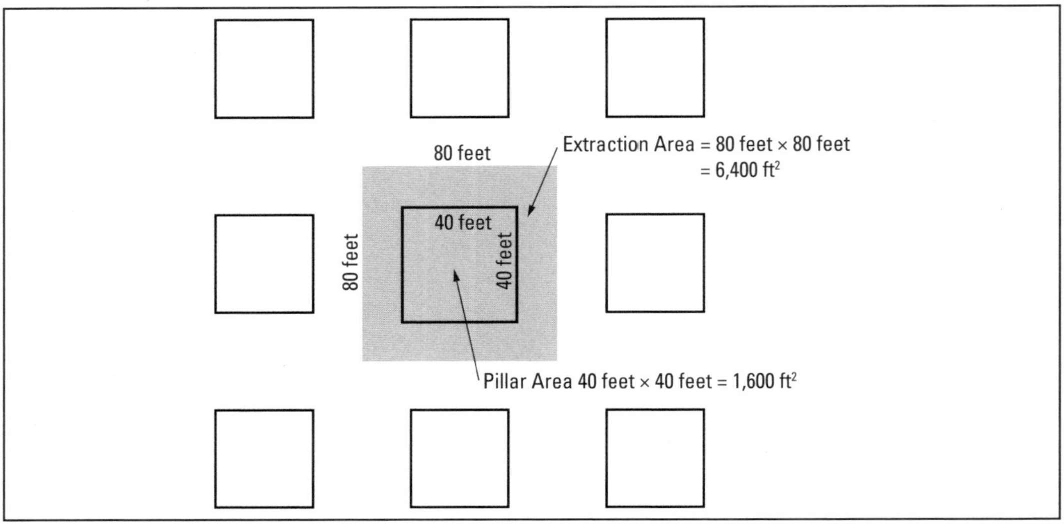

FIGURE 11.7 Extraction (recovery) ratio

$$\text{ER} = 1 - \left(\frac{A_p}{A_e}\right)$$

where

ER = extraction ratio
A_p = area of the pillar
A_e = area of extraction

Using the example shown in Figure 11.7:

$$\text{ER} = 1 - \left(\frac{1{,}600}{6{,}400}\right) = 0.75 \text{ (or 75\%)}$$

The vertical stress on a mine pillar can be estimated using the tributary area formula. Premining vertical stress is normally calculated at 1.1 pounds per square inch per foot of overburden (774 kilograms per square meter per 30 centimeters of overburden) overlying the mine roof. The following formula estimates the vertical stress on an underground mine pillar:

$$S_p = S_o \frac{(r + w)(r + l)}{(w + l)}$$

where

S_p = postmining stress
S_o = premining stress
r = room width
w = pillar width
l = pillar length

Using the example shown in Figure 11.7 with an overburden thickness of 500 feet (152 meters):

$$S_p = 550 \text{ psi}(3{,}792 \text{ kilopascals}) \times \frac{(40 + 40)(40 + 40)}{(40 + 40)}$$

$$S_p = \frac{550 \text{ psi } (3{,}792 \text{ kilopascals})[6{,}400]}{1{,}600} = 2{,}200 \text{ psi } (15{,}168 \text{ kilopascals})$$

The ratio of the width to the height (*w*/*h*) of a pillar also affects pillar stability. Pillars that are too narrow or too slender have the potential to fail, especially if any discontinuities exist. A pillar with a *w*/*h* ratio of 0.4 is considered to be slender and one with a *w*/*h* ratio of 1.4 is classified as squat. Pillar widths should be somewhat overdesigned initially because, in reality, pillar spalling, mechanical scaling, and cleanup inevitably results in the narrowing of pillars over time. Steep, high-angle joints and slips greater than 50–60 degrees can cause pillar failure even in pillars classified as squat.

When mining on multiple underground levels, the pillars must be aligned vertically or superpositioned to allow the pillars on both levels to carry the overburden loads. This requires precision surveys of the existing mining levels prior to developing any other levels. Accurate alignment of the pillars on the next developed level, directly beneath the existing pillars on all other levels, is imperative to equalizing vertical stresses. The area between levels is known as interburden, or a sill pillar. Determination of the thickness of sill pillars is dependent on geology, rock mechanics testing, and pillar superposition.

Although square pillars are the most common, other types of pillars can be utilized to gain additional support. We've already discussed the use of rib pillars, the length of which is considered to be one to three times the width, under certain geological conditions. Barrier pillars (length is more than three times the width) can be used to control ventilation and/or add additional support around critical mine openings like declines and adits. Larger solid blocks of stone (abutments) can be left intact under critical surface facilities or areas with greater overburden thickness or near structural discontinuities. Note the large area in the center right of Figure 1.5 shown as an abutment. This area was directly beneath the processing plant and was left unmined as a safety precaution.

I took over management of an underground mining operation beginning on a Monday morning. The previous Saturday, a substantial section of the mine roof had collapsed. Fortunately, no employees were in the mine at the time and no injuries occurred, but it made for a very exciting first week on the new job. A section of mine roof collapsed in the middle of a heading for a length of about 70 feet (21 meters) including through the center of one crosscut. After consulting a mining engineering firm, the conclusion of our internal investigation into the roof fall was that a "swarm" of seven joints had coalesced at a crosscut intersection. This created a zone of weakness down the middle of the drift. The joint swarm was at just the right angle that it was not supported by any pillars for a length of almost 120 feet (36 meters). After extensive joint mapping both on the surface and underground, we determined that to maintain a stable roof beam we needed to reorient the entire mine layout and utilize 35 × 60 feet (11 × 18 meters) rectangular rib pillars instead of the 40 × 40 feet (12 × 12 meter) square pillars in use at the time. The new angle of orientation was designed so that no predominant joint would be unsupported by a pillar for more than a span of 50 feet (15 meters). This ultimately eliminated the roof problems but the reorientation took over a year of production to accomplish. Because of the location of the roof fall, not far from the entrance portal, we had to physically support the roof to save a crosscut intersection that was necessary to our truck haulage route. We managed to physically support the roof utilizing several truckloads of treated railroad ties to construct four towers of cribbing from the floor to the back.

Underground mines require a stable roof beam for the pillars to support. The roof beam should be composed of thick (massive) beds that have high strength, are weather resistant, and made of durable rock. Planes of weakness (joints, excessive weathering) should be avoided in the rocks of the roof beam, if possible. The roof beam needs to be tested for compressive strength, tensile strength, and modulus of rupture to establish sufficient strength parameters. If the potential roof rock is laminated or contains varying thicknesses of beds, roof bolting will be necessary to laminate the various layers together to form a stable beam. This can be accomplished utilizing resin bolts and/or point anchors to make a competent, solid layer of rock. If the rock in the roof is appreciably softer than the rock in the pillars, the pillars can actually punch into the roof, causing catastrophic roof failure. A well-defined, horizontal parting or bedding plane beneath a massive layer of rock should be located to form a smooth back.

These design considerations are for discussion purposes and should not be utilized for any actual mine design. Because of the critical nature of mine stability and employee safety, an experienced mining engineer should always be consulted in the design process.

UNDERGROUND MINE ACCESS

Several different ways exist to access an underground mine from the surface. An adit is a horizontal shaft going into the side of a hill or into a quarry face (Figure 11.8). Existing surface quarries when faced with excess overburden, special quality requirements, or specific market needs will often utilize adits to access reserves via underground mining. MSHA requires a minimum of two means of ingress and egress for any underground mine in the United States. This is a safety requirement in case one access is blocked by a roof fall, escape and rescue is still possible via the second access.

Vertical mine shafts are another option for accessing underground mine workings. The larger production shafts need to be drilled, blasted, and mucked out using conventional mining methods. These production shafts that range from 24 to 40 feet (7 to 12 meters) in diameter are completed to slightly beyond the depth of the mining horizon and then lined with concrete. A headframe and hoists are constructed above the vertical shaft, which operates a personnel elevator in one section of the shaft (Figure 11.9). The remaining area of the shaft contains a mechanical apparatus called a skip hoist. This hoist supports two buckets called skips; one is lowered while the other is hoisted to the surface (much like a double-grab dredge). At the

FIGURE 11.8 Twin adits in a quarry highwall

Courtesy of Frontier Kemper

FIGURE 11.9 Headframe and hoist house above a vertical shaft

bottom of the mine is a dump hopper and a skip loader mechanism. When a skip breaks the surface, it automatically dumps its load into a chute within the head frame that sends the ore to a conveyor or bin for stockpiling. Skips come in several sizes based on production needs and depth of the mine. The maximum capacity of the skips is directly related to the diameter of the vertical shaft.

A second vertical shaft to provide ventilation and employee escape access in case of an emergency will also need to be developed. These smaller-diameter shafts are usually developed by means of a raise bore operation. A raise bore is a shaft sinking method that begins with drilling a very accurate (±1° deviation) borehole, usually 11 to 12 inches (28–30 centimeters) in diameter, into a previously mined drift inside the mine. After removal of the drill bit, a large reamer attachment is connected to the bottom of the drill string and rotated while the drill string is raised back to the surface. These reamers usually are 15–20 feet (4.5–6 meters) in diameter. The cuttings produced by reaming out the hole fall to the bottom of the mine and are usually removed by a wheel loader. Most raise borings are completed on a 24-hour, 7-day schedule to prevent the reaming device from becoming stuck in the shaft. Upon completion the raise bore is usually lined with concrete or shotcrete to prevent deterioration of the surrounding rock from weathering. A smaller head shaft and emergency escape winch with a four-person escape cage will be installed over the ventilation shaft.

Inclined shafts are a third option to access underground mine workings (Figure 11.10). Incline shafts are also known as slopes or declines. A decline can be combined with a raise-bored ventilation/escape shaft or twin declines can be developed parallel to one another. In the case of twin declines, one is used for production and the other for ventilation and emergency escape access. Depending on reserve requirements, declines can be constructed with switchbacks to enable control of the location of the bottom of the shaft. Declines are extremely versatile, and if large enough, heavy mining equipment can be driven directly into the mine.

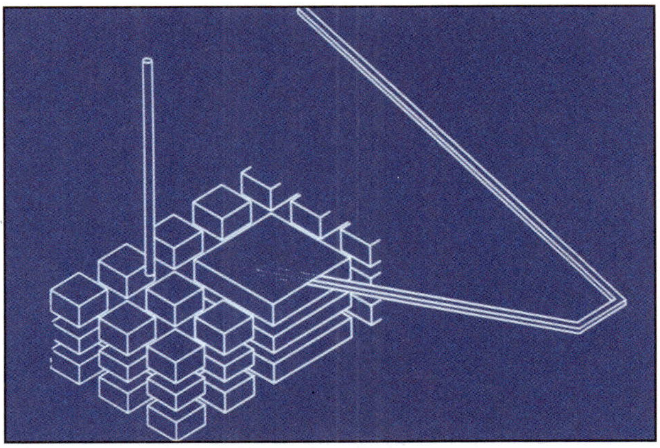

Courtesy of American Aggregates Corporation

FIGURE 11.10 Room-and-pillar mine with access via decline and vertical ventilation shaft

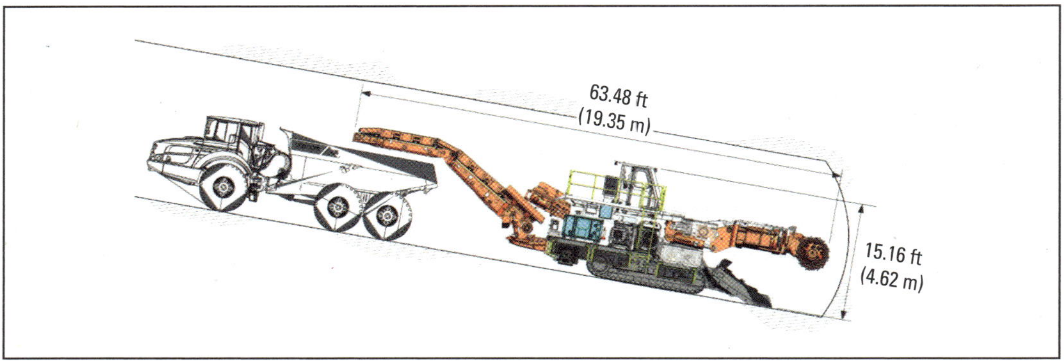

Courtesy of Frontier Kemper

FIGURE 11.11 Roadheader loading a truck in a decline

Most declines are equipped with standard conveyor equipment allowing large production volumes to be conveyed to the surface without the constricting limits of a small-diameter vertical shaft. Declines usually range in size from 10 × 10 feet (3 × 3 meters) up to 20 × 20 feet (6 × 6 meters) in size and are developed at slopes ranging from −12.5% to −17.5%. To allow for heavy equipment access within the decline, the conveyors and utilities are usually suspended from the back (ceiling) of the sloped tunnel.

Decline development can be accomplished by different methods depending on the rock type. The most common method is conventional drill, blast, and muck. Roadheaders, which were essentially designed as tunnel boring machines, are becoming more common in the construction of inclined mine shafts (Figure 11.11). These machines are able to operate at declines as high as −17.5% and are very efficient and usually much faster than conventional mining methods, especially in softer rock formations such as shale. No drilling or blasting is required because the rock breakage is all done mechanically by a rotating head imbedded with an arrangement of replaceable carbide picks. This is generally a less expensive method of constructing inclined slopes when compared to conventional mining methods. As the roadheader progresses downward, secondary support is added to the tunnel behind it, as needed, such as roof bolts, wire mesh, steel sets, and shotcrete.

All underground mines require some form of mechanical ventilation to maintain a healthy working environment for the employees, vent the gases produced by blasting, and dissipate the fumes produced from diesel equipment exhaust. Mine ventilation is an extremely complicated subject because the properties of air change dramatically with variations in temperature, barometric pressure, and humidity. Requirements for ventilating a mine during the cold winter months can fluctuate dramatically from those of the hot, humid summer. Fan size, volume, type, and location must be designed by a competent ventilation engineer specifically experienced in underground mine ventilation. Fans function much like centrifugal pumps, with the primary load on the fan being the friction between the air and the walls of the airway, much like friction head developed in a pipe. But, unlike water in pumps, the density of air changes radically based on thermodynamics.

The total horsepower of all diesel engines and the dilution of diesel exhaust fumes along with the associated diesel particulate matter need to be carefully considered in ventilation design to meet the threshold limit values imposed by MSHA. The gases emitted during blasting operations need to also be considered and vented to prevent employee exposure. Blasting is generally done at the end of a shift with the start of the next shift postponed to give the ventilation system time to remove the blasting-generated gases. The mine plan itself should break down the mine into panels that can be shut off by stopings to eliminate air loss and recirculation when specific panels are mined out. As the mine progresses away from the main ventilation system, additional fans (Figure 11.12) will likely be required to pressurize the air, much like booster pumps in dredging operations. Mine ventilation is a very complex issue that is best left to an experienced mining engineer. Like all mechanical equipment, ventilation fans require regular inspection and maintenance and ventilation plans need to be reviewed and updated regularly.

FIGURE 11.12 Portable ventilation fan

UNDERGROUND DRILLING AND BLASTING

Drilling and blasting underground is much different than initiating surface blasts. When advancing a heading or breast stope, the drill holes have to be drilled horizontally into the rock formation. Underground blasting is done with small-diameter holes ranging from 1 to 2.5 inches (2.54 to 6.35 centimeters) in diameter. Appreciably more drill holes are required because of the confinement and the small-diameter blastholes. The drilling needs to done with extreme accuracy to avoid damaging either the back or the ribs. This type of blast is confined on five sides and is timed similarly to a lift shot in an open pit environment. Two different drilling patterns are generally utilized; one is known as a V-cut and the other a burn cut. The burden in a V-cut is essentially the distance from the bottom of one hole to the bottom of the next (Figure 11.13). These holes are advanced horizontally into the cut at varying angles radiating from the two center rows that form a V. Each successive row of holes is drilled at a slightly smaller angle, maintaining an equal burden at the back of the cut or bottom of the holes. The final row of holes along the ribs are angled slightly (about 2°) outward toward the ribs. The two center rows forming the V are fired instantaneously, removing the center of the cut. Each successive row is delayed to fire into the void created by the V. The variation on this shot design called the baby V has an additional row of angled holes in the center drilled to about half depth to assist in pulling the V.

A burn cut, on the other hand, has all holes drilled horizontally, straight into the face with no angles. One or more holes will be larger in diameter and left unloaded to give the shot some relief. The center of the cut will contain the relief holes and several tightly spaced, loaded holes designed to fire into the relief holes. In theory, the holes all fire in a delayed sequence into the larger empty holes in the center of the cut. My experience underground was that burn cuts were much less efficient in fragmenting the rock and pulled less depth of cut than did V-cuts. At the time we used 18 foot (5.4 meter)–long drill steels drilling a V pattern and routinely pulled 14–16 feet (4.3–4.9 meters) of face depth on most blasts. At one point,

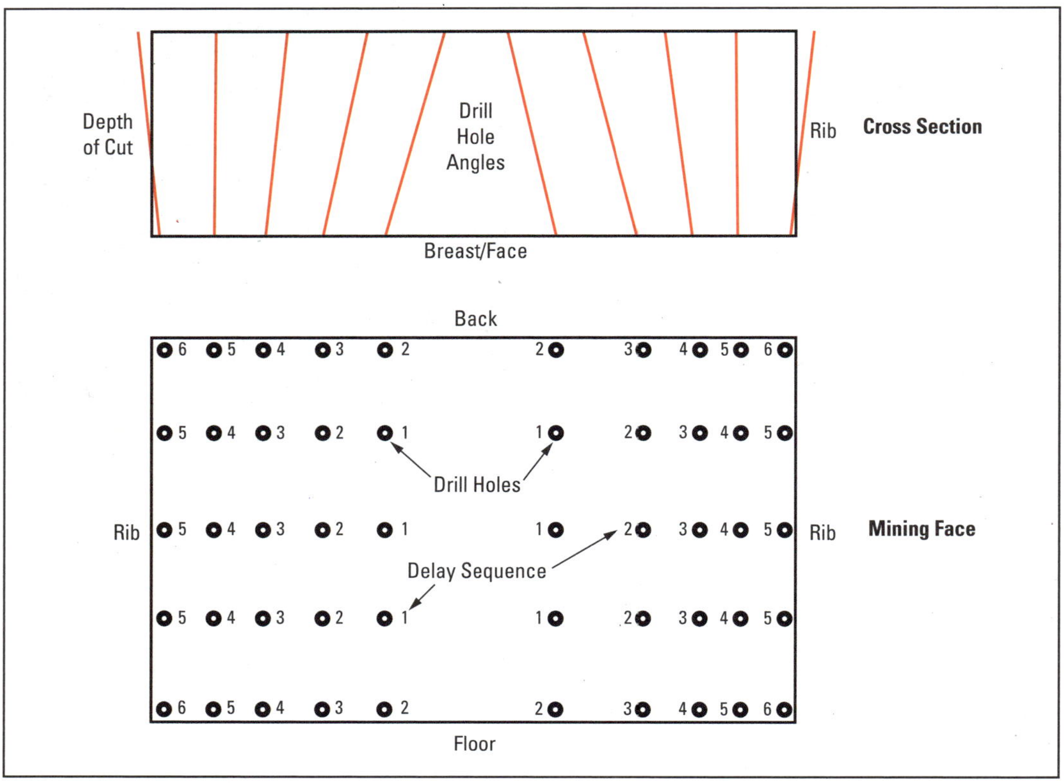

FIGURE 11.13 V-cut blast design

I was approached by an underground mining consultant who said he could improve my blasting results by pulling deeper depths using a burn cut. I told him to "put his money where his mouth was" and agreed to try his burn cut design but, if it didn't pull more than the expected 14–16 feet (4.3–4.9 meters) of depth, I would not pay his consulting fees. He was very confident and readily agreed. We had to buy a larger 6-inch (15-centimeter) drill bit for the relief holes that cost about $600 at the time, which I agreed to pay for. His first shot was an absolute disaster, pulled less than 3 feet (0.9 meters), and left quite a mess behind. On the second shot the driller managed to lose the new drill bit in one of the relief holes and this shot only pulled 6 feet (1.83 meters) of face. I never received an invoice from the mining consultant. Many underground operations successfully utilize burn cuts for their blasting, but no two rock formations are the same.

Drifting, face drilling, and tunneling require very different drilling equipment than bench drilling. Instead of drilling vertical holes, the holes need to be advanced horizontally into the breast face. In certain mines with very tight tunnels, drilling is done using a jackleg drill, which is hand-operated by a single miner. This is not the case with large openings in room-and-pillar mines. The type of underground drills operating in larger headings have become known as drill jumbos, which have their drills mounted on a mobile chassis. Drill jumbos can have multiple drills operating at the same time. The drills are mounted on articulated arms that raise, lower, and change angles via the use of hydraulic motors and cylinders and the drills themselves are hydraulically powered. A twin-boom jumbo cannot double the production of a single-boom machine because of the time required to move, angle, and set up the holes. One man operating

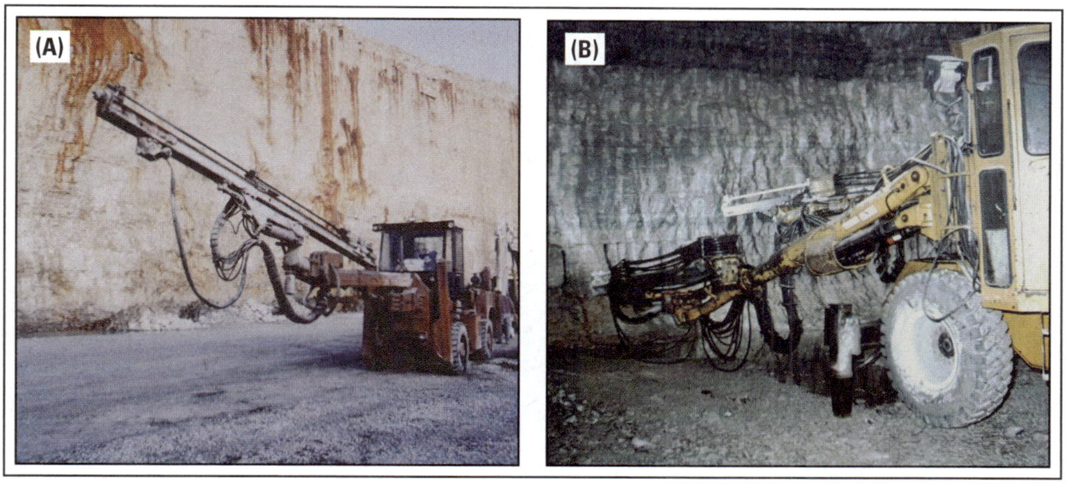

FIGURE 11.14 (A) Single-boom drill jumbo, and (B) twin-boom jumbo drilling a breast face

a twin-boom machine can outperform a single-boom machine by an additional 60%–65% production (Figure 11.14).

Underground room-and-pillar mining requires a specific, choreographed sequence of events and enough available open headings to keep the sequence moving with no interruption. The choreographic sequence consists of:

1. Drilling
2. Loading the blastholes with explosives
3. Blasting at shift's end
4. Mucking out the ore
5. Scaling the roof and ribs
6. Cleaning up scaled rock
7. Roof bolting (when required)

The first step in the dance is the drilling process. The drill jumbo (Figure 11.14) approaches a clean face and begins drilling the specified pattern of drill holes at the precise angles for the V-cut. Upon completion of the drilling, there must be another cleaned, scaled, and bolted face available for the next drill setup or the dance comes to an abrupt end. Step 2 is charging the blastholes with explosives. After loading explosives, the heading will be blocked off to personnel access until such time as multiple blasts are ready to be initiated. Multiple shots are usually detonated at the end of the shift after most personnel have departed to allow the fumes from the explosive gases to be removed by the ventilation system. Step 3 is loading and transporting the blasted shot rock to the primary crusher, which is known as mucking. When the shot rock has been mucked out, the back, ribs, and face need to be scaled of loose rock (Step 4) to prevent any falling material from injuring the miners. Mines requiring additional support will, in Step 5, be bolted on a specific pattern designed to add additional roof support. This needs to be completed prior to the drill arriving for the dance to begin once again. Enough headings must be in the development process so that each step in the choreography is kept moving at all times. If the objective is three blasts per shift, at least 12–15 working headings must be available for production.

The blastholes are loaded with similar types of explosives as in surface blasting but in much smaller quantities because of the smaller-diameter holes and the tight confinement of the headings (Figure 11.15). Where a typical surface blast may produce 30,000–50,000 metric tons of shot rock, a breast shot will produce in the range of 1,000–1,200 metric tons. A high-explosive primer containing a millisecond-delayed detonating cap is inserted into the bottom of each hole and ANFO is pneumatically blown into the hole on top of the primer. A specially designed machine called an ANFO pot is used in this process. The individual holes are usually wired to detonating cord and one cap added at the time of initiation to start the process. Depending on the size of the face, a lift truck of some type is needed to access the drill holes on the upper half of the breast. We always had a safety rule that the blaster had to be at least two crosscuts in front of the shot and an additional two cross cuts at 90° to the shot before initiating the blast. It's a thrilling experience initiating a blast underground, feeling the pressure wave rapidly advancing down the tunnel toward you at the speed of sound.

FIGURE 11.15 Loading a breast (face) shot

Scaling the back, ribs, and face can be accomplished in a number of ways. This can be carried out manually with a long pry bar, but this is slow and dangerous work. Mechanical scaling is necessary in high-production mining operations. The purpose of scaling is to remove any loose rock that could possibly fall onto miners or equipment working in the heading. Excavators equipped with a hydraulic hammer attachment are used for scaling in some mines, but this is usually an overkill situation where too much material is removed. The more the hammer operates, the more material becomes loosened, the more time is spent hammering, and the more cleanup is necessary. A hydraulically operated pick or finger mounted on an excavator or Gradall boom is usually all that is needed to scrape off the blast-loosened rock and prevent injury. At one point, we utilized an excavator/hammer combination as a scaling device, but we were constantly facing structural cracking of the excavator boom and the chassis under the intense stress of the hammering. A rubber-tired machine generally works best because the tires help absorb a portion of the stress.

There are many different applications and different styles of roof bolts. Mechanical bolts with expansion wedges are used for point anchoring, and split set types that are expanded after insertion into the hole are used mostly in shifting ground applications. The most common form of roof bolts used in this type of application are 6 to 8 foot (1.8 to 2.4 meter)–long fully grouted resin bolts. The bolts, fabricated from rebar, are available in several diameters and lengths depending on the roof control requirements. A 4 inch (10 centimeter)–square roof plate is inserted around the bolt like a washer. Two-part cartridges containing reinforced polyester resin in one compartment and a catalyst in another are inserted into the hole, followed by the roof bolt, which is spun to mix the cartridges and then held in place while the resin hardens. The resin cartridges are designed for the specific hole/bolt size combination and the resin fills the annulus between the bolt and the walls of the hole. Resins with various mixing times from 3 to 15 seconds and hold times from 3 up to 245 seconds are commercially available.

According to a study by NIOSH, "a fully grouted bolt anchors itself by frictional interlock between the resin and the rock. The effectiveness of the interlock is measured by the 'Grip Factor,' which is defined as the bolt's resistance to pullout per inch of bolt length" (Mark et al. 2002). The competence of the rock being anchored and the installation procedures play a large role in the effectiveness of resin bolting. The results of several different studies are summarized in Table 11.1.

Roof bolting is done by very specialized equipment designed specifically for this purpose (Figure 11.16). An automated roof bolting machine contains a carousel holding 10–12 roof bolts, a drill, a bolt spinning wrench, and a pneumatic device that shoots the resin cartridges into the drill hole. The operator is positioned back on the chassis under a protective cover for safety. Through a complicated sequencing system, a hole is drilled into the back, then two tubes of resin are pneumatically inserted in the hole. In the following step, the bolt is inserted and spun to mix the resin. After several seconds of the bolt spinning, the bolt is held in place until the resin hardens. This process forms a tight bond between the rock formation and the bolt laminating together the layers of rock above the back into a solid beam much like plywood.

TABLE 11.1 Grip factors of resin-grouted rock bolts

Rock Type	Country	Grip Factor (metric tons/inch)
Coal/shale	Australia	0.7–2.1
Hard sandstone and limestone	Australia	1.4–3.6
Soft rock	United States	0.5
Hard rock	United States	2.0

Data from NIOSH

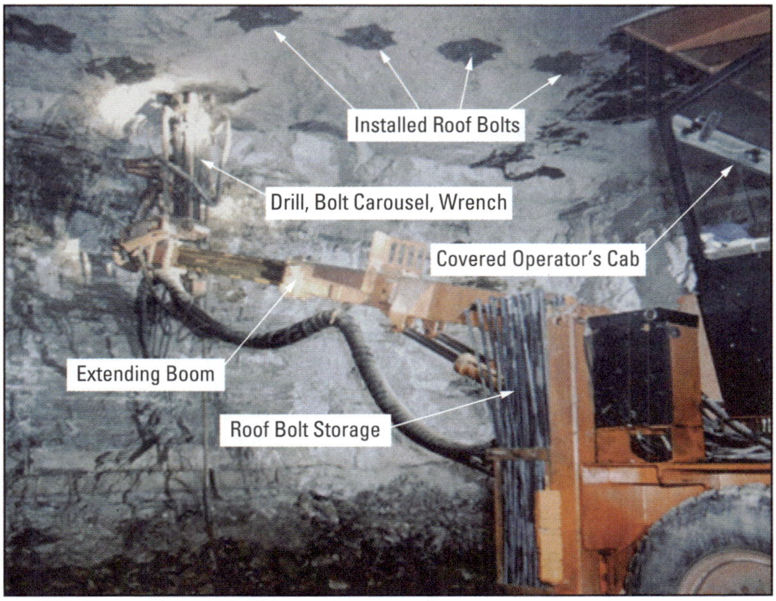

FIGURE 11.16 Bolting the back

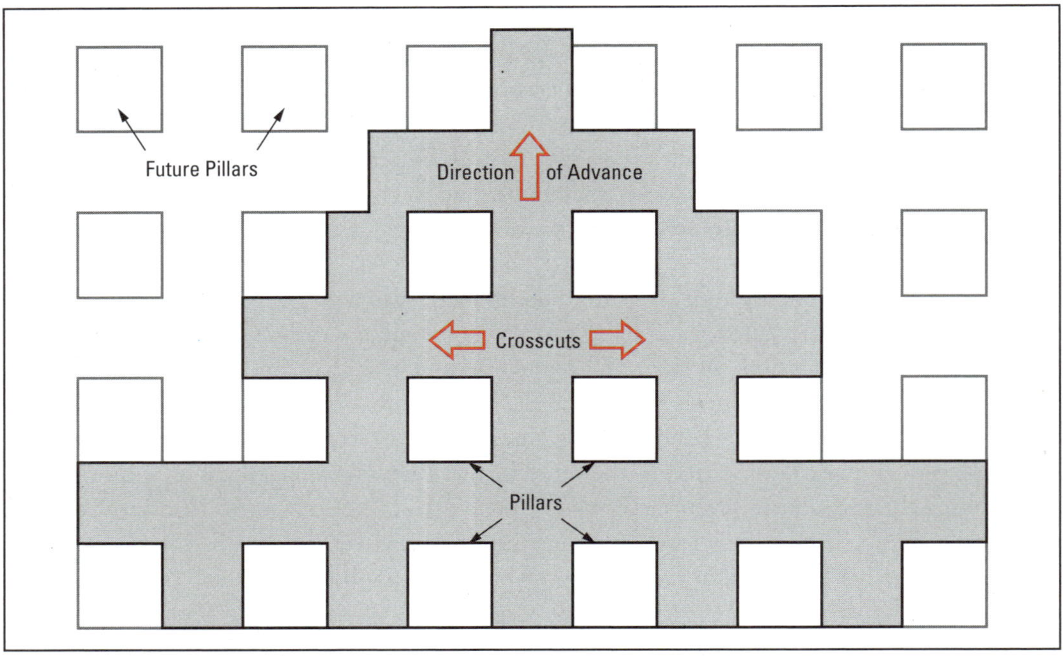

FIGURE 11.17 Christmas tree advance

Depending on the size and shape of the property to be mined, the mining sequence will need to progress in a preferred direction. For roof stability, it is usually a good policy to advance the mining direction in a wedge shape or Christmas tree fashion (Figure 11.17). Pushing one or two faces ahead of the others helps relieve the initial stresses resulting from the stoping process. This type of advance also facilitates opening the large number of faces required to prevent equipment from stalling because of a lack of available headings to work.

In thick deposits, upon completion of a sufficient number of headings and crosscuts, development can begin on a bench or benches to recover the additional reserve. The final height of the openings after benching will be dependent on the rock mechanics testing mentioned previously and the final size of the pillars. Some underground limestone mines have developed openings as large as 90 feet (27.4 meter) tall, but these require very large pillars to support such openings. Final heading heights in the range of 40–60 feet (12–18 meters) tall are not uncommon. Most benches are developed by drilling vertical blastholes much like surface mining, although much smaller diameter blastholes and tighter patterns are necessary (Figure 11.18). Care needs to be taken in locating the outer rows of holes so as not to burn or damage the ribs when blasting the benches. Offsetting the side rows of holes a foot or more away from the ribs is a good idea. Ribs that are wider on the bottom than on the top are much more stable over the long run. Ribs have a tendency to spall in the center as they take on weight and ultimately form an hourglass shape, which should be avoided if possible. Some mines use horizontal drilling in the bench cut instead of vertical drilling, which is known as front benching. In my experience this method is considerably more costly than standard vertical drilling, although front benching can be accomplished with the same drill jumbo, eliminating the need for another drill. In front benching, the bench height will be limited to the reach of the drill jumbo while a vertical drill has a much greater depth capacity (Figure 11.19).

Loading (mucking) equipment is dependent on the size of the headings being developed. In large headings, standard surface mining equipment like rigid-frame haul trucks and wheel loaders can be utilized to muck out the headings and the trucks haul the shot rock to the primary crusher. Low-profile mining equipment, such as haul trucks, loaders, drills, and roof bolters specifically designed for use in smaller headings, are available from several different manufacturers. Load-haul-dumps, also known as scooptrams, are specifically designed to both dig the muck pile and haul to the crusher with the same machine (Figure 11.20). These machines are low profile, articulated, very maneuverable, and able to efficiently deliver rock to the primary crusher much like a load-and-carry operation using a wheel loader in a surface quarry (Figure 11.21).

FIGURE 11.18 Drilling an underground bench shot

Underground room-and-pillar mines require meticulous

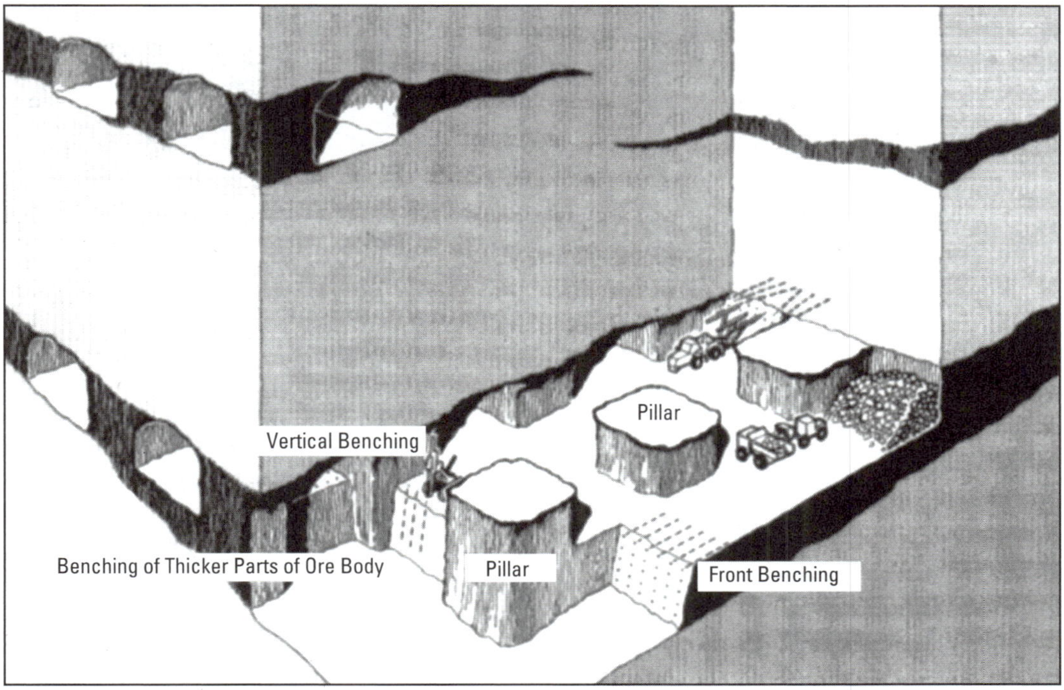

Original source unknown

FIGURE 11.19 Multilevel mine and the heading and bench concept

Courtesy of Epiroc

FIGURE 11.20 Load-haul-dump

FIGURE 11.21 Large wheel loader loading a 35-metric-ton haul truck from an underground bench cut

surveying techniques to maintain proper pillar size. It is very easy to turn a crosscut and be facing 86° instead of the required 90°. Initially, this may not be a problem, but 10 pillars away the two nonparallel drifts will be either merging together or veering apart, and the pillar in between will no longer be the correct dimensions. Eventually, as the drifts progress, you will end up with a too small, triangular, pillar unable to support the vertical load, which has the potential to cause a roof fall. It is somewhat problematic to install typical survey points in an underground mine, especially if a bench will ultimately be developed. In my experience, we would bring in a trained survey crew about once a month and have the crew set up survey pins in the back (mine roof) at the center of every crosscut intersection and hang reflective tape from the pins. When it came time to turn a crosscut at an intersection, the underground mine supervisor would set up a laser level on a tripod in the center of the intersection, line up on at least two reflective survey markers previously installed in the back, and turn the level exactly 90°. This would shine the laser light at the exact center line of the next crosscut to be shot and enable the blaster to accurately mark off the drill pattern. Special care must be taken in multilevel mines to accurately locate the pillars so they are aligned vertically to maintain the proper support.

MSHA requires that a regularly updated mine map be available to the employees at all times. MSHA also requires signage within the mine showing evacuation routes. For evacuation purposes, a method of numbering all the ribs (pillars) should be developed so miners are easily aware of their location within the mine at all times. This can be done in reference to the base of the decline or an emergency exit. Numbering the pillars numerically and directionally is a proven system. As an example, a miner might be at the intersection of 6 N and 11 W, meaning that the miner is six pillars north and 11 pillars west of the exit. The pillars can easily be marked on all four sides using reflective spray paint, although periodic renewal of the paint will be necessary. Generally, escape arrows are also painted on the pillars for miners to quickly and easily identify escape routes.

Although production is a priority in any mining operation, the safety of the employees is the first and highest priority. Mining is one of the more dangerous occupations in the United States, and underground mining more so than surface mining. MSHA has separate regulations for the different types of mines. Underground metal and nonmetal mining have different regulations than underground coal mining. MSHA requires twice the number of inspections in subsurface mines as in surface mines, additional personal protective equipment, stricter ground control standards, more daily inspections, sign-in/sign-out requirements, trained mine rescue teams, and twice the amount of annual training for underground employees.

Underground mining is a viable alternative to surface mining under the right economic conditions. In today's environment of NIMBYism and increasing environmental restrictions, being underground and out of sight can be advantageous. As deposits of quality aggregate materials on the surface become depleted, more and more underground mines will eventually come online. Subsurface mining of aggregates is the future of the industry.

CHAPTER 12

Postmining Reclamation

Mining, whether on the surface or underground, is an interim or temporary land use during the time necessary to remove and market the mineral reserve underlying the specific property in question and reclaiming the property afterward. In most cases, this interim period lasts 30–50 years. Postmining reclamation should be an integral part of the mine planning process and be accomplished contemporaneously with the mining process.

Only a few states had regulations that focused on the reclamation of surface mines prior to the 1970s. Coal mining began in the United States in the mid-1700s and accelerated during the late 1800s as the Industrial Revolution came into full swing. The demand for coal expanded exponentially during World War II and the postwar years, leaving behind tens of thousands of acres of abandoned mine lands, a portion of which were massively polluted by acid mine drainage. The Surface Mining Control and Reclamation Act of 1977 was enacted by Congress for the sole purpose of regulating the reclamation of surface coal mines and also funding reclamation of abandoned coal-mined acreage. The act created the Office of Surface Mining, an agency within the U.S. Department of the Interior with the intent of funding and ensuring consistency among state regulatory agencies while regulating the adverse environmental effects of coal mining.

This act focused on surface coal mines, and there are still no federal regulations that pertain specifically to the reclamation of other types of mining activity. Most states enacted their own mining and reclamation regulations that apply directly to mining activities other than coal starting in the mid-1970s. The majority of these state laws are fairly similar in nature and initially require the operator of a surface mine to obtain a mining permit. These permits require detailed maps showing the location of the operation and what county, township, or municipality has jurisdiction. State permits are usually interdependent with the local land use regulations discussed in Chapter 6, requiring proof of zoning approval for the specific operation.

Following the actual legislation, rules are promulgated that outline the reclamation methods, environmental requirements, and what is deemed necessary to protect the adjacent properties. These rules include some or all of the following:

- Setbacks (mining limits) from property lines, highways, easements, buildings, streams, and so forth
- Limits on the placement and storage overburden
- Restrictions on groundwater removal
- Control of surface drainage
- Sediment and erosion control
- Management of slopes and highwalls

- Control of fugitive dust emissions
- Restrictions on flyrock and vibration from blasting
- Certification of blasters
- Storage of explosives
- Limits on the slopes of final reclaimed land
- Permissible above- and below-water slopes in water impoundments
- Height restrictions of highwalls in hard-rock quarries
- Requirements and growing time limits for replanting vegetation upon reclamation

An application and approval process is required prior to mining. Most mining permit applications require an extensive application map, made to a specified scale, with details that include the following:

- Location coordinates or latitude and longitude
- North arrow and map scale
- Number of acres owned and acres to be affected
- Topography and surface features
- Property lines and easements
- Identification of adjoining property owners
- Buildings, roads, and highways
- Drainage details including streams, ponds, and wetlands
- Utility lines, pipelines, and transmission lines
- Geological information:
    - Geological formation to be mined
    - Thickness of the deposit
    - Thickness and character of overburden
    - Depth to groundwater
- Cross sections showing:
    - Premining topography
    - Postmining topography
    - Depth of proposed excavation
    - Limits of proposed excavation
- Sequence and direction of mining
- Areas of proposed overburden storage
- Proposed water impoundments
- Proposed sumps, settling ponds, and outfalls from dewatering
- Proposed postmining land use

A map of the reclaimed property is required by some states as well. In addition to the detailed maps, an application form is required asking for information on the mine operator like corporate structure, management, ownership, directors, statutory agent, and a listing of the applicant's other mining permits. The application generally requires a statement of intended future use of the property post-reclamation. Many applications ask what methods will be utilized to protect the adjoining property owners. Liability insurance requirements are outlined in these documents and an insurance certificate is required. Most laws require that the acreage affected be bonded at a set rate per acre affected. Bonds can normally be supplied as a cash bond to be held by the state until reclamation is completed or a surety bond from a recognized bonding agency. After submission of the required fees, insurance certificates, bonds, and approval of the application, the mining process can finally begin. Depending on the exact requirements, this process normally takes several months from application submission to final approval. This time frame does not take into account the time required for the maps and application to be completed before submission.

Annual progress reports are required by many states and inspections are mandated once or twice yearly by inspectors employed by the state for that specific purpose. Most mining permits will be limited to a specified period of time (10 or 20 years) and have a detailed renewal process after that time. If the operation is sold to another producer, the permit must be legally transferred and approved by the state agency. There is usually a method to amend these permits if mining circumstances change or additional property is acquired. Upon completion of mining and reclamation, a final report and bond release application is required. Most permits allow for interim bond releases if an operator can prove that a certain portion of the property has been adequately reclaimed and vegetation has been re-established. Each step of the process (application, annual report, permit renewal, amendment, bond release, and final report) generally requires additional fees be paid to the regulator. Most state reclamation laws have the authority to permanently shut down an operation and revoke the bond for flagrant violations of the law. Some state legislation limits the amount of acreage that can be affected during any one time period to encourage simultaneous mining and reclamation.

These types of regulations provide an operator with a definite incentive to reclaim its mining properties and mitigate any environmental issues. A much better incentive should be the residual value of the reclaimed real estate. When we discuss interim land uses, the property in question had a premining use (farming, forestry, pastureland, etc.), and it will also have a postmining use. There is no financial incentive in leaving a polluted wasteland behind like the coal mining companies of the early 1900s. Their philosophy was that the reserve was depleted, the land served its purpose, let it be, walk away! Think of the potential value of the tens of thousands of acres of land that was mined, abandoned, and never reclaimed. What a monumental waste!

When developing a new mine in a rural area, it can be very difficult to foresee what the future land use of a property will be 50 years into the future. A property located close to a metropolitan area has a much brighter future use than many rural properties. The potential for postmining reclamation is only limited by the imagination of the mine planner. Early in the mine planning process, it can be beneficial to consult a landscape architect to assist in developing a conceptual reclamation plan. Properly planned and reclaimed properties can be developed in the future for postmining uses such as:

- Residential lakefront properties
- Public parklands
- Golf courses
- Commercial developments
- Water-based amusement parks
- Industrial sites
- Wildlife refuges/natural areas
- Wetland banks
- Private recreation complexes
- Fishing clubs

FIGURE 12.1 Google Earth image of early sand and gravel residential reclamation in Wayne Lakes, Ohio

- Public water supply reservoirs
- Irrigated pasture or cropland

One of the earliest innovators of reclamation in the aggregates industry was American Aggregates Corporation of Greenville, Ohio. It began reclaiming sand and gravel operations into lakefront residential properties as early as the 1930s. Its reclamation activities over the next 70 years included numerous lakefront residential developments (Figure 12.1), commercial properties, office parks, and community recreational parks. American Aggregates created two lakes in a reclaimed gravel pit near Dayton, Ohio, each more than 1 mile (1.6 kilometers) long, that have been used for the multiple purposes of parkland, hydrofoils racing, and water skiing competitions, and simultaneously as a source of recharge for the Dayton municipal well field since the early 1970s. These types of developments and real estate sales can add a significant level of profit to an aggregates company's bottom line. When considering mine lives of 30–50 years, the property values upon completion of mining will have increased substantially. Just because your reserves are depleted does not mean that the future value of your property is depleted.

Most surface mines eventually intersect the water table and with proper planning are capable of producing attractive lakefront properties that are very valuable. This can be advantageous in many ways. A donation of reclaimed park property can be offered as an incentive to the local, township, county, or municipality in an effort to gain zoning approval on another property. What elected official wouldn't want to provide additional parkland for the future enjoyment of their constituents? (Why not appeal to their egos and name the park after the head of the commission?) Donating parkland to a municipality will eventually result in a sizable tax deduction when the property is assessed, at a much higher value, 30 years in the future. An added value of favorable publicity is a desirable outcome as well.

The options for parkland developments are endless (Figure 12.2). Many sand and gravel operations, by necessity, are located in floodplains where buildings are not permitted because of flooding issues. Seasonal flooding has little effect on parkland as long as any permanent structures are located at the higher elevations above the floodway. It may make economic sense

FIGURE 12.2 Contemporaneous mining and reclamation: an active quarry on the left and an operating city park on the right on the same property

to leave a small percentage of the reserves in situ to have a viable development at the end of the mine's life. Strategies for overburden removal and management during mining will be necessary to shape and slope the lakeshore with material that would otherwise have been waste. Determine the total volume of overburden early in the mine planning procedure and where it can be best utilized in the reclamation process before selling massive volumes of overburden as fill and not having a sufficient amount to complete the final reclamation. If overburden volumes are sufficient, utilize it to fill areas of the floodplain above the 100-year flood elevation and manufacture usable, salable ground (Figure 12.3). Excess, unsalable sand can be pumped or flumed to fill areas for future development. Hydraulically placed sand makes a very stable, buildable fill.

FIGURE 12.3 Sand and gravel slope reclamation in progress

In Chapter 8 on mine planning, I emphasized that reclamation should always be performed contemporaneously with mining to minimize the cost of moving the overburden (Figure 12.2). This requires meticulous planning prior to excavating the first shovel of dirt. Moving overburden using a contractor costs in the range of \$1.50–\$2.50 per cubic yard depending on the length of the haul. If the overburden is stockpiled and moved a second time that cost doubles. When initially opening a surface mine, it may be necessary to stockpile the overburden or use it in site barriers. Site barriers along the periphery of the property should be made up of topsoil

FIGURE 12.4 Recently planted site barrier adjacent to a residential roadway

so that when mining ceases the site barriers can be graded on top of the reclaimed slopes to aid in encouraging the growth of grasses, trees, and such (Figure 12.4).

More than one surface mining operation has been developed into an award-winning golf course. Two of the top 100 rated golf courses in 2021 were formerly quarries: The Quarry at Grant's Ridge in Minnesota (rated No. 35) and Black Diamond Ranch in Florida (rated No. 39) (Duncan 2021). Numerous other quarries and gravel pits have been reclaimed into golf courses as well. I googled "the Quarry Golf Course" and found two in California, one in Florida, one in Indiana, one in Minnesota, one in Missouri, one in Ohio, one in Texas, and one in the Caribbean in Curaçao. All of them have some very interesting water hazards.

The Butchart Gardens is located on Vancouver Island and is a National Historic Site of Canada. The gardens are a world-renowned tourist attraction with millions of bedding plants in over 900 varieties as well as the remnants of an abandoned cement plant. Development of the famous gardens began in 1904 after the limestone reserve was depleted. These gardens are visited by thousands of people annually and host outdoor concerts, night lighting in the summer, choreographed fireworks displays, a children's pavilion, and the Magic of Christmas celebration every December. All of this at a National Historic Site developed within an abandoned limestone quarry.

Quality waterfront residential property demands premium lot prices, especially in suburban areas (Figure 12.5). A beautifully landscaped lake with lakefront lots, boat docks, beaches, and playgrounds is very appealing to many families. There are several alternatives for development. The property can be sold directly to a residential real estate developer who will put in the necessary infrastructure (roads, water, sewers, power, and natural gas), the producer can install the infrastructure itself and market the individual lots, or a partnership between the owner and a developer can be negotiated. Residential developments require a lot of planning and the lots need to be marketed after the development takes place, either to individual buyers or various home builders. Some developers do their own construction and others sell one or a block of lots to various home builders. Depending on the market area, single-family homes or multifamily buildings or some combination thereof can be developed.

Commercial and office park developments around scenic lakes are also valuable in areas where residential developments may not be compatible with the surrounding area or the proper zoning does not exist. Developments with a combination of residential and commercial

FIGURE 12.5 Reclaimed sand and gravel pit in Indianapolis

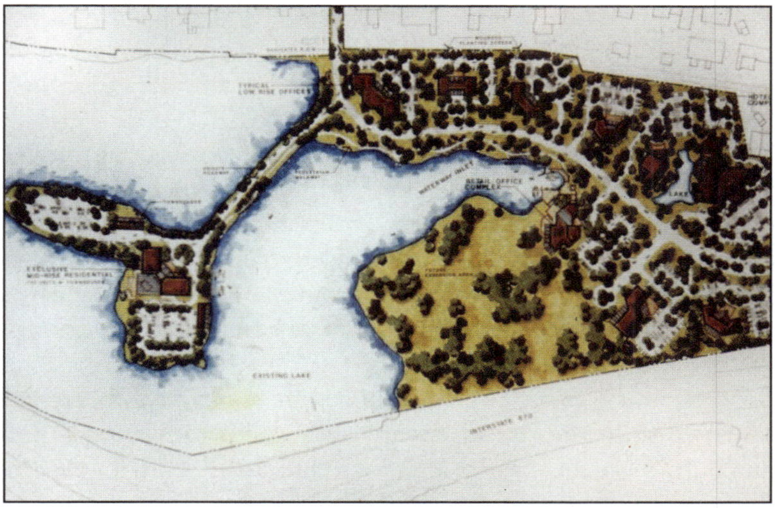

Courtesy of American Aggregates Corporation

FIGURE 12.6 Plan for a combined residential, retail, and office complex in Columbus, Ohio

properties and offices are also possible (Figure 12.6). Luxury hotels are attracted by the ambiance of waterfront views. Boat dealerships adjacent to lakes allow for in-water demonstrations and test drives.

Under the right circumstances with shallow water lakes and the correct plantings, wetland banks can be developed. Wetland banks are areas containing wetlands where shares (acreage) in the bank can be sold to other developers in lieu of completing their own wetland mitigation on their property. Depending on the location and quality of the wetland affected, the regulatory agency will likely require compensatory mitigation at a ratio of 1.5:1 to 3:1 times the acreage of the wetland destroyed. If you destroy 10 acres (4 hectares) of wetland, you will be required to create 15–30 acres (6–12 hectares) in mitigation. This is an expensive and time

FIGURE 12.7 Available underground space

consuming process and a good incentive for developers to purchase wetland credits from a wetlands bank. The wetland banks usually become wildlife sanctuaries for the enjoyment of outdoor enthusiasts.

Underground mines require little reclamation other than securing the mine entrances from access to protect the public. For the most part, underground mines maintain constant temperature and humidity levels throughout the year. As long as the abandoned underground mine is dry, there are many potential uses for this underground space (Figure 12.7). Mushroom farming is extensively done in abandoned underground mines where the dark environment encourages growth of the fungi. Other mined-out areas have been utilized as parking garages, RV storage facilities, and for waste disposal.

The SubTropolis near Kansas City, Missouri, is an abandoned underground limestone mine that has been developed into a temperature-controlled storage facility. The constant temperature and humidity make these spaces ideal for record storage both for paper and electronic records. The rooms between pillars can easily be walled in to produce secure, separate storage facilities for government, commercial, and industrial customers. The United States Postal Service, the United States Environmental Protection Agency, and the National Archives and Records Administration all lease storage space in the SubTropolis. This one facility boasts of having 14 million square feet of available storage with access to both truck and rail sidings. Other similar storage facilities exist in Pennsylvania and several other states. The surface above these underground storage spaces is then available for other uses. An amusement park complex overlies the SubTropolis.

Reclamation can be much more than just the final step in the mining process. If properly planned and executed, reclaimed mining properties can be marketed at a significant profit. The reclamation process must be an integral part of the mine plan prior to beginning development.

CHAPTER 13

Crushing

Aggregate processing plants are a combination of numerous mechanical systems to efficiently manufacture products that meet the gradation and quality requirements of the customers (see Chapter 7). Much of the quality is determined by the inherent geology of the deposit, but the final product quality depends on the size fractions and particle shapes produced through the mining and plant processes. In previous chapters, we discussed the equipment necessary for the excavation of the raw material in both consolidated and unconsolidated deposits. We also explained how the use of explosives fractures the consolidated rock into transportable sizes and various options for transporting the raw feed to the processing plant. The aggregate processing stage consists of crushing, screening, washing, classifying, and transporting the finished product to stockpiles for later shipment to the customers.

Producing crushed stone or sand and gravel into salable products is an energy-intensive business. Large quantities of fuel and/or electricity are required in the stripping and mining processes. Explosive energy from the rapid expansion of explosive gases is required for the first step in the crushing process when mining consolidated deposits. The blasting process as discussed in Chapters 10 and 11 is the critical first stage in size reduction and has a significant affect further downstream in the production process. Blasting needs to be a precision process aimed at efficiently utilizing the explosive energy to produce the optimum fragmentation at the mining face. Efficient blasting should result in approximately 50% of the muck pile being less than 8 inches (20 centimeters) in size and less than 5% being larger than 24 inches (61 centimeters).

AGGREGATES CRUSHING

Beyond the blasting stage, the oversized material transported to the processing plant needs to be mechanically crushed into salable size ranges. The energy required to crush stone is a function of the material's resistance to being crushed, the quantity to be crushed, and the amount of required size reduction. Breakage is caused through abrasion, fracture, or cleavage and modern crushing technology takes advantage of all three of these mechanisms. Abrasion is a result of localized removal of protruding edges and in essence is the rounding of the particles, which is accomplished by crushing as well as by particle-to-particle abrasion. Cleavage is usually a result of compression forces causing breaks along planes of weakness within the rock. Fracture, on the other hand, results when excessive energy is applied, causing the particle to shatter into a large number of smaller pieces. A considerable amount of energy is absorbed during the crushing process. This absorbed energy is ultimately released as rock particles flying in every direction including crashing into each other. Crushers need to be fully loaded to take advantage of this excess energy release.

Reduction ratio is defined as the maximum size of the material fed to a crusher as compared to the maximum size of the product discharged from the crusher. This is usually defined as the D_{80} of the gradation of the crusher feed compared to the D_{80} of the crusher product. The D_{80} is the diameter (screen size) of 80% passing the applicable gradation:

$$\text{reduction ratio (RR)} = D_{80} \text{ crusher feed} : D_{80} \text{ crusher discharge}$$

High reduction ratios are desirable for production efficiency purposes, but high reduction ratios also generate high volumes of fine and ultrafine materials. These fines, when produced in excess, can be problematic and often end up as unsalable waste products. The objective of the production process is to manufacture specific size ranges (gradations) of products that meet customer specifications. The volume of each product manufactured should correspond to the volume of products sold by the sales department as closely as possible and is known as the production/sales mix. When the sales and production mix do not correspond, excess inventory of certain products results. Inventory has monetary value and unsold inventory negatively affects plant profitability.

To avoid excess fines and inventory imbalances, it is imperative to reduce the product sizes in a series of stages, with the maximum reduction ratios accomplished early in the production sequence. Primary crushers (the first stage) generally produce a reduction ratio of from 7:1 up to 15:1. Secondary-stage crushers have reduction ratios in the range of 4:1 to 6:1 and tertiary crushing ranges from 2:1 to 4:1. Crushing is accomplished by either compression or impact forces with some additional abrasion occurring depending on the crusher type. The following discussion concentrates on the crusher types most common to the aggregates industry. There are many other types of industrial crushing equipment that are specific to other applications (fine grinding, coal crushing, ore milling, glass crushing, scrap metal shredding, etc.) that are not covered. Both compression- and impact-type crushers have applications in the primary, secondary, and tertiary stages of crushing.

COMPRESSION CRUSHING

As the name implies, compression crushing consists of compressing the rock between two surfaces with sufficient energy (pressure) to break the material along its inherent planes of weakness. This is accomplished by using two semi-parallel surfaces, one or both moving toward the other to compress, crush, and then release the material from between the two surfaces. There are a number of methods to accomplish this task. In compression crushers, the crushed material must pass through an adjustable opening known as the crusher setting. The jaw crusher consists of two plates set an acute angle (known as the nip angle), which is generally less than 25°, that compress and release, similar to the jaws of an animal. All jaw crushers have a large, heavy flywheel attached to the drive that stores energy on the release stroke to be delivered on the next crushing stroke. The two jaws are lined with readily replaceable wear liners called jaw dies and the sides of the frame are also lined with wear material known as cheek plates. The replaceable liners are normally made from manganese or Ni-hard (nickel chromium alloyed cast iron) and come in numerous shapes with several surface texture options ranging from smooth to deeply corrugated.

One or both jaws are pivoted toward the other using an eccentric motion to close and open the jaws. Jaw crushers with two movable jaws are available but are rare and generally only used on extremely hard materials. The most common type of jaw crusher in use today

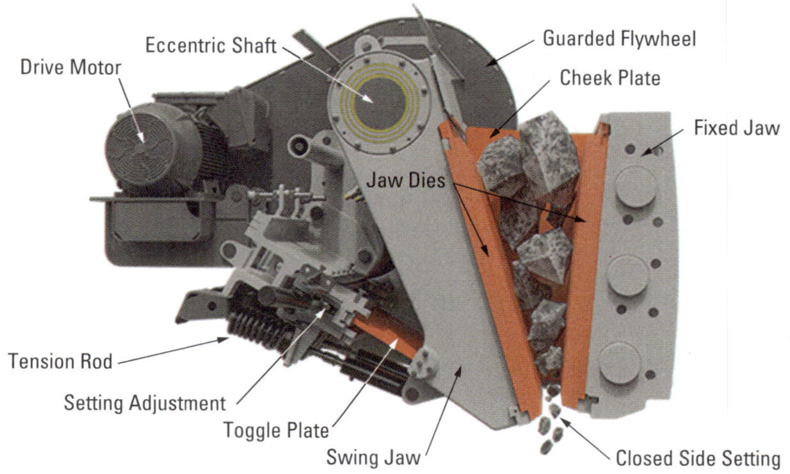

Adapted from Superior Industries

FIGURE 13.1 Overhead eccentric jaw crusher

has one fixed and one movable or "swing" jaw. The three main types of jaw crushers in use are the Blake jaw, the overhead pivot jaw, and the overhead eccentric jaw crusher. In the Blake and overhead pivot types, a separate pitman arm on an eccentric shaft with two toggle plates pushes the swing jaw toward the fixed jaw. The difference in these two types is the location of the drive, length of the pitman arm, and shape of the jaws, all of which affect the stroke of the crusher. The other type, the single-toggle, overhead eccentric jaw crusher, is probably the most common jaw crusher in use today (Figure 13.1). In this type of jaw crusher, the pitman arm is eliminated and the swing jaw is attached directly to an overhead-mounted eccentric shaft. The toggle plate acts as an adjustment mechanism that is either manually shimmed to adjust the setting or can be hydraulically adjusted. The top-mounted eccentric shaft exerts a circular motion at the feed end, which converts to an elliptical motion in the center of the crushing cavity and an almost-horizontal motion at the discharge end. This action encourages the feed into the crusher cavity and the horizontal stroke at the bottom end allows for rapid discharge. A spring-mounted tension rod allows for the passage of noncrushable material (tramp iron). The maximum that the swing jaw moves is known as the throw of the crusher and is usually adjusted by changing the eccentric shaft.

Jaw crushers are typically used in a primary application and are available in numerous sizes. They are normally classified by the size of the intake area, with the width of the jaws the first dimension and the distance between the opening at the intake end or gape as the second dimension. A 22 × 42 inch (56 × 107 centimeter) heavy-duty jaw crusher using 125 horsepower at 275 rpm can produce from 100 metric tons per hour at a closed side setting (CSS) of 2 inches (5 centimeters) up to 500 metric tons per hour at a CSS of 8 inches (20 centimeters). A 42 × 48 inch (107 × 122 centimeter) jaw crusher utilizing 250 horsepower at 210 rpm has a rated capacity of 375 metric tons per hour at a CSS of 3.5 inches (9 centimeters) and larger jaws can produce up to 1,200 metric tons per hour. Much larger models with even higher capacities are also available. Compression crushers are not conducive for use in very dirty deposits containing large percentages of clay. Plastic clays can be easily compacted by the compressive movement and plug the crusher, requiring expensive downtime to remove the plug. Jaw crushers have a tendency to produce elongated versus cubical particles. This is generally not an issue when used as a primary crusher because the material will normally be re-crushed in the secondary and tertiary crushing stages. If the material is scalped off, post primary crushing, for use as a base product, the elongated particles (carrots) can sometimes be problematic to the quality of the finished base product.

When managing a quarry in central Indiana, we had an unusually high demand starting in the following spring because of an airport runway contract. We had acquired this particular plant from another producer the previous year who had not properly maintained all of the equipment. The primary jaw crusher was in dire need of a structural rebuild so we contracted a machine shop in Tennessee to complete the needed work over the winter, prior to spring start-up. We removed the machine and sent it on a flatbed truck to the machine shop. During the rebuild, my superintendent traveled to Tennessee on two different occasions to check on the progress and quality of workmanship. All went well until the day the crusher was scheduled to arrive back at the plant. The rest of the plant maintenance was completed on schedule and we were only awaiting the arrival and installation of the primary crusher to begin the season's production. By 7:00 PM on the scheduled day of arrival, there was still no sign of the crusher and we called the machine shop manager at home for information. His reply was that the crusher had been picked up by the trucking company and left the machine shop on schedule. We next contacted the trucking company, only to be informed that their driver had unexpectedly called, quit his job earlier that day, and walked away from the truck and its load. The problem was that the trucking company had no idea where the driver had abandoned the truck and, unfortunately, our crusher with it. The crusher finally arrived 4 days later after a frantic search of every truck stop between central Indiana and Tennessee. So much for planning!

Another type of compression crushing is the gyratory or cone crusher. Gyratory and cone crushers are very similar crushers but with different internal geometry, crushing chamber configuration, speed, and stroke and are designed to produce different products. Gyratory and cone crushers, instead of using jaws, are essentially an inverted cone within a cone-shaped bowl. The conical head rotates within the bowl in an eccentric motion, causing the head to be closer to the bowl on one side than it is on the other side as the head rotates. The closest dimension of the head to the bowl is the CSS and the widest dimension (180° from the CSS) is called the open side setting (OSS). As the head spins within the bowl, it acts much like a jaw crusher, squeezing the material between the head and the bowl at the CSS. On the opposite side at the OSS, the crushed stone product is allowed to discharge. The material being crushed moves by gravity, assisted by the motion of the head, and discharges from the bottom of the crushing chamber. Both the head and the bowl are lined with replaceable wear materials made of manganese or Ni-hard. In a gyratory crusher, the bowl liners are called concaves and the removable head covering is known as the mantle (Figure 13.2). The machine is driven by a beveled pinion gear on the end of the countershaft that engages with a beveled ring gear at the base of the eccentric.

The long, steep, angled head allows large production volumes to pass through the gyratory crusher and the bowl shape provides clearance for large feed sizes, producing reduction ratios as high as 10:1. Production capacity is a function of the stroke, speed, horsepower, nip angle (angle between the bowl and head at the CSS), and size of the feed material. Gyratory crushers by their nature are very large and heavy and require large structures for support. Quarry blasted

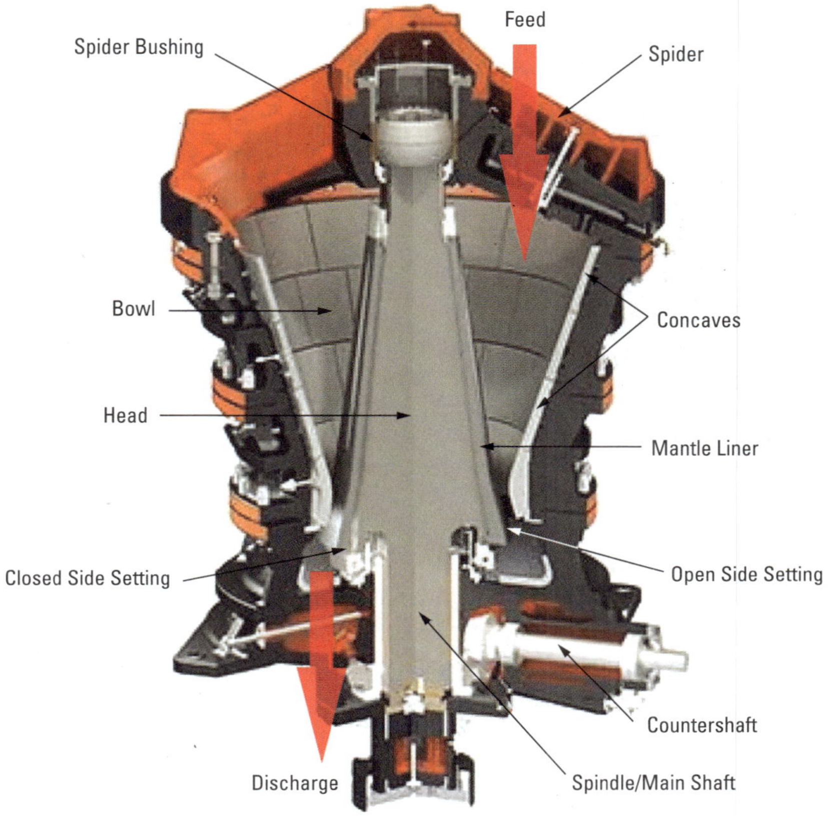

Adapted from Sandvik

FIGURE 13.2 Cross section of a gyratory crusher

rock can be dumped directly on top of gyratory crushers with no need for mechanical feeders or pre-scalping of the fines (Figure 13.3). Gyratory crushers run relatively slowly at between 85 and 150 rpm. Large primary gyratory crushers can have production capacities in excess of 15,000 metric tons per hour. One quarry I managed in the mid-1990s was still operating a primary gyratory crusher that had been built in 1902. Smaller gyratory crushers are sometimes used in secondary crusher applications when feeds range from 4 to 8 inches (10 to 20 centimeters). Gyratory crushers are categorized by the opening dimension (inches) × head diameter (inches), for example, a model 42 × 65 or a model 60 × 89. Gyratory crusher horsepower ranges from 600 up to 2,000 and the crushers can weigh as much as 1.2 million pounds (544,310 kilograms).

The main differences between gyratory and cone crushers are the shape of the head, the bowl shape, speed, throw, and, on the gyratory crushers, the spindle is suspended from the top by the spider. The cone crusher has a shorter spindle that is not suspended by the spider but is supported by a curved bearing beneath the head. Eccentric crushers have a tapered offset bore that allows the main shaft to follow an eccentric path within the bowl on each rotation. Cone crushers come in various sizes and types, some driven with an eccentric bushing and some with spherical roller bearings, which are known as roller cones. The roller bearings do not generally

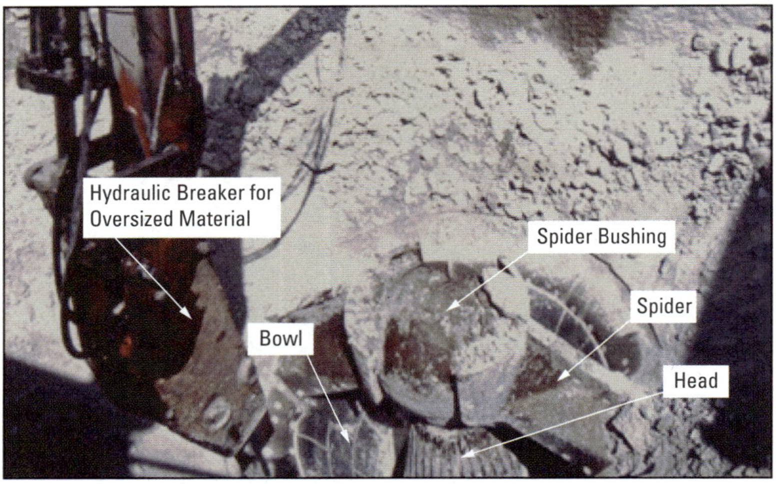

FIGURE 13.3 Gyratory crusher and hydraulic breaker in the bottom of a dump hopper

last as long as the bushing design between rebuilds. Cone crushers have a highly sophisticated lubrication system including an oil reservoir, pump, filters, oil cooler, and plumbing. The lubrication system requires routine maintenance, including oil changes, filter changes, and cooler inspections. Oil pressure and oil flowmeters with alarm systems need to be installed and regularly monitored on all cone and gyratory crushers. Many expensive cone crushers have been extensively damaged because of a lubrication failure caused by a lack of or overheated oil.

The standard cone crusher is used in a secondary crushing application, has a reduction ratio of approximately 6:1, and is generally used to make products in the 1–4 inch (2.5–10 centimeter) size range. These machines operate at more than twice the speed of a gyratory crusher and have a larger throw. Instead of the steeply angled head geometry of the gyratory crusher, the standard cone crusher has a larger, much flatter head, a large feed opening, and a smaller, straighter, more parallel bowl geometry. Cone crushers have shorter spindles and do not require a spider and spider bushing (Figure 13.4). The crusher feed falls directly onto the top of the head or onto a feed distribution plate mounted on top of the head. Numerous combinations of bowl and mantle liners are available for customization of the geometry of the crusher cavity based on the individual feed and product requirements. Older models have spring-activated tramp metal relief (protection from damage by uncrushable material), while excess pressure relief in the newer models is accomplished utilizing nitrogen accumulators.

Careful feed control is necessary on cone crushers. Oversize particles that would normally be retained in the crusher by other particles can pass through the crusher if the machine is not choke fed. For the best possible size control and cubical particle shape, a cone crusher should be fed via a feeder from a small surge bin. Cone crusher settings are adjusted by screwing the bowl toward or away from the head. Crusher settings must be monitored regularly and adjusted to make up for the ever-changing wear on the liners. Older models adjusted the setting utilizing a capstan and chain; later came hydraulic cylinders, followed by a ring gear with an adjusting motor. Many of today's processing plants have automatically adjusting crushers that are computer controlled using proximity switches to determine the crusher setting.

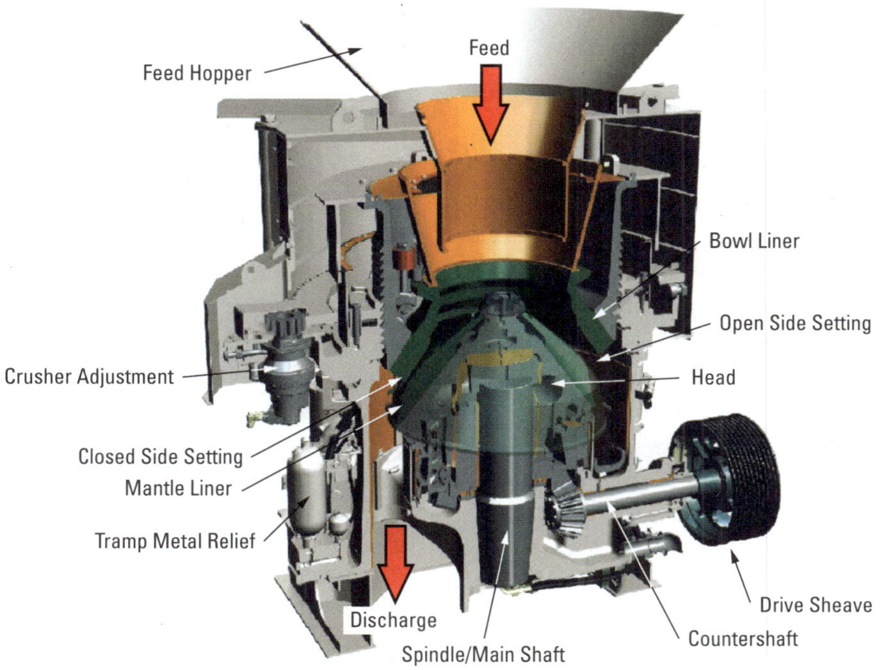

Adapted from Metso Outotec

FIGURE 13.4 Cross section of a cone crusher

Although the presence of some moisture in the crusher feed is expected, excessively wet feed can substantially increase liner wear in most applications.

Tertiary crushing can be accomplished with standard cone crushers set to a tighter CSS than the secondary crushers or by another variety of cone crusher known as the short-head cone crusher. As the name implies, the short-head crusher has a shorter head with a somewhat steeper head angle, a narrower feed opening, and a much longer parallel section in the liners than the standard cone to prevent choking from the finer material that is being crushed. Interparticle energy transfer is required to achieve the best performance in fine crushing applications. The finer the material being crushed, the more horsepower is required. Products in the size range of ⅛ inch (3.4 millimeters) up to 1 inch (25.4 millimeters) are generally produced by the short-head style cone crusher at a 4:1 reduction ratio. Cone crushers are normally identified by the diameter of the head, such as 4 ¼ foot (51 inch) (129.5 centimeter) short-head or a 5 foot (60 inch) (152.4 centimeter) standard model. Typical cone crushers used in the aggregates industry have head diameters ranging from 29 up to 70 inches (74 up to 178 centimeters), utilizing from 125 to 900 horsepower, with 250 up to 2,000 metric tons per hour capacities and weighing from 14,000 up to 170,000 pounds (6,350 up to 77,110 kilograms).

A third type of cone crusher is known as an attrition cone or a Gyradisc crusher (Figure 13.5). The Gyradisc is a specialized cone crusher designed to produce very fine material like manufactured sand. A Gyradisc uses a combination of compression, interparticle comminution, and attrition between layers of particles. This is accomplished by using short liners and a very flat head angle in the crusher retaining the material longer within the crushing chamber to enable the attrition crushing. These crushers produce a very cubical product but have a high recirculating load. The reduction ratio varies between 2:1 and 5:1 and the crusher should be choke fed at all times. Wet, sticky, or plastic feed tends to "pancake" and can easily plug a Gyradisc crusher.

FIGURE 13.5 Nordberg cone crushing limestone

Roll crushers are another type of compression crushing that can also include shear forces if the shells on the rolls have teeth added. The most common roll crushers have two horizontal cylinders rotating toward one another. One roll is in a fixed position, while the other is spring loaded and the gap in between is adjustable on the spring-loaded side. Roll crushers come with one to four rolls and are capable of multiple stages of reduction. A typical roll crusher has a reduction ratio of 3:1 to 4:1. The shells on the outside of the rolls are the wearing surface and are generally built up by welding as the wear progresses. Because of the welding required, they are very maintenance intensive and not suitable for crushing highly abrasive materials. The advantage of roll crushers is that they produce a high percentage of a single-sized product close to the crusher setting and generate very few fines.

The material to be crushed is fed directly in the center of the two rolls rotating toward one another and is crushed in between (Figure 13.6). Because of the low coefficient of friction between steel and stone, a nip angle of less than 30° is required to avoid particle slippage. The coarser the feed, the slower the rolls need to be turned to crush efficiently. The speed can be increased when feeding finer material allowing for increased capacity. Roll speeds range from 300 to 2,500 feet per minute (1.5 to 12.7 meters per second) depending on material, feed size, and nip angle. Shell designs vary considerably based on the application. Some shells have knob-like protrusions, some have shear vanes, and some are smooth. Those with knobs and shear vanes tend to add a shearing action to the compressive forces.

IMPACT CRUSHING

Impact crushers have high reduction ratios and create very cubical products, both of which are desirable properties. Some beneficiation of the product takes place as planes of weakness within the stone particles tend to be eliminated by the impact as well as some softer layers (e.g., shale seams). At the same time, impact crushing also creates more fines than other types of crushers. The energy generated by impact crushers varies with the square of the rotational speed; therefore, crusher sped and feed size are critical to impact crusher performance. Larger feed size is more easily broken by impact crushing than are the smaller particles. From a marketing standpoint, excess fines can become problematic if the sales mix does not match the plant production. The cost of wear materials needs also be considered when specifying impact crushers. Very abrasive stone can cause excess wear in impact crushing, which can add significant operating costs because of increased replacement of blow bars, shoes, apron liners, side

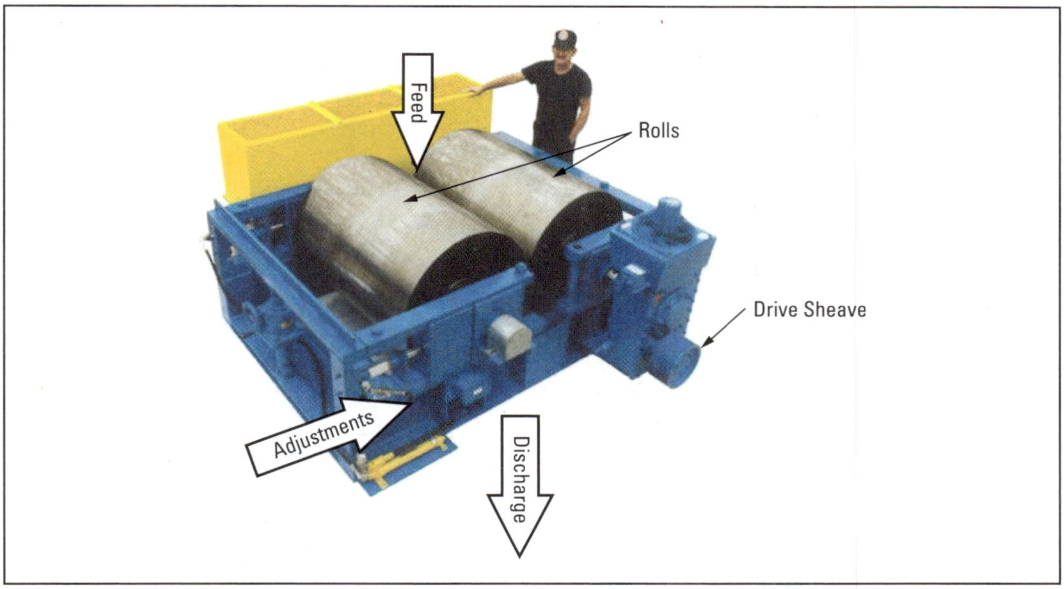

Adapted from Williams Crushers

FIGURE 13.6 Double roll crusher

liners, and so on. Abrasive deposits containing high percentages of silica are best suited to jaw or cone crusher applications because of the potential for inflated wear costs.

HORIZONTAL IMPACT CRUSHERS

There are two basic types of impact crushers depending on the orientation of their drive shafts: horizontal shaft impact (HSI) crushers or vertical shaft impact (VSI) crushers. Several variations exist for both geometries. HSIs are commonly used in the primary application and can have reduction ratios as high as 15:1. The New Holland–style HSI has a single rotor, which spins on a horizontal shaft with a series of replaceable blow bars (two to six) attached to the outside of the rotor. For additional wear, the blow bars can be reversed or flipped up to three times. A series of breaker bars with replaceable wear sleeves line the top and front of the crusher spaced at regular intervals. As the rotor spins, the blow bars impact and break the rock, which is thrown violently against the breaker bars, adding additional impact and rebound forces. Particle-to-particle impact occurs as the rock particles crash into to one another, adding even more impact forces. The distance between the lowest breaker bar and the rotor blow bars is adjustable and controls the size of material discharged from the New Holland–style HSI.

HSIs tend to produce very cubical products, but they also produce a large percentage of fine material because of the impact. A double-rotor HSI has two rotors spinning away from each other, with a similar type of breaker bars as the New Holland, but produces even more fines than a single-rotor machine. The newest style of HSI is known as an Andreas style and was named after the original designer (Figure 13.7). In lieu of sleeved breaker bars, these machines have two or three adjustable impact aprons that are lined with manganese or Ni-hard replaceable wear liners. The aprons are adjusted at different distances from the spinning blow bars, providing multiple stages of crushing; the lower apron, which is set closest to the spinning rotor, controls the CSS. HSIs are lined top, bottom, and sides with replaceable wear liners. The

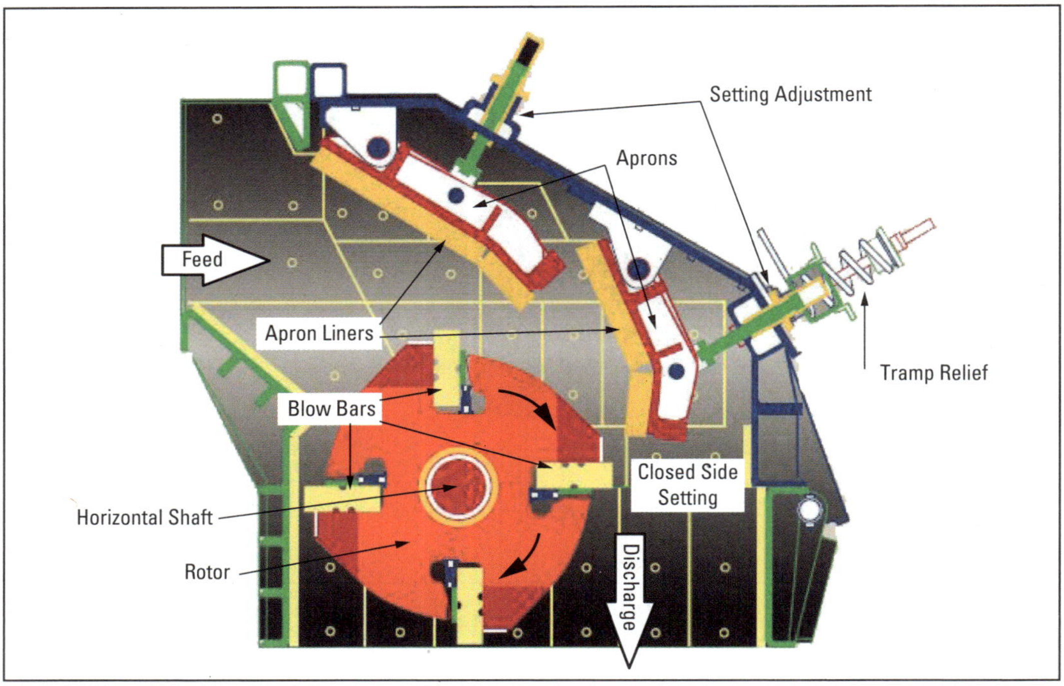

Adapted from Lippman Industries

FIGURE 13.7 Four-bar Andreas-style horizontal shaft impact

aprons are adjusted either manually or by hydraulic cylinders and usually have spring-actuated tramp relief systems. The ability to easily adjust the apron settings allows better control of the reduction ratio and generally lower wear material costs. Like all crushers, HSIs require regular lubrication, maintenance, and inspection of the wear part replacement.

Another variation of the HSI is the hammer mill. A hammer mill has a horizontal rotor, which operates over a series of closely spaced wear bars curved to match the outside diameter of the blow bars, forming a grate. The crushed material has to pass the spacing between grate bars before it can discharge from the crusher, thus providing the CSS for this type of HSI. The distance between the blow bars and the grate is minimal, which adds attrition crushing to the impact as the particles are dragged between the blow bars and the grating. The larger particles are swept up and re-crushed until they are small enough to exit through the grating. Another variation on the HSI is a single-rotor model where the direction of rotation can be reversed when the blow bars are worn in one direction preventing the need to flip the blow bars as they wear.

The cage mill is another type of impact crusher. Cage mill capacity is a function of mill speed, size, and specific gravity of the feed, the number of cages, and the size of the product produced. Cage mills have either one, two, four, or six cages depending on desired reduction ratios. The cages are concentric or circular with a series of pins evenly arranged around each cage. Each successive cage is larger in diameter then the previous one in the series. The material is fed into the center of the inner cage and impacts the rotating cage pins. The cages in multi-row mills spin in opposite directions, causing multiple stages of impact crushing (Figure 13.8). Impact with the spinning pins along with centrifugal force causes the particles to pass through the first cage to impact the next set of pins (cage) spinning in the opposite direction, with each

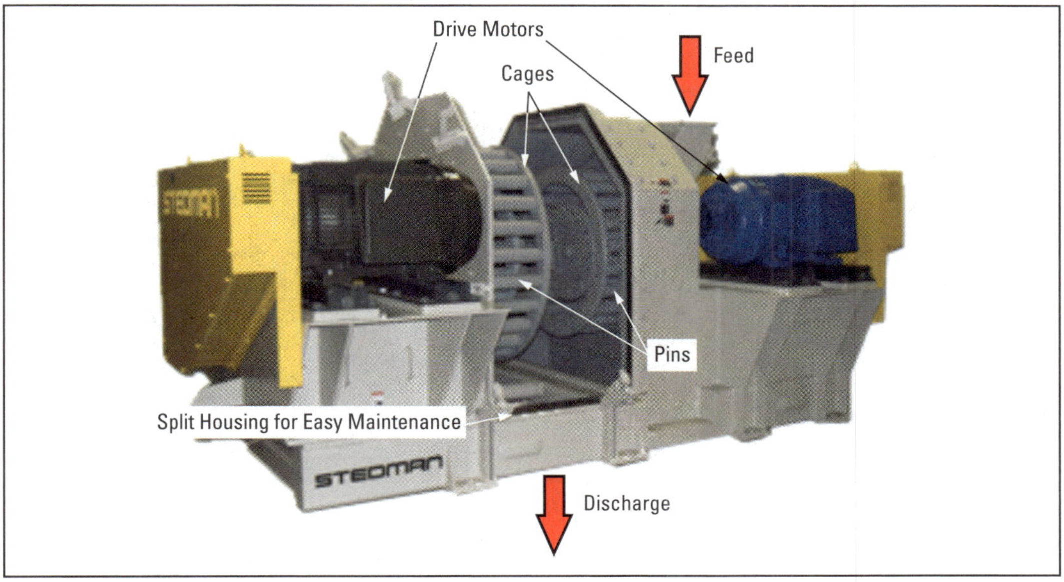

Adapted from Stedman Machine

FIGURE 13.8 Multirow cage mill

successive cage spinning faster than the previous cage. The farther the particles travel from the center of the cage, the more the impact velocity increases and the smaller the product. Reduction ratios are high and range from 8:1 up to 20:1.

The mills are lined with abrasion-resistant liners and the pins are a high chrome alloy steel for maximum wear life. Cages require rebuilds or replacement on a regular maintenance schedule that is dependent on the abrasiveness of the material crushed and it's a good idea to have two or more sets of cages; one to operate and one to rebuild. Many cage mills are reversible so maximum cage wear life on the cages can be achieved. Cage mills are available with production capacities from ½ up to 250 metric tons per hour and feed sizes up to 2.5 inches (6.3 centimeter). A four-row cage mill being fed a 2-inch (5 centimeter) minus feed of high-calcium limestone, operating with 300 horsepower on the small cages and 400 horsepower on the large cages will produce 85–90 metric tons per hour of No. 8 minus material usable as agricultural lime (Figure 13.9). These are ideal machines to produce agricultural limestone for pH adjustments for soil in farm fields and have many other industrial applications crushing nonabrasive materials.

VERTICAL SHAFT IMPACT CRUSHERS

As the name implies, VSIs have a vertical shaft versus a horizontal shaft, and like most crusher species the VSI has evolved into several types. The open-rotor VSI consists of a lined, round table on a pedestal with manganese or high chrome hammers called impact shoes attached vertically around the outside diameter. The shoes function much like the breaker bars in an HSI. To a larger-diameter (anvil) ring on the outside of the table, mounted at the level of impact shoes, are attached a series of large heavy castings called anvils. The material is fed into the center of the spinning table through a feed tube and is thrown centrifugally into the anvils by the shoes at very high speed. The anvils are angled to receive the material accelerated by the impact shoes at a 90°

FIGURE 13.9 New agricultural lime plant under construction

angle for maximum breakage efficiency. The feed tube is a wear item and is adjustable downward to act as a crusher setting device and prevent too large of material from entering the crusher.

A combination of rock-to-steel impact, rock-to-rock impact, attrition, and abrasion results in approximately a 4:1 reduction ratio. As in other types of crushers, VSIs are lined with replaceable wear liners to prevent wear on the table and crusher housing. However, I have seen many VSI crushers over the years with patches welded to the outside of the crusher housing where some inattentive supervisor let the liners wear out. This is inexcusable behavior and would have led to an early termination in any organization that I managed.

Other variations of the VSI have an enclosed rotor on the top and bottom with the shoes attached between in lieu of the open table style. In this configuration, the feed enters through a hole in the top of the rotor. Like any type of impact crushing, highly abrasive materials will result in high wear costs and excess downtime to replace the wear items, such as shoes, anvils, feed tubes, and liners. Several varieties of autogenous or semi-autogenous VSIs have been developed over the years with a rock ledge replacing the anvils. The ledge is designed to fill up with crushed material and provide stone-to-stone impact as opposed to stone-to-steel impact in an effort to save on wear materials. Crusher shoes have also been designed with stone pockets for the same purpose. These designs result in wear savings but also have a tendency to lower the reduction ratio to about 3:1.

Vertical impactors produce a desirable cubical product but also, because of the impact crushing, produce a high percentage of fines (Figure 13.10). Reducing the tip speed of the VSI can reduce fines production but will also affect the reduction ratio of the machine. The product size from VSIs is controlled by the feed tube adjustment and the tip speed at which the shoes are traveling. Tip speed is defined as the circumference of the table times the speed of the shaft (rpm) and is usually expressed in feet per minute. Most crusher manufacturers have crushing laboratories where they can test samples of your stone at various tip speeds to determine the necessary parameters for maximizing reduction ratio, minimizing wear, and minimizing fines production. Crusher manufacturers publish typical gradations produced from their crushers for various different materials at different CSSs, tip speeds, and so on; these are your best resource when specifying crushers.

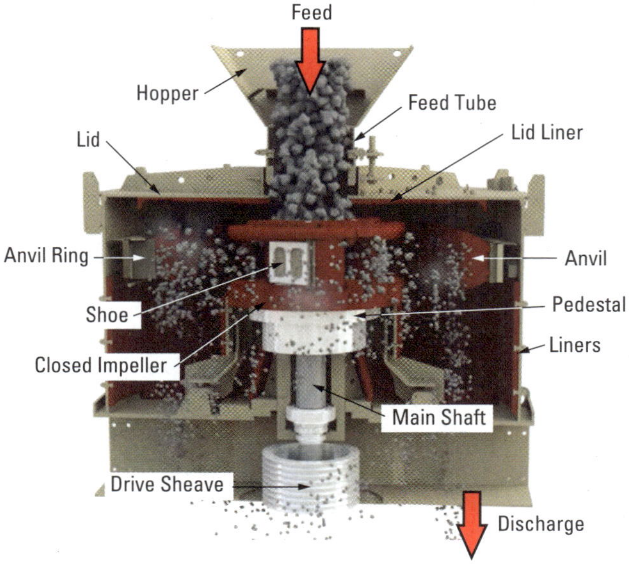

Adapted from KPI-JCL, Astec Industries

FIGURE 13.10 Vertical shaft impact with closed rotor

VSIs are generally utilized in the secondary stage of crushing, but some producers attempt to use them as both secondary and tertiary crushers at the same time. They do this by purchasing an oversized crusher and recirculating the oversize stone back into the same machine rather than purchasing a tertiary crusher. This generally ends up being penny wise and pound foolish because the additional feed produces additional wear costs and the excess volume of fine and ultrafine materials generated are not always salable. VSIs are less expensive than cone crushers and have similar reduction ratios. More crushing stages will alleviate the production of excess fines, but comes at the cost of increased capital expenditure. Many plant designs use a mixture of compression and impact crushing in an effort to minimize fines production and maximize particle shape. VSIs range in capacity from 70 up to 1,000 metric tons per hour plus, powered by 50 up to 1,000 horsepower. As in cone crushers, maintenance of the lubrication system on VSIs is critical to preventing premature failure of the machine components. Oil pressure, temperature, and flow should be monitored at all times. Many manufacturers provide variable sensitivity vibration switches to detect excessive vibration that automatically shut down the machine if necessary.

Crushers are the heart of the aggregate processing plant and represent a significant portion of the capital expenditure. Greater production volume and fewest fines will be generated if the greatest reduction is achieved early in the production cycle. Crushers should be operated as close to their maximum feed capacity as possible (choke fed) to take advantage of the kinetic energy generated and to achieve peak efficiency. Crushers are energy-intensive machines and power is an expensive component of the operating cost. Compression crushing tends to produce a more predictably sized product because of the ability to more precisely set the CSS of compression crushers. High-speed impact crushing is more erratic and unpredictable in its gradation results and more dependent on the inherent strength of the stone being crushed.

CHAPTER 14

Screening, Washing, and Classifying

SCREENING

Product sizing of aggregates is accomplished by the use of various types of mechanical screening devices. The screening of aggregate products is accomplished for five principal purposes:

1. Scalping
2. Sizing
3. Grading
4. Dewatering
5. Dedusting

Scalping is the process of removing the coarsest-size fractions from the feed material, generally prior to crushing, to limit the amount of fines entering a crusher. One thousand metric tons per hour of feed to a plant need not require a 1,000 metric tons per hour of primary crushing capacity if a portion of the feed is already smaller than the crusher setting. If 50% of the feed is smaller than the required primary crusher discharge, a primary crusher with 500 metric tons per hour capacity is sufficient as long as the coarse fraction (crusher feed) is separated (or scalped) from the feed stream prior to crushing. Scalping can be accomplished by using a heavy-duty screen or a grizzly scalper. Sizing and classifying is the separation of the feed stream into specific sized products for further downstream processing (secondary or tertiary crushing), blending, or stockpiling. Grading is the process by which a number of different sizes are sorted and combined into a single product gradation. Dewatering is the removal of excess free moisture, usually from a sand slurry, and dedusting is the removal of excess fines or ultrafines from an aggregate product.

Most screens utilized in the aggregates industry are multi-deck vibrating screens mounted on springs with the vibration resulting from an eccentric drive motion. Each screen deck essentially produces two products; the oversize (overs or rejects) flowing over the top of the screen deck without passing through any openings, and the product (throughs), which are the particles that pass through the screen opening. Screen performance is influenced by a number of variables. These include feed size, desired product size, moisture content, screen capacity, screening media, and the adjustable variables of the vibrating screen mechanism itself. These adjustable variables include screen throw, frequency, screen angle (slope), and screen direction. Depending on the drive mechanism, the resulting eccentric motion is circular, elliptical, linear, or some combination thereof. The eccentric motion is usually generated by an offset counterweight on the rotating shaft that can be adjusted for angle and weights added or removed to modify the amplitude of the throw.

The throw is basically the amplitude of the motion created by the drive as it bounces the screen on its springs, with a smaller throw providing less aggressive motion. Screen throw varies inversely with frequency. The frequency is the speed (rpm) that the drive rotates, with the lower frequencies providing more aggressive movement. Most screens run at between 800 and 950 rpm. The screen angle or slope assists the mechanically induced throw by providing additional movement along the screen via the force of gravity. Not all screens are inclined at a slope and at times it can be advantageous to slope a screen backward. Most inclined screens are adjusted between 15° and 30°, but a 20° angle toward the discharge is most common. The flatter the screen, the longer the material is retained on the deck, resulting in a higher likelihood of a particle passing the screen. The direction can be adjusted downhill (toward the discharge), which is most common, or reversed to run uphill to retain the material longer on the screen deck. Efficient screening is achieved when a screen is operating with a vibration force of between three and seven times the acceleration of gravity (3–7*G*)

$$G \text{ force} = \frac{\text{rpm}^2 \times \text{throw}}{70{,}418}$$

Two different phenomena occur when separating materials with vibrating screens: stratification and separation. Stratification is the process where smaller particles (fines) fall to the bottom of the bed of material and larger coarse particles sit on the top. Stratification has to occur before separation can happen. Most screens contain a feed box (flat pan) where the material is fed to the screen. The feed box distributes the material onto the entire width of the screen deck. The first 10%–15% of the screen deck provides little separation because that is where the stratification is occurring. The next 35% of the screen deck does the majority of the separation, with the final 50% adding to the screening efficiency by removing the remaining undersized and near-size material. The bed depth of the material at the discharge end of the screen should never exceed four times the opening size of the screening media. If this ratio is exceeded, a portion of the undersized particles never have an opportunity to hit the screen and are carried over into the reject product; essentially, the screen is overloaded.

The required size (square feet) of the screening area is determined by a complicated formula that includes various factors from tables derived for this purpose. The screening area formula is

$$\text{screening area (square feet)} = \frac{U}{A \times B \times C \times D \times E \times F \times G \times H \times I}$$

where

U = undersize factor
A = capacity factor
B = oversize factor
C = half size factor
D = deck location factor
E = wet screening factor
F = material weight factor
G = screen open area factor
H = shape factor
I = efficiency factor

Luckily, several computer programs are available to calculate this formula. Most screens utilized in aggregate processing have more than one screen deck (two to four) allowing for several different products or size splits. The feed is introduced onto the top deck with the largest screen opening, and each successive (lower) deck has a smaller screen opening. The undersized particles fall through the top screen onto the second deck for additional screening, while the oversize particles from the top deck are directed to a discharge chute. This process flows downward through all successive decks, producing as many as five size splits. Because the bottom screen deck has the smallest screen openings, it determines the capacity of the entire vibrating screen. Screens are rated by their size expressed in width × length, for example, 5 × 12 feet (1.5 × 3.6 meters) or 6 × 16 feet (1.8 × 4.8 meters). The width of a screen determines its capacity, while the length defines the efficiency because the longer the particles are on the screen deck, the more likely it is they will pass the screen opening.

The separation probability is a ratio of the particle size to the size of the screen opening. With a large probability ratio where the particle is appreciably smaller than the screen opening, there is a high probability that the particle will pass the screen opening. Conversely, when the particle size is close to the screen opening size, the probability is low that this particle will pass the screen. The most difficult particles to remove by screening are those closest in size to the openings of the screen cloth, which are known as near-size particles. Coarse material screening requires lower speeds and higher stroke amplitude, while for screening finer particles, lower stroke and higher frequencies are more effective.

Screening efficiency is defined as follows:

$$E_{\text{screen deck}} = \frac{\text{metric tons per hour of undersize actually passing the screen}}{\text{metric tons per hour of underside in the feed}} \times 100$$

To calculate efficiency, the gradation of the feed material must be known as well as the gradation of the oversized product produced and the metric tons per hour of each. The flowing example assumes you are screening over a 1-inch (2.5-centimeter) screen deck and feeding the deck at 500 metric tons per hour. From a gradation analysis of the feed, it is determined that 86% is passing the 1-inch (2.5-centimeter) sieve. Therefore, 0.86 × 500 = 430 metric tons per hour of minus 1-inch (2.5-centimter) material is fed to the screen. Sixty metric tons per hour of oversized particles were produced from screening the 500 metric tons. A gradation analysis of the oversized product showed that 55% of minus 1-inch (2.5-centimeter) material remains in the rejects, which equates to 0.55 × 60 = 33 metric tons per hour. The efficiency calculation is therefore

$$E = \left(\frac{430 \text{ metric tons per hour} - 33 \text{ metric tons per hour}}{430 \text{ metric tons per hour}} \right) \times 100 = 92.3\%$$

The number and location of drives on a screening mechanism will determine the shape of the stroke and thus the action on the screen. A drive located at the exact center of gravity of a screen will produce a circular vibration. Overhead-mounted drives with the shaft mounted above the screen produce an elliptical motion slanted forward at the feed end with a circular motion in the center of the deck and an elliptical motion slanting backward at the discharge end. Double-shaft machines rotating in opposite directions generally produce a linear motion inclined toward the discharge. Triple-shaft horizontal screens are extremely efficient at screening. In this type of mechanism, one shaft is driven and connected to the other two shafts by a series of gears. The two outer shafts spin in one direction with the center shaft spinning in the opposite direction. This provides an elliptical, almost oval, motion slanted toward the

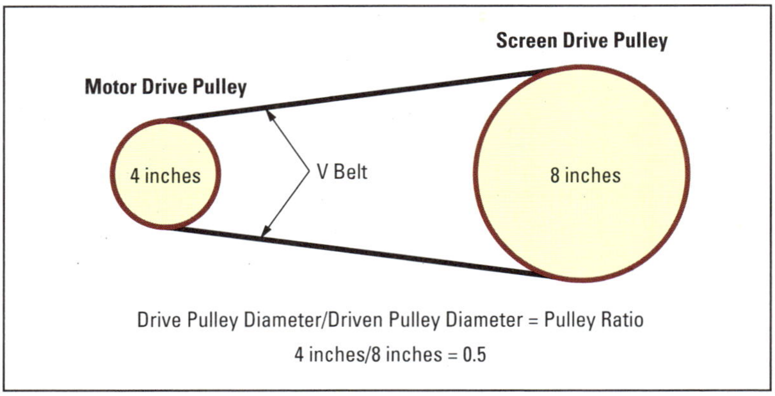

FIGURE 14.1 Pulley ratio

discharge that provides not only bounce but a tumbling action on the particles. The triple-shaft mechanism is extremely efficient but the gearing requires more attention to maintenance, especially oil changes on the gear box.

The throw on most vibrating screens can be easily adjusted by adding or subtracting counterweights on the drive shaft. Frequency or speed is readily adjustable by changing the ratio of the pulleys on the V-belt drive. The direction of flow (upstream or downstream) on a screen can be changed by an experienced electrician by reversing the polarity of the motor. Most screens are operated in the downstream mode to increase production or capacity. Running a screen in the upstream direction holds the material on the screen longer to encourage the near-size particles to pass, but reversing the screen in this manner will also slow the capacity because the material stays on the screen longer. The angle of the screen slope is not as readily adjustable as the other parameters without modifications to the screen's support structure.

The ratio of the diameter of drive sheave to the diameter of the driven sheave times the motor speed is equal to the shaft speed or frequency of the screen (Figure 14.1). The drive sheave is attached to the electric motor and the driven sheave is the one attached to the screen shaft.

$$R(\text{ratio}) = \frac{D_{\text{motor sheave}}}{D_{\text{sceen sheave}}}$$

$$\text{screen speed} = R \times \text{motor speed}$$

For example, a screen with a 4 inch (10 centimeter)–diameter sheave on an 1,800 rpm (1,725 rpm under load) electric-drive motor powering an 8-inch (20-centimeter) sheave on the screen shaft will produce a speed of

$$R = \frac{4 \text{ inches (10 centimeters)}}{8 \text{ inches (20 centimeter)}} = 0.5$$

$$\text{screen speed} = 0.5 \times 1{,}725 \text{ rpm} = 862.5 \text{ rpm}$$

This is an important formula to remember in any processing situation and applies not only to screening and not only to V-belt drives. This ratio also applies to crusher drives, pump drives, conveyor drives, and so on. The same formula applies to gear/sprocket ratios, only instead of diameters, count the number of teeth on the drive gear versus the number on the driven gear to determine the required ratio. Depending on the number of poles in an electric motor, 60-hertz motors come in four different speeds: 900, 1,200, 1,800, and 3,600 rpm (when not under load). The most common 480-volt industrial motors have speeds of either 1,200 or 1,800 rpm, which equates to 1,140 or 1,725 rpm, respectively, under their rated loads. A method of power transmission (PT) to either reduce or increase the speed for the specific applications is then required, such as V-belt drives or gear drives. Some drives have a combination of a V-belt drive, driving a gear box, driving a component (most conveyor drives), and both drive ratios need to be considered when calculating speeds. These PT ratios can be used to either reduce or increase the speed from standard-speed motors.

The velocity of the particle, the incline of the screen, and the thickness of screening media reduce the effective openings in the screen cloth. Depending on the material screened, the moisture content, the plasticity, the screening mechanisms, and the screening media utilized, certain problems can arise. When plugging (pegging) occurs, the near-size particles get stuck within the openings in the screen cloth. This is usually a result of the particle shape, especially if elongated particles are present. Pegging can significantly reduce the open area of the screen, which reduces screening efficiency. If pegging becomes problematic, one or a combination of the following should be attempted: (1) increase the stroke of the screen in an attempt to get additional force to lift the particles, (2) change the shape of the openings in the screening media, (3) change the type of screening media, or (4) tighten the crusher settings prior to the screen in question.

Blinding is another common screening problem that can occur, especially when screening wet material or when materials with high plasticity are present in the feed. Blinding is a result of the slow buildup of wet or plastic material adhering to the screening media and eventually plugging the openings. Blinding can be a combination of both pegging and adhesion. Increasing the stroke on the screen may help with blinding but may also reduce the efficiency of screening the fines. An increase in the frequency of the screen also helps accelerate the particles by reducing the time they have to adhere to one another. Changing the screening media to something more flexible (screen cloth to live wire or polyurethane) may also help prevent blinding. Adding spray bars and nozzles to the screening process is a good solution if the screen is set up for wet operation. If using water is not an option, special screen decks called ball decks are usually effective. Ball decks have special compartments attached to the bottom of the screens that contain rubber balls that bounce along the screen bottoms to add vibration in an effort to remove the plugs.

SCREENING MEDIA

Screening media (screen cloth) is available in a variety of types, sizes, construction, and configurations designed for myriad screening applications. The screening media is attached to the decks in the vibrating screens and accomplishes the actual particle separation. Screen cloth is a wear material that requires regular inspection and replacement. On heavy-duty applications when separating large particles (>8–10 inches [20–25 centimeters]), a 1 to 1.5 inch (2.5 to 4 centimeter)–thick steel plate with round, square, hexagonal, or rectangular perforations, called a punch plate, is often used. Woven wire is the most common screening media used in aggregate production. Woven wire comes in opening sizes from several inches to fine mesh down to the 240 mesh or smaller. These woven wire cloths come in various configurations from square to rectangular (slotted) openings and with different crimp styles: double crimp, locked crimp, corrugated crimp, and flat top crimps for free flow. Wire cloth is available in different alloys to fit almost any need: oil tempered, abrasion resistant, and stainless steel.

Stainless steel is worth the premium charged because it tends to wear longer because of its corrosion resistance and allows sticky materials to pass more readily.

When ordering screen cloth, the size of the wire, the size of the screen opening, the shape of the opening, the size of the screen panel, the crimp style, and the method of attachment need to be specified. Wire size is an important concept to consider when ordering screens. The larger the wire specified, the longer the screen cloth will last before replacement, but less open area will be available for particles to pass the screen, which significantly affects screen efficiency. This is also true on the spacing between holes in a punch plate. Open area can vary from 20% to over 80% of the screen area and will be listed as a specification by screen cloth manufactures. Slotted screens with rectangular openings are sometimes used in an attempt to increase open area. The slots can be oriented either with or perpendicular to the flow of the screen. Slotted screen cloth has an application, but can also allow flat and elongated particles to pass through the slots that would not normally pass a square opening. The smaller the screen opening, the less the open area because many more wires are required on the smaller-sized opening. When specifying screen cloth between screening efficiency and wear life of the media, a serious decision must be made. The more open area (smaller wire), the better for efficiency, the less open area (large wire size), the better for wear life. This becomes a balancing act for the processing manager and the sweet spot can only be determined through experience and experimentation.

Various other materials have been developed as alternatives to woven wire. Rubber, polyurethane, rubber-clad punch plate, and piano wire screens are available alternatives. Rubber-clad punch plate is very durable and has a place in screening large material but tends to add appreciable bounce to the screening process, which can discourage particles from passing through. Polyurethane is a good durable product that screens well because it has a tendency to flex somewhat, but it also has a smaller percentage of open area than wire cloth and comes with a much higher price tag. Many a plant manager, after retrofitting screen decks with polyurethane for additional wear resistance without considering the loss in open area, has been very disappointed with the ensuing capacity limitations. Polyurethane screen decks are a good product as long as the vibrating screen was initially designed with the extra capacity required for the reduced open area of the polyurethane cloth. The structure of the deck frame is different for punch plate, screen cloth, and polyurethane applications and the screening media to be used needs to be specified at the time of the screen purchase (Figure 14.2).

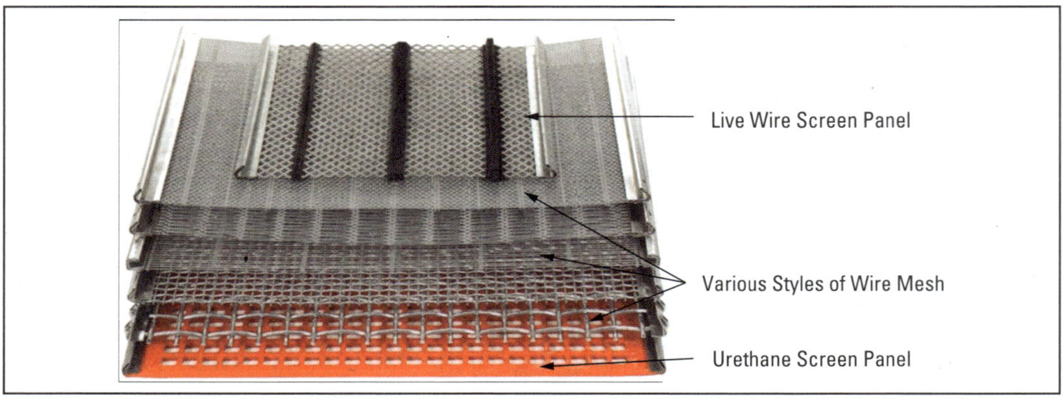

Adapted from Durex Products

FIGURE 14.2 Screening media

Many polyurethane systems are designed to be modular, having individual small panels that can be selectively replaced quickly and easily. This is an advantageous feature where the small panels can be replaced in areas of heavy screen wear on a more frequent basis. A punch plate is generally bolted to the surface of the screen deck. A screen cloth is normally side-tensioned with hooks or hook rails built into the screen and held down by side tensioning bars that are either bolted or held tightly with wedges. Wedge tensioning of side rails, although much faster to change than bolted rails, can be a problem with heavily loaded screens. Multiple pieces of screen cloth will be used to cover the individual decks (four 4-foot [1.2-meter] pieces on a 16 foot [4.9 meter]–long deck) and different screen openings can be attached to the same deck if required to tweak gradations. Piano wire screens are usually tensioned on the ends with a tensioning device built into the screen frame.

Hybrid screening media, also known as self-cleaning wire, is becoming more popular, especially for screening wet or sticky fines. These Livewire screens (a registered trademark of Durex Products) are a combination of stainless steel wire attached to vulcanized rubber or polyurethane strips. The flexible strips allow the wire a small amount of additional bounce, add a vibration effect to the screen cloth, and increase the open area by eliminating some cross wires. These come in various configurations with long slots, diamond-shaped openings, herringbone, and zigzag styles. Most screen cloth manufacturers have developed their own variations of these hybrid screens. The vibration effects caused by the flexible strips between the wires can also assist in the passage of plastic or wet materials that would normally agglomerate.

COMMON SCREEN TYPES

Several types of screening units are utilized within the industry including rotary screens (trommel screens), vibrating grizzly screens, stationary inclined screens, vibrating inclined screens, and vibrating horizontal screens. Trommel screens are one of the oldest types of screening devices and consist of a cylindrical frame with a perforated plate or screen on the surface that rotates on a near-horizontal axis (Figure 14.3). Most rotary screens are inclined slightly at 5°–7° from horizontal for gravity to assist with the discharge of particles. The material to be screened is fed into the center rear of the cylindrical screen and as the screen rotates the particles smaller than the surface perforations fall through. Different size screen panels (perforations) can be mounted to the outside of the screening cylinder to produce different sized products. Trommel screens range from 5 feet (1.5 meters) in diameter × 12 feet (3.7 meters) long up to 7 × 35 feet (2.1 × 10.7 meters).

Vibrating grizzly screens (or grizzly feeders) are normally used to scalp the fines from run-of-mine material prior to the primary crusher. These are heavy-duty, single-deck screening devices, and instead of mounting screen cloth on the screening surface they are equipped with parallel (grizzly) bars facing in the direction of the discharge with a specific spacing in between bars. The spacing of the bars is usually the same or slightly less than the closed side setting of the crusher being fed to remove any material that is already smaller than the crusher setting. Some grizzly bars are tapered from the feed toward the discharge end to allow for ease of discharge and prevent the wedging of larger material between the bars. Grizzly bars are made of abrasion-resistant steel or manganese for longer wear life. Grizzly feeders sometimes have one or more steps (levels) in the deck to encourage the feed to tumble during the feeding process. Other types of grizzly screens have a flat pan that precedes the bar section to help stratify the material prior to screening. The spacing allows for ample open area to remove or scalp the fines

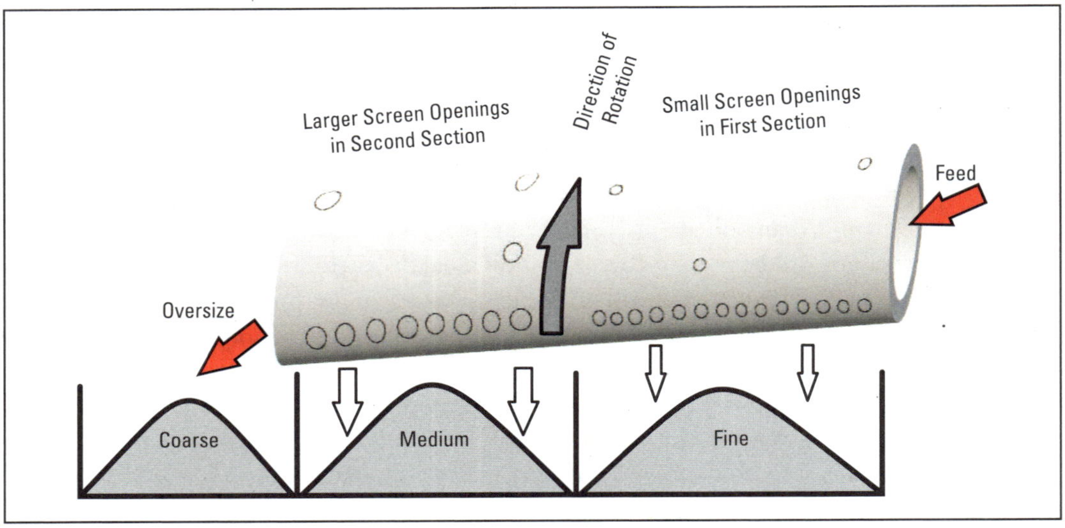

FIGURE 14.3 Trommel screen

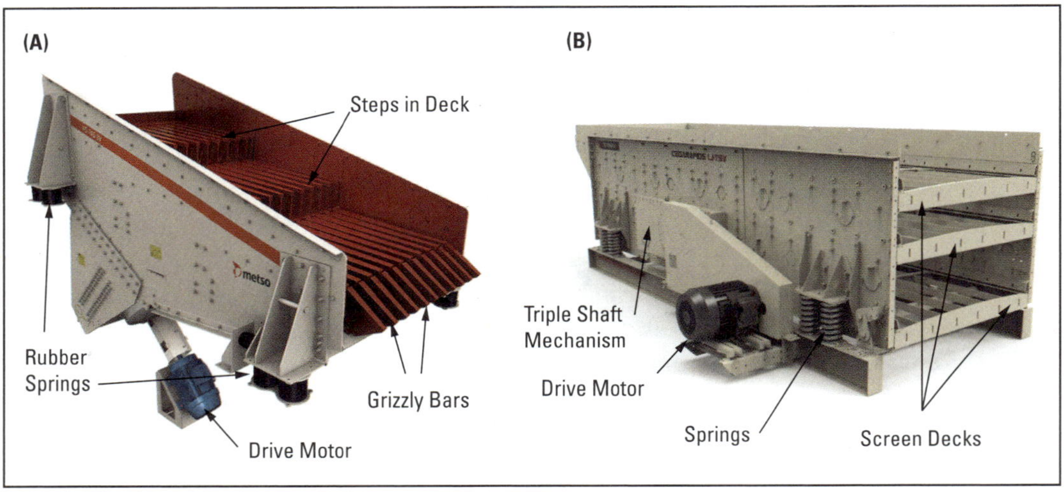

(A) Adapted from Metso Outotec; (B) Adapted from Terex

FIGURE 14.4 (A) Stepped grizzly feeder left, and (B) three-deck triple-shaft horizontal screen

(undersize) and the oversized material is fed to the crusher for further reduction. Vibrating grizzlies are used to feed most types of primary crushers except gyratory crushers, which do not require a feeder. Inclined, stationary grizzly decks are occasionally used preceeding crushers and sometimes used to produce riprap.

Most aggregate screening is accomplished using two- or three-deck inclined screens operating at an angle of 20° (Figure 14.4). These screens can have two or four bearings and are usually driven by an electric motor via a V-belt drive. Screens commonly range in size from 3 × 6 feet (0.9 × 1.8 meters) up to 8 × 24 feet (2.4 × 7.3 meters) depending on required capacity. Four-deck screens are available but are not in common use. As these screens become larger and larger, the additional weight of the screens and the material being screened requires heavier bearings, larger counterweights, and higher horsepower to generate the vibrations.

Heavier structures are also required to support the heavier screens and additional cross bracing to absorb the vibrations.

Horizontal screens were first utilized where low head room within a structure did not allow sufficient space for inclined screens. The mechanisms on horizontal screens usually run through the center of the screen box and come in either two- or three-shaft versions. Twin-shaft horizontal screens produce a linear stroke and tend to bounce the particles across the screen deck, while the triple-shaft mechanism provides an elliptical motion that bounces as well as tumbles the material toward the discharge. Horizontal screens are often found on portable (mobile) screening plants because they reduce the height of the chassis. It has been my experience that the triple-shaft horizontal screens are the most efficient screening mechanism available and should be used where tight gradation specifications are required. Some of the newer horizontal screens are coming with an option of a slight incline (3°–10°) to increase their capacity.

Screening of fine material requires smaller strokes and higher frequencies. Wet material is generally more difficult to effectively screen because moisture adheres to the fine particles by capillary action and makes them stick together. When screening wet or sticky material, a wash screen should be used whenever possible. Wash screens are the same basic vibrating machines with certain options added and the receiving pan beneath the screen modified to accept a slurry instead of dry material. Additional water should be added into the feed box of a wash screen and spray bars located at intervals over the screen deck with individual spray nozzles aimed at 90° to the screen deck. Urethane fans can be added to assist the high-pressure nozzles direct the water to cover the screen deck. The water spray assists in pushing the fines through the screen cloth, holding the material longer on the deck, breaking the capillary bonds, and washing away any plastic material coating the particles. The water required to wash the material through the screen deck generally ranges from 3 to 5 gallons per minute (11 to 23 liters per minute) per metric ton of material being screened. Feeds containing coatings of plastic fines will require additional water.

Multi-slope screens (banana screens) are becoming more popular with aggregate producers. These screens usually begin with a slope of 30°–40° at the feed end, which is incrementally reduced to 0°–15° at the discharge end. The steeper slope on the intake end allows the material to flow faster and reduces the bed depth, allowing the undersize material to then pass more quickly through the screen. The flatter slope toward the discharge end retains the material longer, allowing for the passage of the near-size particles. The mechanisms on these screens typically produce a linear motion. Banana screens provide high capacity as well as good screening efficiency.

High-frequency screens are also utilized for sizing both wet and dry fine materials. As mentioned earlier, lower stroke with higher frequency is more effective at screening finer material. Although most vibrating screens are operated in the range of 700–1,200 rpm, high-frequency screens can operate at as high as 3,600 rpm. This is usually accomplished by using electric eccentric motors, which have eccentric weights mounted on both ends of the motor shaft inside the motor housing. Two motors are generally mounted over the screen box rotating in opposite directions. These motors are capable of creating frequencies of up to 4,000 rpm. Some high-frequency screens utilize variable-speed hydraulic vibrators mounted underneath the screen deck.

Dewatering screens are similar to high-frequency screens and are extensively used on clamshell dredges to dewater the material prior to conveyor systems. Many producers are starting to use dewatering screens after the sand screws in the plant to further dewater their sand product and are obtaining moisture contents in the range of 7%–7.5%. Dewatering screens are normally sloped at about 5° in a counterflow direction (See Figure 9.11). The eccentric motors are mounted overhead toward the discharge end of the screen rotating in opposite directions. This creates a linear motion at approximately 45° to the screen deck, generating about 5*G* of force. When the slurry is fed onto the screen, this linear motion along with a fairly large stroke and aggressive *G* force accelerates the particles forward and upward, forming a thick bed of material. The thick bed of material tends to retain fines and filter the water through the bed to pass the polyurethane screen media. The slurry feeding the dewatering screen is generally preceded by hydrocyclones that have already removed a large percentage of the water.

Static dewatering screens are sometimes utilized to dewater slurries from suction dredges. These large flat surfaces are lined with wedge wire (profile wire) spaced closer than the smallest particles in the feed. This specialized wire is wedge-shaped with the narrower dimension on the bottom, allowing very fast discharge of water. These are usually installed on a small slope of 3°–5° with the wedge wire perpendicular to the water flow. Depending on the volume of slurry to be dewatered, these static screens are usually quite large to handle large volumes of water.

Many other varieties of mechanical screening devices exist including hybrid varieties with different types of mechanisms on individual decks. Flip-flow screens (e.g., BIVITEC) alternately stretch and relax flexible screening media to expand and contract the openings in an attempt to prevent blinding in fine materials. Mogensen sizers, modular screens (Omni-Screener), roller screens, circular tumbler screens, sifting screens, and sieve bends are other types of screens less commonly used by aggregate producers.

WASHING AND SCRUBBING

In the section on screening, we mentioned the use of wash screens with high-pressure nozzles to help break the capillary bond and remove plastic materials while screening. Adding pressurized spray nozzles to the screening process can substantially improve efficiency, especially when screening fine, wet materials. The addition of water into the screen feed box to make a slurry also aids in the screening. Nozzles that produce a flat, fanlike spray are the most effective. The nozzles should be evenly spaced along the spray bars so that the spray fans overlap and cover the entire width of the screen. The nozzles can be specified with various fan angles, with the narrower angles providing greater impact. Depending on the material being screened, spray nozzles can be utilized on all the screen decks or individual decks only. When using wash screens, additional equipment is required downstream to remove the dirty wastewater, further classify the fine materials, and dewater the fines. Three to five gallons per minute (11 to 23 liters per minute) of water per metric ton of material being washed on each individual screen deck is typically required for a wash screen to be effective. The design of the under-screen pan differs from a dry screen because the water has to be collected along with the material passing the bottom screen deck. The water volume to be collected includes the majority of the water from any upper screen decks as well as the bottom deck.

Very dirty materials containing clay will require even more equipment to produce a marketable product. Log washers remove plastic clays from a coarse aggregates feed by interparticle scrubbing with a combination of water and material-on-material attrition. A log washer consists of two parallel, tubular shafts (logs) with intermeshing paddles installed along the length of the shafts that are contained within an inclined tub or trough. The log washers are installed at an upward discharge angle from 5° to 15° with the water flowing toward a level retention area in the rear. Adjustable-height weirs control the discharge of the dirty wastewater, clay, and silt particles. As the shafts rotate in opposite directions, at 25 to 32 rpm, the paddles force the rock material to interact via particle-to-particle attrition and push the material toward the discharge. The clay minerals are broken up, dissolved, and absorbed into the water, which overflows the discharge weirs as wastewater. Log washers are designed to scrub material in the ¾- to 6-inch (1.9- to 15-centimeter) size range and are generally not used on fine materials because of excessive abrasion wear. The solubility of the plastic clays will dictate the amount of water added to the system. Some systems provide optional rising current water to help move the dirty water to the top and most have spray bars at the discharge to help rinse off the surface clays. Generally, added water is minimized to take advantage of the particle-to-particle attrition breaking the bonds of the clay minerals.

Dirtier feed material requires a longer retention time within the log washer for the attrition scrubbing to be effective. This is accomplished by either increasing the slope on the washer, using a longer log washer, or a combination. Log washers come in several sizes and capacities with shaft lengths between 24 and 35 feet (7.3 and 10.7 meters) and log diameters of 36 to 46 inches (91 to 117 centimeters). Capacities range from 50 to 190 metric tons per hour requiring from 75 up to 200 horsepower to operate. Extremely high clay contents or insoluble plastic clays may require multiple log washers to produce a salable product. The discharge from a log washer should always be fed to a rinse screen to remove any residual clay particles still adhering to the aggregates. Wear-resistant replaceable paddles in various alloys are necessary options for these machines to reduce maintenance costs and downtime.

Coarse material washers work in a similar way to log washers and are used to remove deleterious clays from coarse aggregate. They use a combination of paddles and spirals (augers) with weirs to control the discharge of the dirty water. They are available with either a single or a double shaft and are designed for the removal of lesser amounts of clay using less horsepower than required for log washers (Figure 14.5). Coarse material washers can also float out lightweight materials like sticks, twigs, and lignite. As the spirals force the material forward along the trough, the product is cleaned by particle-to-particle attrition and also partially dewatered, enabling the product to be conveyed to a stockpile. Coarse material washers typically process material in the size range of ¼ to 1 inch (0.63 to 2.5 centimeters). A blade mill is another type of similar of machine with alternating paddles and spirals but is mounted at a shallower angle of 0°–5° and has no water discharge. Blade mills are designed to break the bond of clay minerals and deleterious materials to increase the efficiency of wash screens.

Fine material washers (spiral classifiers or sand screws) provide washing, classification, and dewatering functions to material in the sand size range of minus ⅜ inches (0.95 centimeters) or No. 4 feed sizes. The sand size products typically are screened and washed on a wash screen prior to being fed to a sand screw. A slurry of approximately 50% solids is fed into the tub of the sand screw, which has either one or two spiral (auger) shafts inclined at about 13°–15°. The rotation speed of the spiral(s) determines the agitation rate and capacity of the screw. The spiral forces the sand up the inclined trough, providing attrition scrubbing of the particles, while the adjustable weir on the rear of the tub allows the dirty water to discharge. Increasing weir height allows more settling time while also decreasing agitation. Adding additional water will dilute the solution, increasing settling and allowing additional, finer particles to settle. One of the

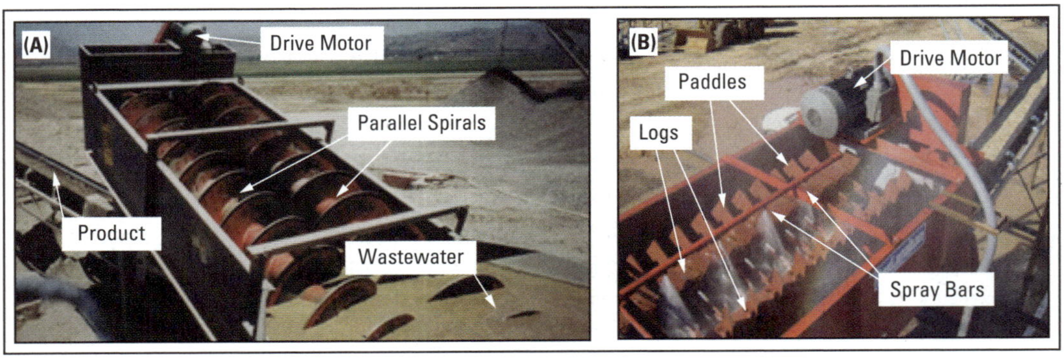

(A) Adapted from Eagle Iron Works; (B) Adapted from McLanahan

FIGURE 14.5 (A) Twin sand screw, and (B) log washer

primary functions of a sand screw is to dewater the sand so it can be transported via conveyor after discharge from the screw.

Sand screws come in numerous sizes and capacities ranging from 30 to 475 metric tons per hour for single screws and 200 to 950 metric tons per hour for double screws. Spiral sizes range from 20 inches (51 centimeters) in diameter by 20 feet (6.1 meters) in length up to 72 inches (183 centimeters) in diameter by 38 feet (11.6 meters) in length with horsepower ranging from 5 to 150. Abrasion-resistant replaceable wear shoes that bolt to the spirals are available as options made from various metallic alloys or urethane products. Other options include support structures, safety covers, and discharge chutes.

CLASSIFICATION

Classification is the separation of various size particles by the use of settling velocity. This method typically applies to sand-sized particles (material minus ⅜ inches [0.95 centimeters] or minus 4 mesh) that need additional processing to meet gradation specifications that screening alone will not accomplish. Bulk specific gravities (SGs) of the sand particles from an individual deposit are usually fairly consistent unless large quantities of deleterious, lightweight materials are present. SGs in the range of 2.65–2.70 are typical of most sands being classified. Based on a uniform SG, individual grains of sand of a specific size will settle in water at the same rate. Classification is therefore a function of the differing rates of settling for different sized particles. These weight differences between particle sizes mean that larger (coarse) particles will settle out early in the tank and lighter (finer) particles settle out progressively further down the tank (Figure 14.6). Horizontal current classifiers are considered to be the free-settling type and are not hindered by the viscosity of the solution. Maximum efficiency occurs when the SG of overflow slurry is in the range of 1.025 to 1.030. If the overflow slurry SG exceeds 1.065, additional water will be required to prevent buildup of fines in the tank.

The density of the slurry (sand-and-water mixture) is also a consideration in classifying sand and is generally expressed as percent by weight of solids or SG of the slurry. The following formulas can be used to calculate any liquid-to-solid mixture:

$$\text{GPM}_{\text{slurry}} = \text{TPH}_{\text{slurry}} \times \frac{4}{\text{SG}_{\text{slurry}}}$$

$$\text{GPM}_{\text{water}} = \text{TPH}_{\text{water}} \times 4$$

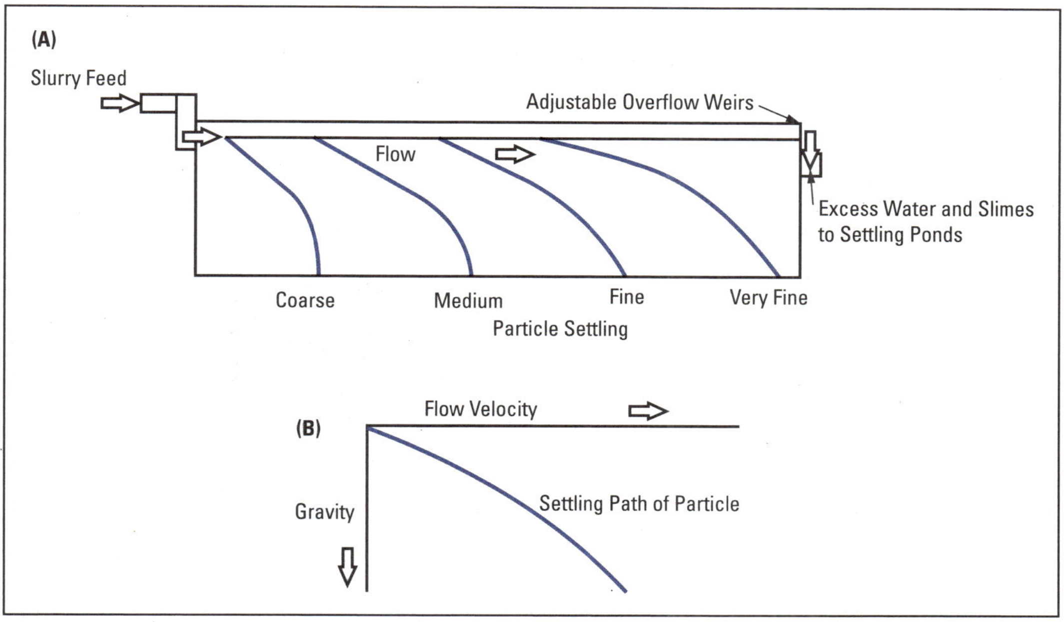

FIGURE 14.6 (A) Horizontal settling tank, and (B) settling forces

$$GPM_{solids} = TPH_{solids} \times \frac{4}{SG_{slurry}}$$

$$SG_{slurry} = TPH_{slurry} \times \frac{4}{GPM_{solids}}$$

$$GPM_{solids} = 1.5 \times TPH_{solids} \text{ (when SG = 2.67)}$$

where

GPM = gallons per minute
TPH = metric tons per hour
SG = bulk specific gravity

The parameters in Table 14.1 can all be calculated utilizing the above formulas if the SG of the solids and any two other parameters are known. The results would apply at an SG of 2.67 for the solids, 100 metric tons per hour sand, and 40% solids by weight in the slurry.

Classifying of sand is generally done to eliminate excess fines in the sand mixture, either minus 100, minus 150, or minus 200 mesh materials, and remove the excess volume of water from the sand slurry. The longer the sand tank, the more efficient the settling of the particles. Classification of the sand is a function of the infeed gradation, material density, rate of feed, volume of water, SG of the slurry, and the required final gradation. Unfortunately, the infeed gradations can change radically, especially in natural sand and gravel deposits. Based on the size of the classifying tank, manufacturers can recommend the water volumes necessary to remove various mesh sizes from the slurry. The general rule for water requirements is 10 gallons per minute (38 liters per minute) for every metric ton of sand feed or 100 gallons per minute (378.5 liters per minute) for every metric ton per hour of minus 200 mesh material, whichever is larger.

TABLE 14.1 An example of slurry density relationship

	Weight (%)	Metric Tons per Hour	Specific Gravity	Gallons per Minute
Water	60	150	1.0	600
Solids	40	100	2.67	150
Slurry	100	250	1.33	750

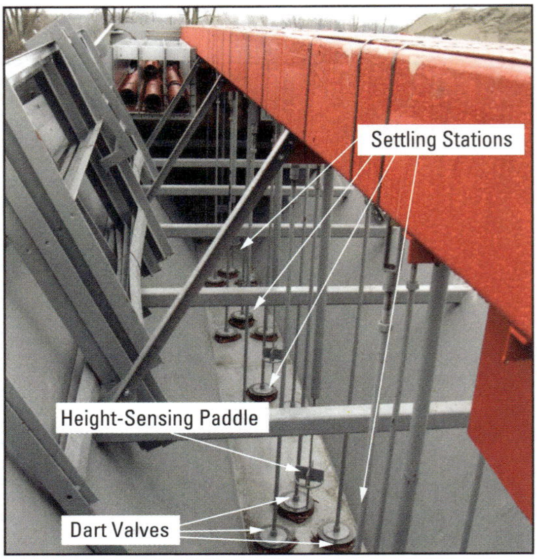

FIGURE 14.7 Interior of a sand classifying tank

Sand classifying tanks (SCTs) consist of an integral system of several components that operate together and usually produce two separate products plus a waste product (Figure 14.7). The products produced can be a combination of concrete sand and asphalt sand or concrete sand and mason sand, and so on. Three separate collecting and blending flumes beneath the tank receive various sized sand particles from different areas along the length of the tank and blend them back together to make the final gradations and waste the unnecessary size fractions. Depending on the length of the tank, 6–11 settling stations are located at intervals along the bottom of the tank. These stations contain three dart valves, each feeding one of the three collecting flumes, and a height-sensing paddle that controls the depth of sand that accumulates at each station. A controller receives a signal from the sensing paddle and controls when each dart valve opens to discharge a sand-and-water mixture into the appropriate collecting flume. An electric or hydraulic mechanism receives signals from the controller and the paddle sensor to activate the dart valves to release the sand into the appropriate trough.

Today's SCT controllers are mostly programmable logic controller (PLC)–controlled, where percentages from each settling station can be reblended into a final required gradation. The more sophisticated controllers allow for the infeed gradations to be entered as well as the required product gradations and assign minimum and maximum timer settings for all the stations to create the necessary product gradations. Large fluctuations in the gradation of the infeed material can be problematic with the timed system. Rising current options can be added to the tanks to enable hindered settling at certain stations if additional settling control is required. SCTs range in size from 8 × 20 feet (2.4 × 6.0 meters) up to 12 × 48 feet (3.6 × 14.6 meters) depending on capacity, water volume, and infeed gradation. Water volumes range from 700 up to 16,000 gallons per minute (2,650 up to 60,566 liters per minute) depending on the cut size required and the fines content of the infeed. The products produced will need to be fed to a sand screw for further dewatering prior to putting the sand product onto a conveyor for stockpiling.

PLC controls on classifiers can at times be a problem. When entering specifications into the PLC, it will attempt to make a product that is right in the center of the percentage range on each specification size split. Many natural sand deposits will have an excess of a certain size range (a belly in the gradation) in the feed to the classifier (Figure 14.8). When starting a newly installed classifier at a sand and gravel plant, my superintendent requested assistance from the manufacturer's representative in setting up the new PLC. The desired specifications for the concrete sand and the mason sand products were entered into the PLC and the classifier did a marvelous job of producing two specification products. The only problem was that sand production was reduced by almost 30% and the waste product increased by a similar amount as the classifier removed the "belly" from the infeed. A reduction in the production was not the outcome we were expecting. We ended up bypassing the automation and manually adjusting the timing and were able to include much of the previously wasted size range (belly) while still producing two specification sand products.

Another method of classifying is the use of hydrocyclones, which are a continuously operating classifier that accelerates the settling rates of particles by the use of centrifugal force (Figure 14.9). As a pressurized slurry is introduced via a tangentially mounted pipe, the swirling motion generates a vortex with a low-pressure zone along the vertical axis. The vortex creates an inward drag while the centrifugal forces push the particles outward toward the outside of the cylinder. The faster-settling (larger) particles are forced toward the wall of the cyclone and fall toward the apex opening, while the drag from the center vortex moves the finer

FIGURE 14.8 Eagle Iron Works sand classifying tank awaiting installation

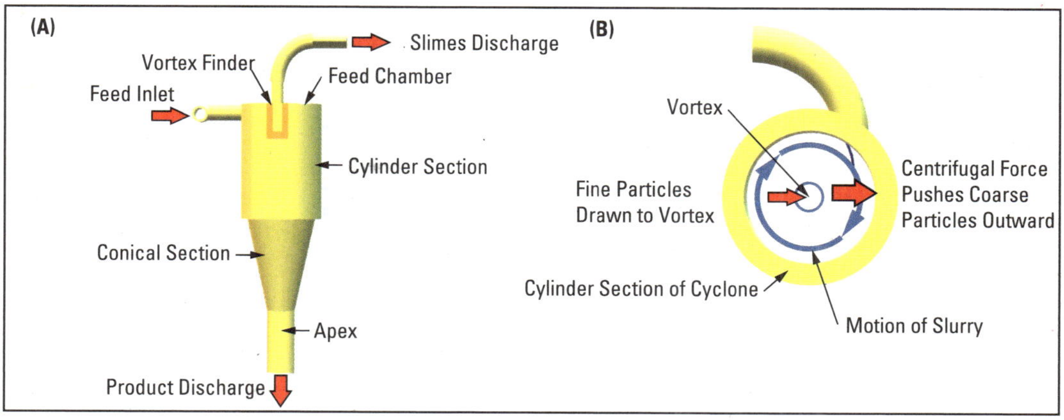

FIGURE 14.9 Hydrocyclone classifier: (A) side view, and (B) cross section

particles upward through the vortex finder toward the discharge. Smaller-diameter cyclones classify finer particles and larger diameters are used for coarser separations. The vortex finder is generally a replaceable wear part. Hydrocyclones by their very nature make excellent dewatering devices and in many cases are utilized solely for that purpose.

The efficiency of classification decreases with the density (viscosity) of the feed and is increased by limiting the water underflow. The split of flow from water to underflow for dewatering purposes increases with a smaller vortex or a larger apex and increases with a higher density or feed viscosity. The cut size is inversely related to recovery of solids and increases with feed density and decreases with the flow rate. Pump controls should be utilized to minimize fluctuations in flow rate to hydrocyclones. Certain types of sand processing plants rely solely on various sized cyclones to produce specification sand products. If multiple similar-sized cyclones are utilized, it is best to mount them in a circular fashion with a common feed from the center of the cyclone array to provide a balanced feed to each cyclone.

Rake classifiers are not commonly used for aggregate products but can be utilized to recycle water in very dry climates. In lieu of a spiral, a rake classifier drags the fine material up an incline to dewater and classify. The rakes are actuated by an eccentric mechanism that drags the material upslope, withdraws at the top of the stroke to move back to the starting position, and the dragging process is repeated. Rake classifiers generally have multiple (up to four) raking mechanisms that alternate raking and withdrawing to keep the process constantly moving forward.

Specialized blending or mixing equipment is sometimes required for certain materials and certain customers. Pug mills are similar to blade mills and include two tubular shafts that rotate in opposite directions with paddles having a spiral pattern overlap. Pug mills are designed to mix two or more dry materials or to add liquid to material such as cold-mixed asphalt. Certain state departments of transportation require base products be to be pugged with a specific percentage of water added. The material(s) to be blended is metered into the pug mill via a hopper(s) and thoroughly mixed by the mill. Spray bars near the intake end are utilized to add liquid at a metered rate. Pug mills are available in various sizes and capacities from 200 to 1,200 metric tons per hour. Replaceable paddles are generally reversible for longer wear and made of abrasion-resistant alloys. The number and spacing of paddles depend on the size and type of materials being mixed and the infeed is limited to 2.5 inches (6.3 centimeters) in top size.

Most aggregates processing plants are custom designed to meet the production requirements and the eccentricities of each deposit. They can utilize an almost unlimited variety of washing and classifying equipment in various combinations to meet the needs of the individual deposit.

CHAPTER 15

Material Transport and Stockpiling

MATERIAL TRANSPORT

Within the processing plant, the movement of material between crushers, screens, and stockpiles is handled in a number of different ways. The easiest and least expensive method is usually gravity. In the past, I worked with a colleague that always said: "Mount your crushers low and your screens high; gravity is your friend." In some cases this is correct and in others . . . questionable. Chutes are a common method of moving material with the assistance of gravity. Chutes can be used to feed crushers, screens, hoppers, and conveyor belts. Special attention must be paid to the angle of repose of the material being moved within the chutes. Chute angles can vary from 30° for coarse material to near vertical for plastic fines. Most chutes today are enclosed to help reduce dust emissions so they essentially consist of a long, rectangular box made from steel plate sloping from one component to another.

The sidewalls of chutes should be a minimum of 12 inches (30 centimeters) tall and up to 24–30 inches (60–76 centimeters) if personnel access is necessary. The tops of chutes should be removable for ease of maintenance access. The width of chutes is dependent on particle size and should be a minimum of four times the size of the largest particle expected to pass. A good general rule is to construct chutes a minimum of 24 inches (60 centimeters) wide to allow for ease of maintenance. Chutes are generally lined with replaceable wear liners. These liners can be made of stainless steel, abrasion-resistant steel, ultrahigh molecular weight (UHMW) polyethylene, or ceramic materials. Ceramics are very good in increasing the life of chute work by resisting abrasion but do not respond well to impact and are an expensive and heavy lining material. As an alternative to chute liners, stairstep chutes, which retain material for particle-on-particle wear, can also be constructed (Figure 15.1). Depending on the length of the stairstep chutes, they can become very heavy, considering the weight of the steel plus the material in the steps, and may require extra support.

Various types of feeders are utilized in material handling applications within aggregate plants in conjunction with chutes. Apron feeders are heavy-duty feeders used to move large, coarse, heavy material, typically in applications feeding shot rock to a primary crusher or static grizzly. Apron feeders are similar to caterpillar tracks and consist of overlapping plates attached to a chain drive and are usually inclined at about 10°–15°. These feeders are sometimes used in conjunction with (feeding onto) a grizzly feeder if fines removal is necessary prior to the crusher. The feeder width corresponds to the opening of the crusher it feeds.

Reciprocating feeders use an eccentric motion to push and retract a flat plate. As the plate moves forward on a series of rollers at a controllable rate and amplitude, the rock on top is forced forward, and as the plate retracts from beneath the load it falls into the crusher.

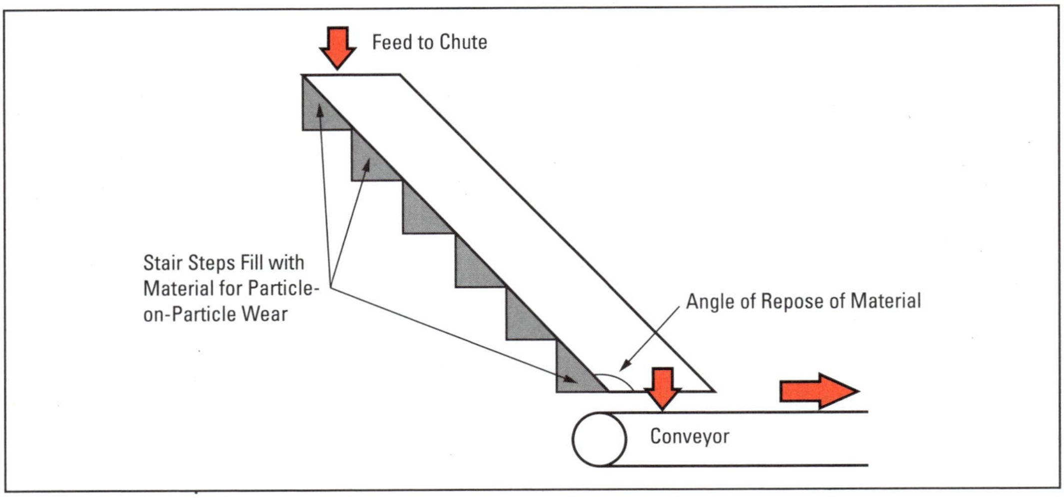

FIGURE 15.1 A stair step chute feeding a conveyor

Reciprocal feeders are usually utilized with damp, plastic, or lumpy materials. I have personally never seen one utilized in aggregate processing, but they are an optional type of feeder in certain situations.

Vibrating pan feeders are probably the most common feeders used in aggregate material handling. These feeders require very little head room and can be mounted in tight locations. They can be used in a multitude of feeding applications such as feed bins, bucket elevators, and conveyors, and are especially useful in keeping cone crushers choke fed for the production of uniform cubical products. They are composed of a shallow-sided sloped trough with a flat bottom and an open end for discharge, are suspended on springs, and are available with either a top- or bottom-mounted mechanism. Bolt-in, abrasion-resistant liners are available in a variety of different alloys. Capacities are a function of the size of the trough, trough angle, material density, moisture content, plasticity, and frequency and amplitude of the vibrations. These feeders are self-cleaning, can start when fully loaded, and can handle material from fine dust to large stone depending on the size and horsepower of the feeder. Two different types of mechanisms are available on pan feeders, either electromechanical or electromagnetic. The vibration for electromechanical feeders is provided by one or more electric eccentric motors. These feeders operate at approximately 1,100 vibrations per minute at 60 hertz with a stroke in the range of 0.25–0.30 inches (6.3–7.6 millimeters). Heavy-duty electromechanical pan feeders come in capacities of from 600 up to 4,000 metric tons per hour.

The vibration in electromagnetic feeders is controlled by a magnetic field being applied to a series of leaf springs within the drive mechanism. This type of feeder has no mechanical parts to wear out, no shafts, no bearings, no eccentric weights, and no belts, and thus requires no lubrication. Electromechanical feeders operate at approximately 3,600 vibrations per minute at 60 hertz with a stroke in the range of 0.06 to 0.08 inches (1.5 to 2 millimeters) (Figure 15.2). This type of feeder comes in models with capacities of between 25 and 1,600 metric tons per hour. Accurate and repeatable flow control levels are accomplished using a rheostat in electromagnetic feeders.

Variable-speed belt feeders are commonly utilized in aggregate processing plants especially when tight conditions are encountered. A belt feeder consists of a short, flat conveyor equipped

Adapted from Syntron MH

FIGURE 15.2 Electromechanical pan feeder

with flat closely spaced idlers. An adjustable feed gate is used to feed the material from a bin or chute onto the conveyor. These belt feeders typically run at 150 feet per minute (76 centimeters per second) or less and the feed rate is controlled by the adjustable feed gate opening along with modifying the belt speed.

Conveyors are the transportation highways within a modern aggregate processing plant. Conveyor capacity is a function of the width of the conveyor belt and the speed at which the belt travels. Conveyor belting comes in standard widths of 12, 18, 24, 36, 42, 48, 60, 72, 84, 96, and 108 inches (30, 45, 61, 91, 107, 122, 152, 183, 213, 244, and 274 centimeters) and can be split to custom widths by your belting supplier. Conveyor belts are a sandwich of multiple layers of multistrand woven fabrics (polyester and/or nylon) and rubber compounds with typically two to six fabric plies. The fabric plies are woven in two different directions with the warp cords (lengthwise) providing tensile strength and transverse tear and impact resistance. The weft cords are at 90° to the warp cords and are designed for holding as well as rip and tear resistance. Each fabric ply is separated by a layer of vulcanized rubber. Certain heavy-duty twill fabric layers designed for high impact and larger particles are also available as an option. Top and bottom surface layers are added to the cord plies, the thickness of which can be specified by the user. Extra heavy-duty conveyor belting has galvanized steel cable (wire rope) in lieu of fabric plies. This type of belting is used on very long, overland conveyors with complex horizontal and vertical curves and is very flexible as well as puncture and abrasion resistant.

Conveyor belting must be spliced together during installation or when a belt tears and there are many belt splicing techniques. Mechanical belt splices consist of hinged steel plates with teeth that cut into the covers and usually require drilling holes through the belt carcass. A technique called skiving removes belt covering material to enable the mechanical fastener installation to be even with the belt surface. Vulcanizing a splice is the process where the conveyor is actually separated at the plies and essentially melted (vulcanized) back together in a similar fashion as the original manufacturing process. Creating a vulcanized splice requires specialized equipment, curing time, and personnel experienced with the process, but will more than pay for itself with added belt life.

Conveyor belting is specified by the width and length, number of fabric plies, top cover thickness, bottom cover thickness, pounds per inch of width (PIW) rating, and the specific rubber compound. The bottom cover of the belt receives much less abrasion and is usually specified at about half the thickness of the top cover, for example, a ⅜-inch (9.5-millimeter) top cover with a 3⁄16-inch (4.7-millimeter) bottom cover. Consulting with the conveyor belting suppliers and/or manufacturers concerning the conveyor location, application, and material to be transported is always recommended when specifying conveyor belt. Various different specialized rubber compounds are also available depending on the application. There are several very reputable conveyor belt manufacturers and also many poorly designed import belts on the market. Be cautious when buying belting on strictly a price per foot basis; it has been my experience that cheap conveyor belting is not the bargain it appears to be on the surface.

Conveyor structures are generally made from steel truss or sometimes channel frame depending on the application and location of the conveyors (Figure 15.3). Truss conveyor is more substantial and will normally outlast the channel frame variety for only a small premium in price. The conveyor system consists of a structure, supports (conveyor bents), a head pulley, tail pulley, pulley shafts, bearings, drive mechanism, belt length adjusting system (take-up), maintenance access (walkways), troughing idlers, return idlers, safety guards, skirted receiving hopper, belt scrapers, and an emergency stop switch running the entire length of the conveyor. Head and tail pulleys, sometimes referred to as drums, are cylinders with a centrally located shaft supported by pillow block bearings that guide and drive the belt.

The conveyor belt rides on a series of idlers (rollers) that run the length of the conveyor usually installed on 4-foot (1.2-meter) centers with return idlers (on 10-foot [3.0-meter] centers)

FIGURE 15.3 Two 40 foot (12 meter)–long conveyor trusses awaiting installation

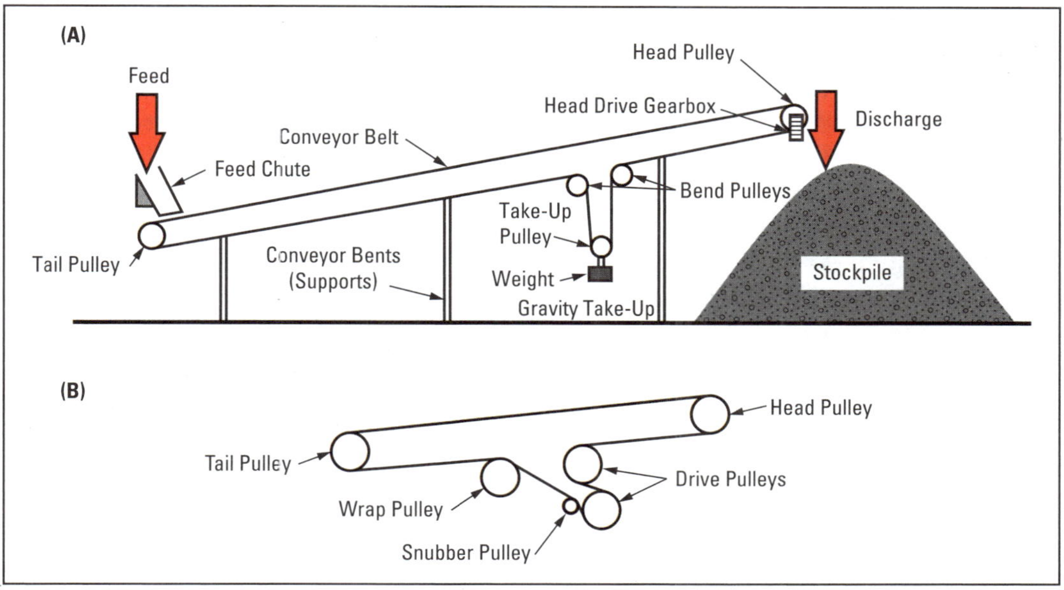

FIGURE 15.4 (A) Typical incline conveyor with a head drive, and (B) optional wrap drive

on the bottom of the structure for the return belt ride. Normally three roll troughing idlers are used for belt support and to contain the load. Idlers come in several varieties: flat (usually for belt feeders), and 20°, 35°, and 45° troughs. Thirty-five-degree troughing idlers are by far the most common in aggregates handling. Idlers are specified by the width of the conveyor belt, diameter of the rollers, angle of trough, and construction. The Conveyor Equipment Manufacturers Association specifies the design of conveyor components, and idlers come in several categories based on application. Idlers classified as A and B are for light duty, C and D are for medium duty, and E and F are for heavy-duty applications. Classes C and D are ample for most plant applications and Class E should be considered for heavy-duty applications at the primary crusher or on field conveyors carrying run-of-mine material. I never specified a can (roller) size less than 5-inch (12.7-centimeter) diameter or a class less then C.

Conveyor inclines for transporting crushed stone are normally limited to 15°, but 18° inclines are possible (Figure 15.4A). Sand and gravel conveyors are typically sloped at a 12° incline or less because the round gravel particles have a tendency to roll backward, down an inclined belt, especially at higher belt speeds. Specially designed conveyor belts with ribbed top covers can be purchased to aid material moving on belts inclined greater than 18°. Idlers come with either greaseable or sealed bearings, but most common are the sealed type. Where a chute or another conveyor feeds onto a moving conveyor belt, an area of impact occurs and specially designed impact idlers or an impact bed made from strips of UHMW polyethylene replace the standard troughing idlers. This is done in an effort to protect the conveyor belt from premature failure owing to impacting particles cutting the top cover of the conveyor belt. Spring-tensioned belt scrapers, mounted on the head pulley, are necessary to remove residue from the conveyor belt prior to the return trip over the return idlers. This is especially true when the material conveyed is moist or contains plastic, claylike particles and always when conveying fine (sand-sized) materials.

Conveyor belting, being rubber, has a tendency to stretch and some method of tensioning is necessary on all conveyor systems. One common tensioning method is the use of an adjustable tail pulley where the tail pulley is mounted in a movable frame that is adjusted by long bolts to maintain tension on the belt. Another tensioning method that keeps a constant tension on the conveyor belt is known as a gravity take-up. This requires three additional pulleys, two bend pulleys attached to the conveyor truss and a take-up pulley suspended by the conveyor belt in a frame that has a counterweight attached. Weights added to the counterweight frame keep constant even tension on the belt even as it continuously stretches.

Conveyor drives come in several designs, the most common of which is the head drive. On this type of drive a gear box with a specific reduction ratio is mounted to the rotational shaft of the head pulley with the gear box driven by a V-belt drive to an electric motor. The speed of the conveyor can be changed by changing the pulley ratio on the gear box (as shown in Figure 14.1), but the ratio of the gear box must also be accounted for in the speed calculation. Another common conveyor drive is a wrap drive and is usually specified when additional power is required on a conveyor subject to heavy loads. Wrap drives are more expensive because they require an addition of at least three more pulleys. Wrap drives (Figure 15.4B) can be mounted anywhere along the conveyor structure but are generally located near the center of the conveyor. Conveyor pulleys are normally lagged to prevent slipping and provide additional start-up torque. Lagging is a rubber lining on the outside of the pulley or drum that adds additional friction between the conveyor belt and the pulley surface. Lagging can be smooth rubber or have a number of tread patterns depending on the application.

Conveyor belt speed can be calculated utilizing the following formula (which utilizes the pulley or sheave ratio mentioned in Chapter 14):

$$R = \frac{A \times B}{C}$$

$$P = \frac{R}{D}$$

$$S = P \times E \times 0.2618$$

where

A = motor revolutions per minute
B = motor sheave diameter
C = reducer sheave diameter
D = speed reducer ratio
E = diameter of head pulley
R = reducer input speed
P = drive pulley revolutions per minute
S = conveyor belt speed

Conveyors normally operate at 300–350 feet per minute (152–178 centimeters per second) in most plant operations, although it is not uncommon to see conveyors running at 600 feet per minute (305 centimeters per second) or faster. Best practice is to keep the conveyors running fairly slowly for a number of reasons. The slower the speed, the greater the longevity of the conveyor components; bearings, idlers, belt scrapers, pulleys, and so on. Another good reason for slower conveyor speeds is spillage. Faster belt speeds create many more opportunities for rocks to bounce with greater velocity, creating potential spillage at transfer points, backups in chutes, and messes requiring additional labor to clean up. Labor costs to clean up constantly

FIGURE 15.5 Wobble feeder feeding a horizontal shaft impact crusher

accumulating piles of rock are wasteful and extremely frustrating for employees. I once worked for a large international company that put on a quarry managers class every August in Georgia. A part of the course included giving the managers shovels and requiring them to clean up the spillage around the processing plant under the very hot August sun. The point of the exercise was to convince the managers to find and repair the cause of the spillage and not expect their employees to clean up the same mess day after day.

Specialized conveyors that can convey vertically at 90° are available but come with a large capital cost. Vertical conveyors use either two conveyor belts that sandwich the material between them or specialized belting that incorporates specialized pockets or rubber buckets vulcanized to the conveyor belt to lift the material. Bucket elevators were common in the past to elevate material vertically between plant levels but are no longer in common usage. They consist of a series of enclosed, small steel buckets attached to a chain that is driven vertically by sprockets. Screw conveyors, which consist of an auger rotating in a near-horizontal trough, are also uncommon in aggregate processing but can be useful in moving very fine materials like agricultural lime for short distances. Wobble feeders consist of long shafts with an elliptical or triangular cross section equally spaced that are mounted horizontally with the ellipses at 90° to one another. As the shafts rotate, the elongated side of the ellipse pushes the material forward along the feeder path and the finer particles bypass through spacing between shafts (Figure 15.5). Wobble feeders are normally utilized when sticky plastic feeds are encountered and, because of the timing on the shaft drive, are rather maintenance intensive.

STOCKPILING

Once the final products are made, they will need to be stockpiled for final load-out to delivery trucks. Most aggregate plants put stock into inventory in outside stockpiles made by inclined conveyors, although some producers utilize storage bin systems or a combination of bins and stockpiles. Storage bins can be fabricated in almost any size, but the angle of repose of the product to be stored needs to be considered for the side slopes of the bins to facilitate fast load-out and prevent buildup inside. Some processing plants have 24-hour load-out for the convenience of their customers where certain products are stored in bins that the customers can drive under for automated loading, weighing, and ticketing around the clock. Railcar loading is commonly accomplished using very large bins straddling the railroad tracks. Material is conveyed into the bins and the railcars are loaded via gravity as they pass underneath (Figure 15.6). Some of these systems are very sophisticated utilizing programmable logic controller (PLC) computers for control. I once toured a coal load-out facility that loaded, weighed, struck off the top and

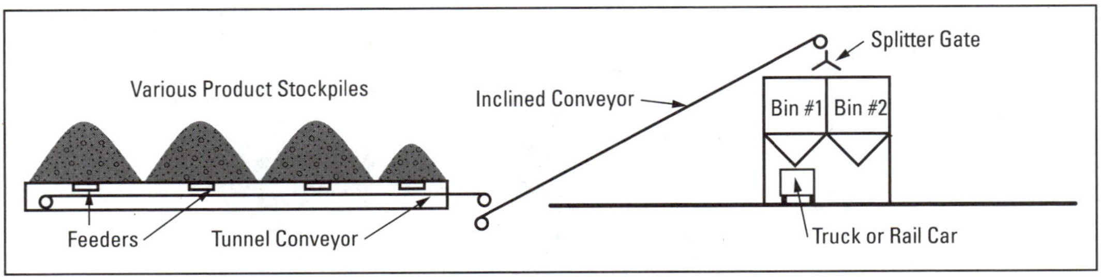

FIGURE 15.6 Bin load-out system fed from a tunnel conveyor

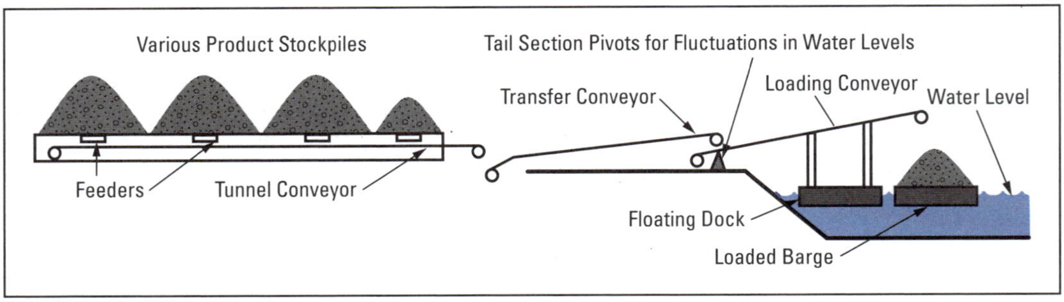

FIGURE 15.7 Barge loading via conveyor system

sprayed a dust suppressant on the coal in an entire 100 car unit train while it traveled at 5 miles per hour (8 kilometers per hour). Other aggregate plants have truck load-out bins that are fed computer-blended aggregates from several stockpiles via a conveyor in a tunnel underneath the stockpiles. When the truck enters the plant gate, the order is sent to the load-out computer and a custom-blended product is waiting in its assigned bin when the truck gets to the load-out bin. These are very sophisticated systems and are not common in most aggregates facilities.

Barges can be loaded directly by rear dump trucks using a correctly engineered heavy-duty dock securely attached to the bank. When loading barges via conveyor, the conveyor is generally in a fixed position with the head section mounted on a barge moored to the shore (floating dock). The tail section is mounted on a vertical pivot, which allows the conveyor to float up and down with changes in the water level in the river (Figure 15.7). As the loading conveyor discharges, the barge is moved forward using a winch or a tow boat to distribute the load evenly on the barge. Telescoping stacking conveyors are ideally suited for ship load-out. The telestacker is positioned on the shore and telescopes outward over the various holds in the ship.

For the most part, aggregates are stockpiled using inclined conveyors from the processing end point to the storage yard. The aggregate piles are then loaded into customer trucks utilizing the industry standard wheel loader. Fixed conveyors will yield a cone-shaped pile limited in size by the discharge height of the conveyor feeding them. For larger stockpiles, radial stacking conveyors have the ability to make kidney-shaped piles up to approximately 270° of arc. These radial stackers have large stockpiling capacities but also take up a considerable amount of real estate (Figure 15.8). The conveyor has a pivot point under the tail pulley and the undercarriage of the conveyor has wheels set at an acute angle to the conveyor, allowing it to travel in a circular motion around the pivot while conveying material onto the ground. The wheeled undercarriage is usually powered by a low-horsepower electric-drive motor, allowing the conveyor to move under its own power. The conveyor movement can be manually operated by the

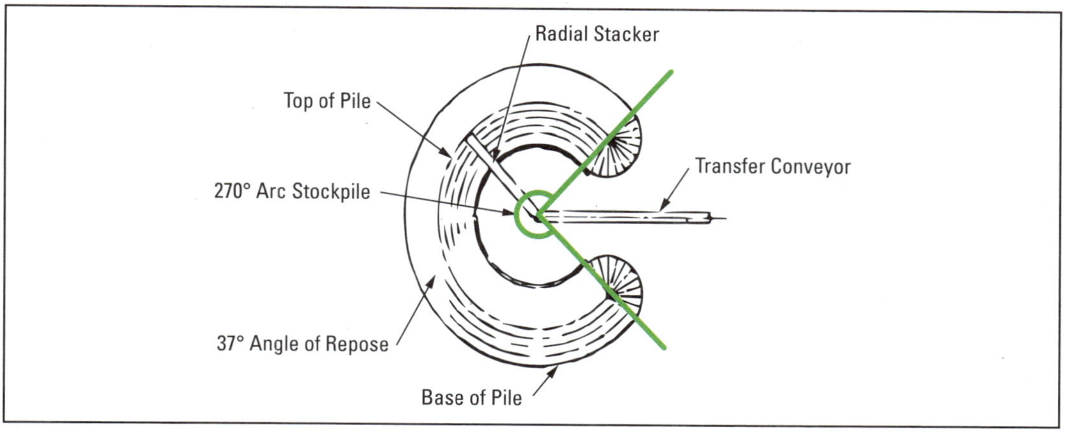

Original source unknown

FIGURE 15.8 Radial stockpile

plant operator or automated using a tilt switch that detects when the pile at a particular point is full and activates the drive motor to move the conveyor several feet forward or backward to continue filling the kidney-shaped pile.

The following formulas will enable you to estimate the volumes of stockpiles. Volumes of conical stockpiles in cubic yards can be calculated as follows:

When the height and the radius of the base are known

$$V_1 = r^2 \times H \times 0.039$$

When the height of the pile and angle of repose are known

$$V_1 = \frac{H^3 \times 0.039}{\tan\theta}$$

When the height and the slope length are known

$$V_1 = H(S^2 - H^2) \times 0.039$$

For radial stockpiles, the volume in cubic yards for 1° of arc of the pile with a 37° angle of repose can be calculated as follows:

$$V_2 = \frac{H^2 \times R_P}{1,160} = \text{volume of a } 1^\circ \text{ arc section}$$

$$\text{volume of the entire pile} = (V_2 \times \text{actual degrees of arc}) + \text{volume of conical ends of the pile}$$

where

V_1 = volume of a conical pile, cubic yards (use the unit weight of the material in question to convert cubic yards to tons)
r = radius of pile base
H = height of pile
S = slope length
θ = angle of repose of material
V_2 = volume of a 1° arc segment of a radial pile
R_P = radius of the pile arc

FIGURE 15.9 (A) Telescoping radial stacker producing a nonsegregated stockpile, and (B) telescoping stacker folded for transport

In Chapter 7 on quality control, we discussed the problems associated with material segregation caused by falls from high conveyor discharge heights. The newest innovation in radial stacking is the telescoping radial stacking conveyor, also known as a TeleStacker (registered by Superior Industries) (Figure 15.9). Telescoping conveyors consist of two individual conveyors contained within the main conveyor truss; the main conveyor in the rear is fixed at the pivot point and the stinger conveyor moves in and out on rollers, which extends the length of the conveyor. The main conveyor feeds the stinger conveyor as it extends and retracts. The discharge height of the conveyor adjusts hydraulically from about 9.5 feet (2.9 meters) above the ground to as much as 72 feet (22 meters) depending on the total length of the conveyor with the stinger extended. Telescoping stackers are also equipped with radial drive on the wheels, much like standard radial stackers. By controlling the stinger movement, the height adjustment, and the radial movement of the entire system with PLC controls, a number of stockpiling configurations are available.

Nonsegregated piles can be constructed by lowering the conveyor discharge to eliminate the gravity effect of the long fall of the material from large heights. The PLC can be programmed to make small, short stockpiles (6–15 feet [1.8–4.6 meters] tall) for the entire arc or a portion thereof; the stinger is then moved out and another low pile created at a longer distance from the pivot point. After filling the entire pile footprint, the discharge is raised a few feet and the process repeated a number of times until the full discharge height of the conveyor is achieved. This essentially combines hundreds of small, short piles into a large radial stockpile of material that is much less likely to be segregated than dropping the material from a fixed discharge height of 50–70 feet (15–21 meters). Products with a large range of sizes like base products (2.0 inches [51 millimeters] down to minus 200 mesh) are the most prone to segregation.

The onboard computer controls the entire process and provides numerous other programmable stockpiling options as well. Telescoping stackers come in lengths from 110 (34 meters) up to 210 feet (64 meters) and have the capability to make various geometries of radial stockpiles with volumes up to 400,000 metric tons. The capacities of nonsegregated piles will be less than the total capacity available. Telescoping conveyors are available with 36- to 72-inch (91- to 183-centimeter) conveyor belt widths and have capacities of up to 5,000 metric tons per hour. Optional features include road portable models, different undercarriage styles, four-wheel drive models for operation on uneven ground, track-mounted models, wireless remote PLC control, galvanized steel structures, and models having self-contained generator power. When folded for roadway transport, the machines range in height from 12.5 to 15 feet (3.8 to 4.6 meters) and in length from 60 up to 113 feet (18 up to 34 meters). When having a 150 foot (46 meter)–long (100 feet [30 meters] folded) model delivered to a site in central Ohio, we kept one intersection (and the local police constable) in a small town blocked for 90 minutes until the truck driver finally managed to negotiate the turn between buildings, power poles, sidewalks, traffic signals, and pedestrians.

Telescoping radial stacking conveyors are extremely versatile, very sophisticated machines with lots of stockpiling and loading capacity. The multitude of electronic sensors, limit switches, and controls require additional inspections and maintenance compared with normal conveyors. Several manufacturers make telescoping radial stacking conveyors that differ somewhat in their designs, undercarriages, stinger suspension, stinger drives, and computer capacities, but they are all very capable machines. The methods of advancing and retracting the stingers vary from cable winch drives to chain drive systems. The telescoping radial stacking system was originally designed in Canada in 1992 by THOR Global.

Installing tunnels containing feeders and conveyors underneath stockpiles or surge piles provides a controlled system to withdraw products from these piles. This system can be used to blend products from different piles (single-sized aggregates) or feed different stages of the plant independently from other stages (surge piles). Surge piles are constructed primarily to assure consistent feed rates to the processing plant to increase efficiency and consistency, maintain product shape, and increase uptime. Feed rates from the quarry to the primary crusher tend to be somewhat erratic and vary because of truck timing, changes in feed gradation from the different sections of the shot, weather conditions, haul road conditions, and other unforeseen quarry issues. It is always a good idea to have a significant surge pile (two to three shifts of secondary production) between the pit/primary crushing operations and the secondary processing. Surge piles provide a large degree of flexibility in the processing plant. Primary and pit production can be done on first shift during daylight hours while maintenance is being performed on the remainder of the plant. There is a definite advantage to performing maintenance when it is light outside and parts are more readily available then during overnight hours. Secondary processing can then be accomplished on another shift. This also provides a production buffer if a maintenance issue arises requiring the primary crusher or the secondary plant to be out of service for a period of time.

When loading via tunnel conveyor, only a portion of the total stock or surge pile above the tunnel is readily available to convey. A portion of the stockpile will remain as dead storage. If the angle of repose of the material in a conical pile is 37.5° (typical for most aggregates), then the live draw (live storage) will be an inverted cone within the pile, above the feeder, with the

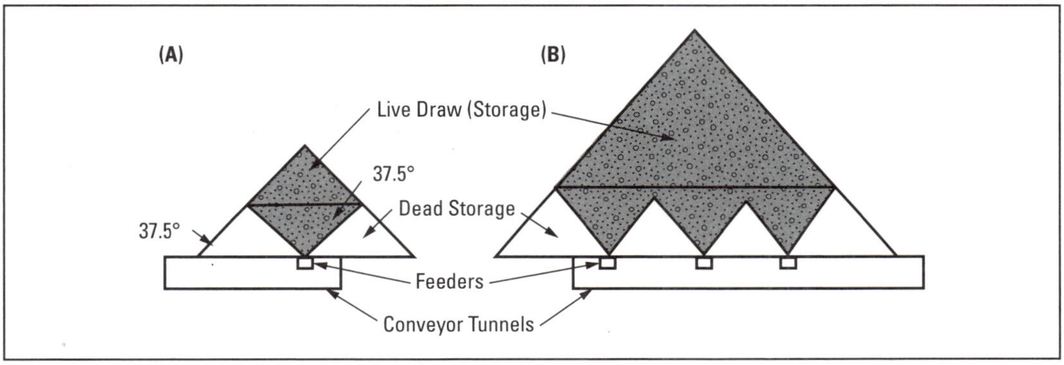

FIGURE 15.10 Live draw from stockpiles: (A) single feeder, and (B) three feeders

FIGURE 15.11 Conveyor replenishing live feed in a surge pile

same slope as the angle of repose of the pile (Figure 15.10). Large piles, especially surge piles, usually contain multiple feeders to increase the amount of live draw available to be pulled from the pile (Figure 15.11). If additional feed is required from a pile, a wheel loader or bulldozer can be utilized to push the dead storage to the feeder(s).

When planning a new processing facility, it will soon become obvious that a considerable amount of acreage will be necessary to encompass the processing plant, material conveyors, truck access, roadways, and especially the stockpiles. One 270° arc radial stockpile produced by an 80 foot (26 meter)–long stacking conveyor will occupy over an acre of ground. A similar pile from a 150 foot (46 meter)–long radial stacker occupies 4+ acres (1.6+ hectares). Piles produced via telescoping radial stackers can be appreciably larger, requiring even more area. Not all products will require the same size stockpiles because this is dependent on the plant's capacity to produce each individual product and the anticipated inventory needs based on the sales/production mix.

CHAPTER 16

Plant Design and Portable Plants

PROCESSING PLANT DESIGN

Now that we've discussed the basic components of a processing plant, the various pieces need to be put together to make an efficient, cost-effective, processing facility. Final plant design is a field that should be left up to the professional engineers that have specific experience in the intricacies of aggregates processing. But prior to the detailed engineering, a basic plant flow needs to be established. At this point we come back to the idea of the sales/production mix. To start the process, we need to know the total projected annual sales, the anticipated sales volume of each individual product needed, and the time frame of the specific product sales. Processing plant design needs to engineer flexibility into the system to allow for making many different products. All plants have different modes where base products are made more efficiently or concrete aggregate is prioritized. Product gradations can be altered by changing screen cloth size on one or more screen decks, which is labor-intensive and time consuming, but the same result can be achieved by designing blending gates into the plant system to blend percentages of material from one deck into another product either downstream or sometimes upstream. High percentages of finer products in the sales mix will ultimately dictate the total plant capacity. If considerable tonnages of No. 8 (⅜ inches [0.95 centimeters] × 8 mesh) for asphalt surface mix are sold, additional crushing capacity will be necessary to crush more material down to this size range.

Some processing plants are designed to produce graded products directly from the screens, while other producers prefer to make single-sized aggregates (1.5 × 1 inches [3.8 × 2.5 centimeters], 1 × ¾ inches [2.5 × 1.9 centimeters], ¾ × ⅝ inches [1.9 × 1.6 centimeters], etc.) using fractionated plants. In fractionated plants, single-sized materials are then blended back together to meet the specific gradation requirements. Single-sized aggregate plants require a higher capital expenditure because more screens are necessary, as are tunnels under stockpiles, tunnel feeders, additional conveyors, and computerized reblending systems. The advantages of fractionated plants are improved quality control by minimizing segregation, producing more consistent product gradations, and increased flexibility to respond to changing customer specifications. Graded versus fractionated plant design is a function of available capital, specifications, quality considerations, and management philosophy. Base products requiring fine crusher dust for compaction are generally made shortly following the primary crushing to take advantage of the fines generated in the quarry blasting. The oversized product scalped from the base product is further crushed and screened to make the clean (dust-free) aggregates products.

The first step in the plant design process is the selection of the primary crusher type and size and is based on the geology, rock type, most efficient reduction ratio, production rate requirements, and initial feed gradation (Table 16.1). Once this has been determined and the primary reduction ratio is established, the secondary and tertiary crushing types and sizes can be determined. An understanding of open versus closed circuit crushing and recirculating load is a necessity at this stage in the design (Figure 16.1). Open circuit crushing is when the oversize material remaining in the crusher discharge is sent onward in the process to another screen, a stockpile, or different crusher. In a closed circuit situation, the crusher discharge is sent to a screen with the top deck screen opening similar to the crusher setting. The oversized material (rejects) from the top deck of the screen is then sent back to the same crusher to be re-crushed. This is known as the recirculating load.

Primary crushers should always be operated in an open circuit arrangement so the recirculating load does not inhibit the total production infeed to the plant. Closed circuits are normally restricted to the final (tertiary) stages of crushing where no oversized material is wanted in the finished products. Both screening efficiency and crusher settings will affect the amount of recirculating load. As the recirculating load increases, the crusher becomes less efficient because it can no longer pass the same amount of virgin crusher

TABLE 16.1 Typical shot rock gradations

	Percent Passing	
Sieve Size	**Limestone Quarry Shot**	**Granite Quarry Shot**
24 inches	100	94.0
17 inches	93.7	—
12 inches	93.1	54.0
8.5 inches	58.8	—
6 inches	48.2	33.0
4.5 inches	37.9	—
3 inches	29.2	28.0
1.5 inches	14.8	15.0
1 inch	10.0	12.0
¾ inch	7.6	10.0
⅜ inch	4.0	8.0

Source: USACE

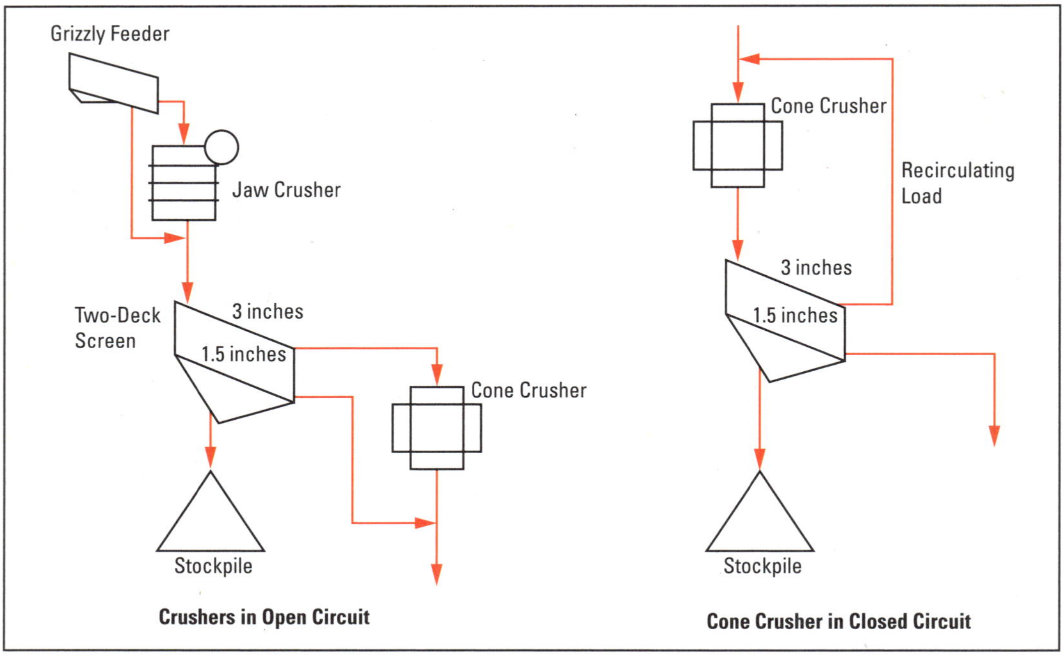

FIGURE 16.1 Open versus closed crushing circuit

feed. Large or increasing recirculating loads are indicative of the need to tighten the crusher setting. If the screen efficiency and the percentage of oversized material in the crusher product are known, the percentage of recirculating load can be calculated

$$RL = \frac{100}{(SE/OS) - 1}$$

where

RL = recirculating load
SE = screen efficiency, %
OS = percent of oversized material in crusher discharge

Assuming SE = 92% and OS = 18%, then

$$RL = \frac{100}{(92/18) - 1} = \frac{100}{4.1} = 24.4\%$$

Because crushers have both an open side setting and a closed side setting, no crushers are capable of producing a product without a certain percentage of oversized material in their discharge. Although, choke feeding a crusher will help to increase this efficiency. If this oversized material is not needed in a finished product, it must be sent either to another crusher or be recirculated back to the same crusher for further size reduction.

Any screen running at or near capacity will not be 100% efficient because of bed depth, particle adhesion, moisture content, and stratification. These efficiencies will need to be taken into account during the plant design process. Screens should be sized to achieve efficiency levels in the range of 90% whenever possible. The bottom deck of the screen dictates the efficiency and capacity of the entire screening unit because it will always have the smallest screen openings and least amount of open area, and thus will pass fewer particles. The capacity of a screen is dependent on the width of the screen deck, while efficiency is a function of the screen length. The longer the deck dimension, the longer the material will remain on the screen and the more opportunities for the particles to fall through the screen openings. Screening efficiency was discussed earlier in the Chapter 14 on screening, washing, and classifying.

Most equipment manufacturers publish theoretical product gradations produced from their crushing equipment for various different rock types, at different capacities, and different crusher settings. This type of data is very helpful in planning and some manufacturers also have computer software available to assist in the plant design process. An excellent resource for plant flow simulations and design purposes is a computer software package called AggFlow. AggFlow contains an extensive database of equipment from many different manufacturers and can calculate required screen sizes, provide theoretical product gradations, estimate conveyor sizes, run multiple different production modes, and calculate production volumes of individual products. This is an excellent tool to evaluate different combinations of equipment, and do what-if scenarios by changing screen types and sizes, crusher types and sizes, crusher settings, screen cloth, and so on. If planning a new plant installation or alteration of an existing plant, an internal analysis of the desired plant flow can be beneficial prior to meeting with plant design and structural engineers.

If you have a basic knowledge of aggregates processing, the AggFlow program is very intuitive and easy to use, and gives accurate estimates of capacities, production rates, and expected product gradations (Figure 16.2). The software allows you to choose individual crusher types

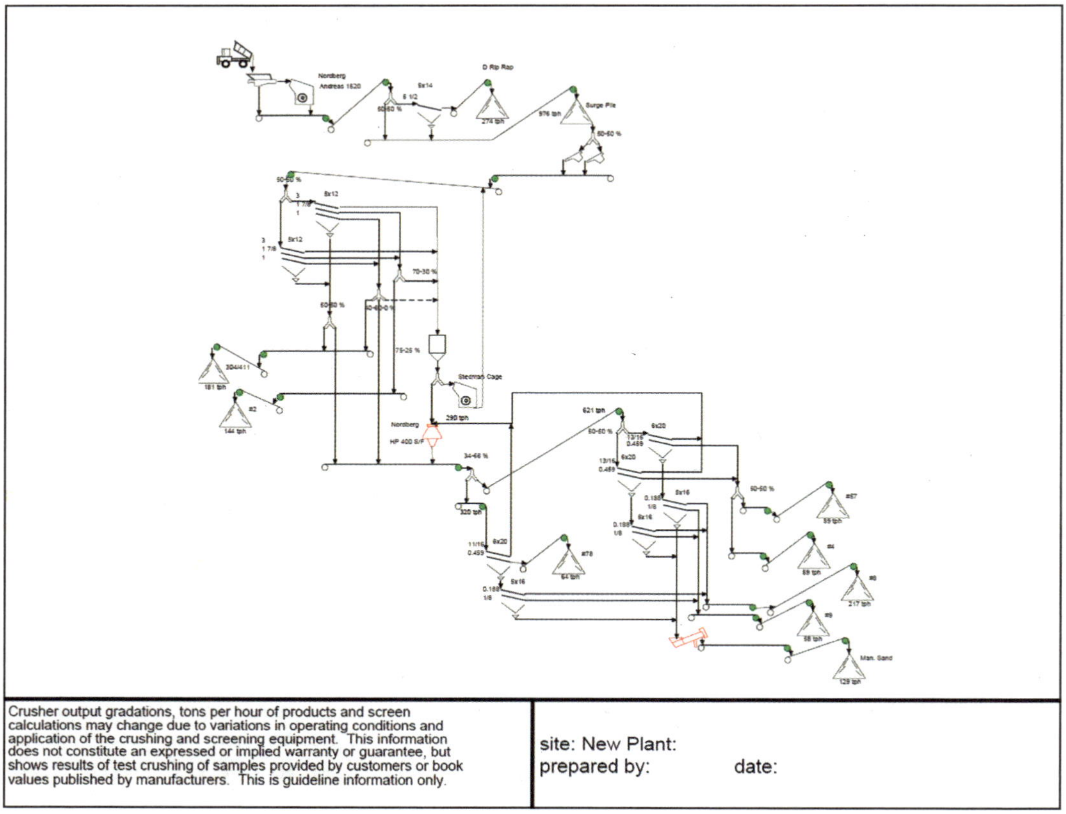

FIGURE 16.2 Plant flow for a 1,000 metric tons per hour limestone plant generated on AggFlow software

and sizes, crusher settings, feeders, vibrating screen types, screen media sizes, proportional feed splitters, washing equipment, classifying equipment, and conveyors, and links the chosen components together to create a plant flow. If the capacity of a specific crusher is overloaded or the bed depth on a screen deck exceeded, the software provides an immediate warning when you push the "run plant" button. Equipment components in the flow will turn red if an overload situation arises. AggFlow is a subscription-based software that requires an annual fee.

There is no substitute for processing experience in the plant design process. A good plant design must always integrate equipment maintenance considerations into the process engineering. Ease of maintaining equipment is essential to achieving optimal running time, productivity, and efficiency. I took over management of a limestone quarry in the early 1990s only to find that the bearings on the primary crusher needed to be replaced (a vital necessity to keep the plant operational). The only problem was that to remove the rotor from the horizontal shaft impact (HSI) crusher, the entire overhead hopper had to be removed. This turned a 1-day maintenance project into a 3-day ordeal with 2 days of unnecessary downtime. The design of the overall plant process was very good, but the design engineer failed to consider the need to remove the HSI rotor for maintenance. Little design flaws without consideration of maintenance needs can be very costly over the life of a processing plant. Raising the tail pulleys of all conveyors above the ground by 2 feet (0.6 meters) costs very little when constructing a new plant but saves countless hours in cleanup over the life of the plant. Only direct processing experience can integrate the many small maintenance time-saving issues into the final plant design. Unnecessary unplanned downtime, no matter the reason, is disastrous to plant profitability. Processing plant design needs to focus on the required product mix, screening efficiency, productivity, and minimization of waste (nonsalable) products.

ELECTRICAL CONSIDERATIONS

Up to 30% of the cost of a new processing plant will be tied up in the electrical system. This estimate does not include bringing three-phase high-voltage power to the plant location if it is not already available. Power can be generated using diesel generators, which may be the only option in certain areas where three-phase power is not available within a reasonable distance. The cost to generate power is usually about 1.5 to 2 times greater than the cost to purchase power from a local power company depending on the location and the power utility. Generator power cost is dependent to a large extent on the price of diesel fuel, but the generators also require additional capital investment, fuel tanks, regular maintenance, and periodic rebuilds.

When purchasing electricity from a power company, an industrial power contract is typically required. Purchasing primary power is usually the most cost-effective but requires that the producer provide its own power transformers, which adds additional capital cost. Secondary power contracts normally come at a higher cost per kilowatt hour but the transformers are usually provided by the power company, lowering the capital investment. Most power contracts come with demand charges, which can ultimately be the most expensive portion of the bill. The demand charge is based on the largest monthly power demand (highest instantaneous usage) incurred by the producer during a 1-year period. Demand is measured in kilowatts and is designed to recover the cost of the utilities' infrastructure (such as power transformers, wires, and power plants) required to provide the maximum amount of electricity the aggregates producer may require at any time. Power factor is also a component cost in many power contracts. Power factor is a calculation of how efficiently a producer utilizes the power provided by the utility and is expressed as a percentage, with 100% being the best. Many power contracts have a penalty if the producer's power factor falls below a certain level (usually 70%). Low power factor can be corrected by installing capacitor banks into the electrical system to increase the efficient usage of the power provided by the utility company.

Many power companies offer cost-saving incentives (lower cost per kilowatt hour) if the producer uses off-peak power, which typically requires operating the processing plant during the overnight hours when normal power demand lessens. Interruptible power contracts with lower rates per kilowatt hour are sometimes available as well. Interruptible contracts require the aggregates producer to cease production (all power usage) during periods of extremely high power demand, at the discretion of the utility company. This usually happens during the hot summer months when air conditioning usage is at its peak, which also corresponds to the time frame when demand for aggregates product is the highest. This becomes a trade-off between power cost savings and a willingness to cease production at a moment's notice during times of peak aggregates demand. A cost–benefit analysis is necessary to determine the cost-effectiveness of interruptible power contracts.

A good deal of the power usage by an aggregates plant comes from starting large-horsepower motors across the line. Direct-on-line starters apply power instantaneously to the motor with a surge of inrushing current accelerating the motor to nominal speed. Across-the-line starting of alternating current induction motors causes a current surge (instantaneous demand) 7–10 times the normal running current and a starting torque of up to three times greater than the

FIGURE 16.3 Electrician wiring a complex motor control center

normal operating torque. This not only raises power demand, it is hard on the motor windings, the motor shaft, and the wiring, can cause premature motor failure, and adds to the electrical maintenance costs. When designing motor control centers (MCCs), soft starts should be incorporated on all motors 50 horsepower or larger. Mechanical soft starts are available that include various types of clutches, fluid couplers, magnetic couplers, and steel shot couplers, which are generally mounted to the equipment drive shaft. Electrical soft starts contain thyristors, also known as silicon-controlled rectifiers (SCRs), which apply current to the motors at a controlled ramp-up rate, reducing the surge of excess current, the torque load on the motor shaft, and the electrodynamic stress on the power cables. Variable-frequency drives (VFDs) are a type of programmable soft start where the speed and torque of the motor are controlled by varying the incoming frequencies and voltage usually via programmable logic controller (PLC). VFDs also sense the operating load on the motors and can vary the motor speed based on the load, reducing power consumption. Any combination of these soft starts will significantly lower power demand (costs) and increase the life of electrical components. When consulting an electrical engineer on MCC design, a producer should stress electrical safety, electrical component life, power demand reduction, and total power cost management (Figure 16.3). Power cost can be a significant factor in the total cost of producing a ton of aggregates.

AUTOMATION

Automation is an important component of any new aggregate plant and there are companies that specialize in automating aggregates processing plants. Automation is a very sophisticated process that requires extensive knowledge of the mechanical systems, as well as sensing devices, proximity devices, electrical controls, and computerization. Automation can be limited to certain components (crushers, conveyors, etc.) or applied to the entire plant including the opening and closing of blending gates. There are actually a couple aggregate plants in the United States that are totally automated and require no personnel to operate. One midwestern producer operates the quarry and primary crusher during the daylight hours, filling a surge pile. The night shift is fully automated, with the secondary plant crushing, screening, classifying, and stockpiling, until no more surge feed is available. This is all accomplished with no operators on site. If a mechanical failure occurs during the automated shift, the plant shuts down and calls the plant supervisor to notify the supervisor of the problem. I personally would not feel comfortable without someone on site during an automated shift just to monitor the processing situation and be able to troubleshoot any problems that might arise.

As with any option, automation comes with a significant price tag and a cost–benefit analysis should accompany any automation plan. The benefit of eliminating the labor of two or three people may not outweigh the capital cost of total automation. At a minimum, automated start-up and shutdown sequences, electrical interlocking of components, control of crusher settings, and controlled infeed rates are necessary automation parameters. Plant operators have a tendency to run the processing plant at some psychological comfort level where they feel they are in control. This comfort level is generally 10%–20% less than the actual plant capacity. Processing plants need to be operated at peak capacity to maintain consistent product gradations, optimum product shape, and most of all minimize the cost per metric ton produced. Well-planned automation systems can eliminate the human error factors and comfort level issues associated with human operators.

I once had a plant operator named Eugene at a 1,000 metric tons per hour crushing plant. Eugene had a comfort level where he wanted to operate the plant at 850–900 metric tons per hour because he didn't like the idea of cleaning material spillage if something in the process went awry. Every time I drove past the plant, I'd call Eugene on the radio and ask: "Eugene what do the belt scales read?" He would reply: "900 tons per hour boss!" My retort was always the same: "Crank up the feed to 1,000 tons!" I would wait until I saw the level of feed on the belt increase and go about my other business, only to return an hour later to repeat the process. Once while attending a managers meeting in Macon, Georgia, a fellow superintendent was demonstrating a new automation system recently installed on his plant in Columbus, Ohio, that he was able to access remotely on his laptop computer. We found the plant operator running the plant at his comfort level about 15% under the optimum production rate. So, without the operator's knowledge, we increased the infeed (via the laptop) to the proper rate using the computer interface. You could almost visualize from 650 miles away the operator's mind working when he realized the feed rate had increased. He would back down the feed rate and a few minutes later we would increase it once again. We played with the operator clandestinely for about 15–20 minutes and he probably never knew what was causing his frustration with the fluctuations in the feed rate. At that meeting we came to the conclusion that we should probably install manual feed rate controls that did not actually function but allowed the plant operators to believe they controlled the feed at their own predetermined comfort level, while the actual feed rate was being computer controlled.

Existing processing plants can be retrofitted with automation, but this is much more difficult to accomplish and more costly than automating a newly designed facility. None of the control (low-voltage signal) wiring or sensors and proximity devices are available on older plants and can be a challenge to retrofit. New processing facilities should always include budgetary capital for at least a minimum of automation. When our company built a new pulverized limestone facility a few years ago, we totally automated the process. We were utilizing enclosed roller mill systems consisting of a feeder, roller mill, flash dryer, system fan, cyclone, and bag house that were controlled by differential pressures within the system. We were able to enter one of several different recipes that produced the various product requirements. At the start of the shift, the operator enters a predetermined recipe for each roller mill, specifies which silo the product is destined for, pushes the start button, and walks away from the controls until the end of the shift or a product change is required. The automation produced very consistent product gradation and sizes that were sent to the correct storage silo for future load-out. The pneumatic truck loading system was totally automated and the bagging operation was partially automated with an automatic bag feeder, palletizing robot, and shrink-wrap system. This was a very successful automation project that saved substantial labor costs but came with a significant capital investment.

Well-thought-out plant design incorporating ease of maintenance is crucial for efficient processing. Plant designs should be based on product mix, processing efficiency, waste (fines) minimization, and flexibility for future expansion. To minimize the production of excess fines, crushers with high reduction ratios should be used early in the process, with smaller reduction crushers in the secondary and tertiary stages. Automation should be designed to prevent the human error factors from inhibiting the productivity. A well-thought-through and engineered crushing and screening plant can reduce the costs of maintenance, power usage, and labor, providing increased profitably.

PORTABLE (MOBILE) PLANTS

Two types of portable processing plants exist, in-pit portable and road portable (mobile) processing plants. Mobile plants are limited in size to maintain ease of portability and meet the requirements of highway transportation limitations; height, width, and weight restrictions. In-pit portable plants are unlimited as to crusher size, total weight, and exterior dimensions (See Figure 10.6B). The only limitation to the size of the in-pit components would be the weight-bearing capacity of the rock beneath the quarry floor. Normally only the primary crushing function needs to be portable if the entire processing plant does not need to move from location to location (Figure 16.4). In-pit portable primary crushers can eliminate and/or limit the amount of hauling with off-highway trucks, which can provide a substantial savings in operating costs. Any typical type of primary crusher (jaw, gyratory, HSI) can be mounted onto a portable chassis. The largest types of in-pit portable crushers are mounted on either crawlers or walking mechanisms similar to large draglines.

FIGURE 16.4 Mobile primary feeding an inclined field conveyor

Mobile portable plants will require multiple component plant sections tied together by road portable conveyors to be transported via highway from location to location. Road portable conveyors generally have folding head and tail sections to reduce the length and will have highway-approved braking and safety light systems. Highway restrictions as to weight, height, and width vary from state to state and many mobile plants will require escort vehicles to be highway legal. Numerous mobile aggregates processing plants are available for purchase and their various components need to be matched to fit the product mix required and the deposit to be processed. By their very nature, mobile plants are very maintenance intensive. While fixed processing plants can be designed for ease of maintenance with large chutes, conveyors raised for ease of cleanup, walkways added around conveyors for access, and so on, mobile plants have to be engineered for compactness. This compaction makes maintenance much more difficult because access to crushers, chutes, feeders, electrical parts, motors, speed reducers, conveyor pulleys, and such is more problematic.

Most new mobile processing plants come with PLC automation controlling crusher feed, crusher settings, and so forth. Parts for mobile plants are often proprietary to the manufacturer, harder to obtain, and generally more expensive. The size limitations also come with capacity limitations; there is only so large a primary crusher that can be mounted on a mobile chassis and still meet the highway portability and weight limitations. Some mobile plants are mounted on a wheeled chassis with multiple axles and a fifth wheel hitch pin designed to be towed by a semi-tractor; others can be crawler-mounted and hauled from site to site on low-boy trailers.

Mobile plants definitely have their place in today's aggregates business. Operations in rural areas usually have limited markets because of smaller populations. One properly designed mobile plant can service multiple smaller operations by moving between the various sites, putting 100,000–150,000 metric tons of aggregates products into inventory and then moving again. Knowing exactly what products will be sold when and producing the correct product mix can sometimes be challenging in this type of mobile scenario. Many aggregate producers also recycle concrete and asphalt pavement into salable products utilizing mobile processing plants. Recycling is well suited to the use of mobile crushing and screening plants because the volumes needed to be recycled at any one location are usually somewhat limited.

Normally, a mobile primary crusher with a grizzly feeder and discharge conveyor will be mounted on one chassis (Figure 16.5A). A second chassis may contain a feeder, a cone crusher, an under crusher conveyor, and a screen. Large screens with under-screen and side discharge conveyors may require their own chassis (Figure 16.5B). Many different mobile processing plant combination component options are available from several different manufacturers. Do your homework before mixing and matching components from different manufacturers. Make sure things like conveyor discharge heights from one component to another are compatible. I was once caught in this dilemma when purchasing a new mobile secondary cone crusher from a different manufacturer and found that the discharge from the existing primary screen was several feet too high to meet the receiving hopper on the new crushing plant. Luckily, the new cone plant was on a crawler chassis and we could build a small pad to drive the machine onto in order to meet the feed conveyor.

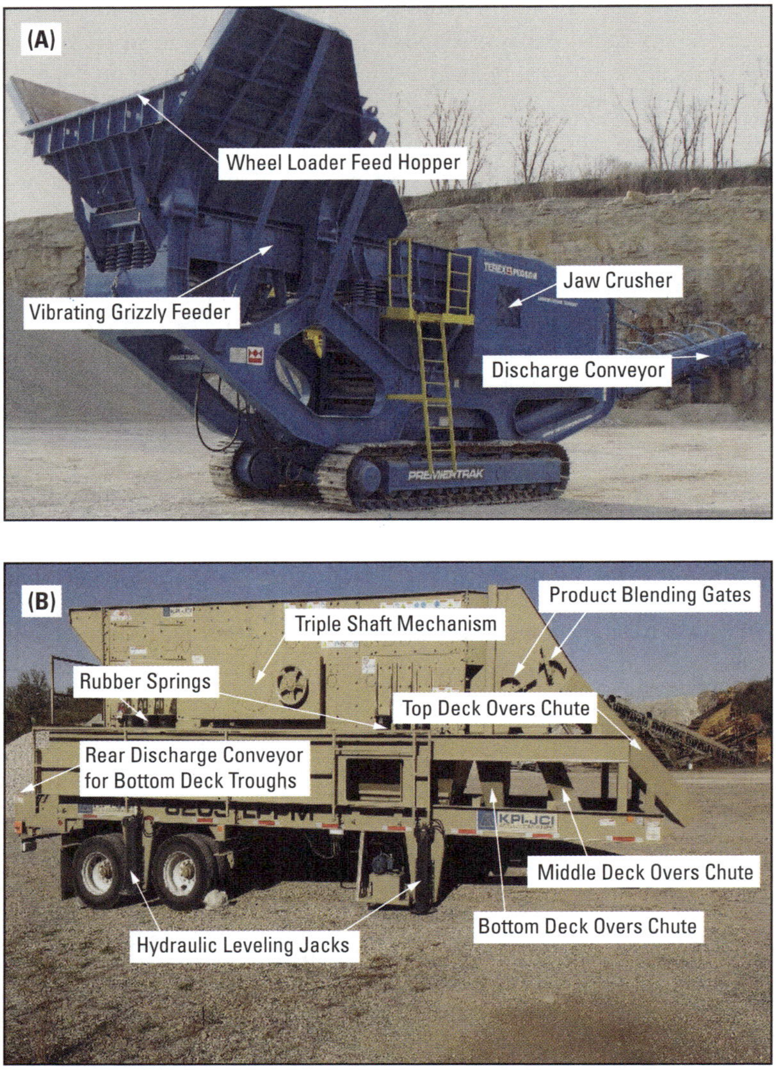

FIGURE 16.5 (A) Mobile primary crusher, (B) mobile horizontal screening plant

CHAPTER 17

Safety Management

The greatest responsibility that managers in the aggregates industry have is to the health and safety of their employees. Safety and health have been the subject of a number of books over the years and the *Aggregates Handbook* published by the National Stone, Sand and Gravel Association (NSSGA) (2013) has a detailed, comprehensive chapter on industry safety and health that is well worth reading. In the late 19th and early 20th centuries, the mining industry had a reputation of being a very dangerous one. When mining fatality statistics were first tracked by the Bureau of Mines (BOM) in 1910, there were more than 3,000 deaths annually in the United States related to mining. Although aggregates constitute only a portion of the entire mining industry, only nine fatalities were reported for the aggregates industry in 2010, the best year on record. The metal and nonmetal mining (MNM) category, in which aggregate safety statistics are tracked, reported an average of 17.6 fatalities annually over the last 10 years. Even nine fatalities out of 97,000-plus industry employees are unacceptable! Since the late 1800s, work conditions, employee training, equipment, processing, and work cultures have changed considerably (Figure 17.1). The National Safety Council now considers the aggregates industry to be one of the safest in the United States. As a comparison, in China, "independent reports continue to be significantly different from official figures. Unofficial data provided by a senior work-safety bureaucrat suggests that as many as 20,000 miners actually die each year. That count does not include the tens of thousands more miners who die of lung afflictions and other mining-related diseases every year. A further obstacle to accurate information comes from the mine operators and owners, who routinely falsify death counts in order to avoid mine closures or fines" (Tu 2007). The U.S. coal industry reported a total of 12 fatalities for the same period.

Courtesy of Special Collections and University Archives at Indiana University of Pennsylvania

FIGURE 17.1 Early miners with no safety equipment

The U.S. Department of Labor has established the Occupational Safety and Health Administration (OSHA) as well as the Mine Safety and Health Administration

(MSHA) to aid industry and business in identifying and controlling industrial hazards and accidents. MSHA was formed as an entity separate from OSHA in recognition of the fact that mining is inherently different than most other industries. Early federal legislation was primarily geared toward regulating coal mining. The first federal mine safety legislation was passed 1891 and prohibited the hiring of children under the age of 12 and established minimum ventilation standards for underground coal mines. The BOM was established in 1910 as part of the Department of the Interior to aid in the development of mineral resources but had no authority to do safety inspections. Safety inspection of underground MNM was first authorized in 1966. In 1973, the Mine Enforcement and Safety Administration was formed as a new department, assuming the safety and health enforcement functions, with the BOM concentrating solely on mineral resource issues.

MSHA, as we know it today, was established by the Mine Safety and Health Act of 1977, which came to be known as the Mine Act. Enforcement responsibilities were transferred to the Department of Labor from the Department of the Interior at that time. The Mine Act included the following:

- All federal health and safety regulations of the mining industry were consolidated under a single statutory scheme
- Four annual inspections of all underground mines were made mandatory
- Two annual inspections of all surface mines were made mandatory
- The rights of miners were strengthened and expanded
- Protection of miners from retaliation for exercising such rights was enhanced
- Mine rescue teams were required for all underground mines

In 2006, the Mine Improvement and New Emergency Response (MINER) Act amended the Mine Act principally to enhance the civil penalties outlined in the 1977 legislation. Aggregate operations are regulated under the MNM section of the MSHA regulations.

MSHA regulations are found under Title 30 of the Code of Federal Regulations (CFR) which can be found at MSHA (n.d.) and are updated annually every July. Part 56 outlines the safety regulations for MNM surface mines, Part 57 outlines the regulations for MNM underground mines, and Part 58 outlines the health standards for MNM. MSHA serves a valuable function in writing and enforcing safety and health regulations and also provides regular safety inspections for both surface and underground MNM. MSHA regulations recognize and incorporate industry consensus standards from the following organizations:

- American Standards Institute (ANSI)
- ASTM International
- SAE International
- National Fire Protection Association (NFPA)
- American Society of Mechanical Engineers (ASME)
- Institute of Makers of Explosives (IME)
- American Welding Society (AWS)
- Institute of Electrical and Electronic Engineers (IEEE)
- National Electrical Code (NFPA)
- Underwriters Laboratories (UL)
- American Conference of Governmental Industrial Hygienists (ACGIH)

TABLE 17.1 Mine Safety and Health Administration metal and nonmetal statistics, 2014–2019

Calendar Year	2014	2015	2016	2017	2018	2019
Number of mines	11,992	11,865	11,823	11,899	11,887	11,853
Number of miners	250,805	247,526	237,408	238,642	249,427	250,404
Fatalities	30	17	17	13	16	15
Fatal injury rate	0.0147	0.0084	0.0088	0.0066	0.0077	0.0072
All injury rate	2.11	2.03	1.94	1.79	1.75	1.72
Total mining area inspections hours/mine	23	23	23	24	23	23
Citations and orders issued	58,500	58,371	56,525	57,843	50,769	55,718
Significant and substantial citations and orders (%)	24	24	22	22	20	20
Dollar amount assessed (millions)	33.8	28.1	26.1	28.7	23.1	26.3

Courtesy of MSHA

These standards along with MSHA regulations should be complied with by all MNM operators. Many MSHA regulations are open ended with multiple interpretations available (Table 17.1). Although many operators find the MSHA inspections intrusive because of their punitive nature (the possibilities of fines and criminal penalties), they do provide regular, biannual safety and health audits.

OSHA has a much larger jurisdictional responsibility by volume of businesses regulated than MSHA. OSHA has safety and health authority over all industrial and commercial facilities in the United States, but has no jurisdiction over any mining operation. An MSHA inspector has the right to enter any mine site in the United States at any time without a warrant. MSHA is the only federal agency that has that authority, indicating the level of importance considered by Congress when the Mine Act was promulgated in 1977. If an aggregate producer were to refuse an MSHA inspector entrance onto its property, it would soon be confronted with a force of U.S. marshals at the front gate. This is the seriousness with which MSHA takes its responsibility.

The MSHA standards also require various levels of training for employees under 30 CFR Parts 46 and 48. Producers must develop their own training plans that need to be approved by MSHA. Various state aggregate associations and NSSGA have developed examples of training plans that can be adapted for use by producers. New, inexperienced miners added to the payroll must go through a combination of 40 hours of classroom and on-the-job training. When operating a new piece of equipment or doing an unfamiliar job function, task training for the inexperienced employee by a person experienced with that equipment or job is required and needs to be documented in the employee's personnel file. A minimum of 8 hours of annual safety and health refresher training is another requirement for all employees, new or experienced, both hourly and salaried. Site-specific hazard awareness training is required for any individual entering a mine site, including vendors, truck drivers, visitors, delivery personnel, and family members. It is the responsibility of the mine operator to provide site-specific hazard training and document the training of any subcontractor employees on the site, as well. Subcontractor employees need to have MSHA-approved miner training and training plans, all of which needs to be documented by the producer before allowing any work to proceed.

MSHA regulations require that all personnel on a mine site be equipped with personal protective equipment (PPE). Required PPE includes hard hats, safety glasses, steel-toed shoes, and hearing protection. Many producers go a step beyond and require high-visibility reflective vests even during daylight hours. Other PPE is required when doing specific tasks, like face masks when grinding, life vests near water, welding helmets, gloves, and special protective clothing when working on electrical circuits to protect employees from arc flash. All installed safety devices on mobile equipment must be kept in working order at all times. Mobile equipment needs to be equipped with backup alarms to warn other vehicles of rearward movement. All machinery parts must be provided with guarding to prevent any possible physical contact with moving parts such as pulleys, belts, drive shafts, and drive chains. Handrails with toe boards must be installed and maintained along any elevated walkways, platforms, and stairways, as well as ladder hoops to prevent falls from installed access ladders. Fall protection (personnel harness and lanyard) must be utilized when it is necessary for an employee to be outside of a handrailing in an elevated area. Emergency stop devices must be installed along all conveyor systems.

Lock-out, tag-out systems must be utilized when working on any mechanical equipment. The energy source of any piece of equipment must turned off and locked out prior to any maintenance activities. Each employee working on the equipment must individually lock out the power source with their own lock to assure the equipment cannot be energized. No employee should attempt to operate any equipment unless the employee has been tasked trained on that particular equipment. Unless properly trained, no employee should handle explosives or attempt to work on electrical circuits. Seat belts must be worn while operating any mobile equipment, and all warning signs and traffic regulations obeyed. All haul roads need to be bermed and barricaded to prevent mobile equipment from accidently leaving the roadways. Procedures and warning signals need to be developed to barricade and prevent accidental access to active blasting areas. These are the type of regulations that are found within 30 CFR Parts 56 and 57. The regulations are too numerous to reiterate here and an interested person should consult MSHA (n.d.) for the entire set of extensive regulations. Part 57 (underground) requires two additional annual inspections beyond the two required by Part 56 and many additional ground and roof control regulations.

MSHA routinely tracks citation statistics and will periodically focus enforcement on specific issues if the frequency of specific citations and/or accidents become excessive (Tables 17.2–17.4). These special enforcement initiatives will require producers to focus their inspections on the same specific areas, such as guarding, fall protection, or berms. MSHA citations are classified as significant and substantial (S&S) or non-S&S at the time of issuance depending on the circumstances, negligence, and potential danger perceived by the inspector. S&S citations come with larger fine assessments from MSHA than do non-S&S citations. Citations and fines can be appealed by the producer within a specified time limit and any questionable citations should be appealed immediately. Although somewhat rare, MSHA has the power to file criminal charges and impose monetary fines against individual employees if the inspector and/or an investigator feels that intentional negligence existed at the time of violation.

In 2003, MSHA and NSSGA entered into MSHA's first alliance with an industry association to promote mining safety and the terms of the alliance have been renewed five times since then. The alliance page on the msha.gov site says "Both groups will work to promote a national dialogue on mine safety and health by sharing information on best practices or effective approaches to eliminating mining hazards, as jointly determined by MSHA and NSSGA. The alliance will conduct this dialogue with others in the mining industry through individual or joint outreach" (USDOL 2019). Various training materials and several accident

TABLE 17.2 Top 10 Mine Safety and Health Administration violations for surface stone operations, 2018

30 CFR	Subtitle	Violations	Percent of Total
56.14107(a)	Moving machine parts	2,010	1.02
56.12004	Electrical conductors	1,799	0.91
56.14100(b)	Safety defects; examination, correction and records	1,631	0.83
50.30(a)	Preparation and submission of MSHA Form 7000-2—Quarterly Employment and Coal Production Report	1,340	0.68
56.20003(a)	Housekeeping	1,327	0.67
56.14132(a)	Horns and back-up alarms	1,296	0.66
56.11001	Safe access	971	0.49
56.12032	Inspection and cover plates	937	0.48
56.14100(c)	Safety defects; examination, correction and records	936	0.47
56.14112(b)	Construction and maintenance of guards	833	0.42

Courtesy of MSHA

TABLE 17.3 Top 10 Mine Safety and Health Administration violations for underground stone operations, 2018

30 CFR	Subtitle	Violations	Percent of Total
57.3200	Correction of hazardous conditions	122	0.12
57.20003(a)	Housekeeping	119	0.12
57.12004	Electrical conductors	110	0.11
57.14107(a)	Moving machine parts	82	0.08
57.14100(c)	Safety defects; examination, correction and records	68	0.07
57.14100(b)	Safety defects; examination, correction and records	59	0.06
57.12032	Inspection and cover plates	57	0.06
57.11001	Safe access	49	0.05
57.14112(b)	Construction and maintenance of guards	47	0.05
57.14207	Parking procedures for unattended equipment	45	0.05

Courtesy of MSHA

TABLE 17.4 Top 10 Mine Safety and Health Administration violations for surface sand and gravel operations, 2018

30 CFR	Subtitle	Violations	Percent of Total
56.14107(a)	Moving machine parts	1,345	1.38
56.12004	Electrical conductors	949	0.97
50.30(a)	Preparation and submission of MSHA Form 7000-2 —Quarterly Employment and Coal Production Report	826	0.85
56.14132(a)	Horns and back-up alarms	699	0.72
56.14100(b)	Safety defects; examination, correction and records	594	0.61
56.12032	Inspection and cover plates	474	0.49
56.12028	Testing grounding systems	459	0.47
56.14112(b)	Construction and maintenance of guards	430	0.44
56.4201(a)(2)	Inspection	414	0.42
56.11001	Safe access	408	0.42

Courtesy of MSHA

alerts have been generated through this alliance. NSSGA says: "NSSGA's commitment to safe and healthful workplaces, as well as full compliance with related standards, is illustrated by the association's leadership in the first alliance ever entered into by the U.S. Mine Safety and Health Administration (MSHA). The MSHA-NSSGA Alliance was established to forge a public/private partnership to encourage training and education that helps operators do the best possible job of safeguarding workers" (NSSGA, n.d.). NSSGA at its annual meeting recognizes producers with excellent safety records via safety awards and one individual receives the Safety Professional of the Year Award.

The BOM was moved in 1997 from the Department of the Interior and incorporated into the National Institute for Occupational Safety and Health (NIOSH), which is a part of the Centers for Disease Control and Prevention (CDC). (That's a lot of government acronyms!) The BOM stills does a considerable amount of research on mining health subjects, as well as safety issues, especially pertaining to underground mines. CDC (n.d.-b) contains volumes of publications, research papers, and statistics on mining-related safety and health issues. It has a generous amount of research on pillar and roof stability in underground room-and-pillar mining, noise suppression, dust suppression, and ventilation. The CDC's website states: "The NIOSH Mining Program's mission is to eliminate mining fatalities, injuries, and illnesses through relevant research and impactful solutions. We value relevance, impact, excellence, integrity, innovation, and collaboration as we join our stakeholders in fostering an industry of safe mines and healthy workers" (CDC, n.d.-a).

SAFETY CULTURE

The safety and health of a producer's employees is not the sole responsibility of MSHA or the federal government. The responsibility lies entirely with the management of the company. A culture of safety needs to be established within the company and come directly from the top of the management structure. This safety culture should begin by defining the goals of the safety and health program, the responsibilities of the employees, and the responsibility of management, and provide the resources necessary (both personnel and dollars) to achieve these goals. The goals need to be re-emphasized on a regular basis, and an example of safety awareness set by all managers at all times. The CEO or company owner should never set foot at a mine site without donning the appropriate PPE employees are expected to wear. "Safety first" must be a mindset instilled into all employees, not just a slogan on a poster on the employee bulletin board. The issue of personal responsibility should be stressed and instilled into all employees. Individuals need to be personally responsible for their own safety and the safety of fellow employees.

We've stressed the fact that good employees are a company's most valuable asset and a culture of safety awareness emphasizes and reinforces that value. Like all successful company cultures, every employee in the organization is responsible for the safety and health of not only themselves, but every other employee as well. This requires open communication between all levels of the organization concerning safety and health. It is the responsibility of every employee to make sure their supervisor is aware of any potential unsafe practice, any mechanical problem with equipment, any tripping hazard, any broken guarding, and any electrical issue. If an employee sees a coworker performing an unsafe act, the employee should feel comfortable enough to approach the coworker and suggest a better, safer method of doing the job. Looking out for your fellow employees is the key to a successful safety culture. If employees are asked to do something by their supervisor that they feel is unsafe, they should have the right to refuse with no fear of retribution until a safer method is agreed upon. Prior to beginning projects not routinely done on a daily basis, the supervisor should take the time to hold a quick safety meeting to discuss the procedures, processes, and tools required, and outline each individual's responsibility, as well as the hazards and how to avoid them. Input from the employees should be sought on the safest method to complete the project.

As additional training and safety re-enforcement, tailgate safety talks should be held with the employees. In my last job, we required our site management to hold these meetings on a weekly basis prior to each shift. The safety department provided 52 suggested safety subjects with outlined bullet points to discuss with employees, one for every week of the year. These were designed to be short, 10- to 15-minute meetings, held in the field, thus the term tailgate meeting. We required a sign-up sheet to have an ongoing record of each employee's training. Some of the more innovative managers would assign various hourly employees the task of presenting the subject matter. Pertinent issues were discussed at pertinent times, like avoiding heat stress at the beginning of summer or winter hazards early in that season. MSHA will allow the use of short tailgate meetings to suffice as the 8-hour refresher training requirement, but we chose to do these meetings in addition to an annual 1-day refresher training session. Safety awareness needs to be in the forefront of all employee's consciousness.

An unsafe work environment is not only bad for the employees, it is also expensive for the company involved. It is only good business to eliminate workplace accidents. Much like the management information system we discuss in Chapter 18 on operations management, a safety information system that tracks and documents MSHA violations, MSHA fines, accidents, and health-related issues on an individual employee and plant operation basis needs to be established. This should be a prime function of the safety department. The average industrial accident can cost in the range of $250,000. This includes medical expenses, disruption of work, loss of income for the employee, property damage, and increased insurance and worker's compensation premiums. This cost does not include the pain, suffering, and psychological effects experienced by the injured employee involved and their family members. Some government-sponsored and large construction projects refuse to do business with companies that have inferior safety records. The corporate reputation with investors in publicly traded companies can also suffer if safety records fall beneath shareholders' expectations.

I regret to say this, but I believe that some people are actually accident prone. Certain employees have a tendency to have a repetitive problem with accidents. These individuals, who are generally very good employees, tend to act before thinking about the danger in which they are placing themselves and their fellow employees. Most employees that I've seen exhibiting this propensity for accidents do it with all the good intentions of getting back into production quickly, not wanting to take the extra time to get the proper PPE, or getting the job completed in a timely manner. An employee exhibiting a trend of having multiple accidents needs to be taken aside, counseled, and educated on methods of slowing down, focusing, and evaluating the hazards of their potential actions prior to undertaking a job. This type of counseling is not always heeded by the individual employee and may as a result require severing the employee's service before the employee severs an arm or some other extremity. I have personally demoted one supervisor and one manager, both of whom were good at their jobs, back to hourly positions because they could not prevent themselves from having accidents. Accident-prone supervisors are extremely bad for employee morale and the success of a safety culture.

Aggregates Division

Safety Manual

Table of Contents

FIGURE 17.2 Safety manual table of contents

Like all good performance, outstanding safety should be recognized and rewarded by management at every chance possible. The last company I worked at had a policy of providing a steak luncheon for any operation that completed a biannual MSHA inspection without any citations being issued. The corporate safety manager would bring a grill on-site and personally grill steaks, pork chops, and swordfish, with all the fixings for the entire crew and the operation would shut down for the festivities. The employees enthusiastically participated, looked forward to these lunches, and strived to repeat the citation-free record every 6 months. The positive MSHA safety records were also published in the company newsletter. As operations manager (OM), I attempted to make as many of these luncheons as possible and pass on my congratulations on a job well done (with a nice lunch included). Many of these same employees had a bonus program that was partially dependent on their individual safety record. At a previous employer, employees were recognized for good safety records annually with a gift of their choice from a gift catalog; the more years with no accidents, the nicer the gifts became. Recognition for safety achievements is another integral part of creating a culture of safety.

Step 1 of developing a safety culture begins with writing a comprehensive employee safety manual via a collaborative effort between management and the corporate safety department (Figure 17.2). Upon completion, the manual should be reviewed by all plant managers for comment prior to distribution. The safety manual should then be published and copies distributed to all employees of the aggregates division, and also included in the new employee pre-employment paperwork. This publication provides employees with management's commitment to safety, expectations on a safe working environment, and employee safety responsibilities and housekeeping expectations. Each employee should be required to sign an acknowledgment of receiving and reading the safety manual.

We also established a safety committee within the aggregates division. This committee met every other month and consisted of four hourly employees, one foreman, two superintendents, the safety manager, the OM (myself), and usually the president of the aggregates division. The objective and bylaws of the committee were defined as follows:

Objective

To ensure a safe and healthy environment for the employees, customers, vendors, and visitors at the operating sites of the Aggregates Division through the development of training procedures, employee awareness, site inspections, and a general commitment to safety.

Bylaws

1. The Aggregate Division Safety Committee will consist of seven members; four hourly employees, one foreman, and two plant superintendents.
2. At least one member of the Corporate Safety Department and the operations manager will attend all committee meetings in an advisory capacity.
3. The purpose of the Safety Committee is to promote general safety awareness among the employees, act as an advisory group to management on health/safety/training issues, identify and eliminate unsafe equipment and work practices, and provide incentive to the employees to work safely.
4. Each committee member will serve a two-year alternating term. On alternating years, one plant superintendent, the foreman, and two hourly employees will be replaced by other employees in similar positions. The following year the other plant manager and two hourly employees will be replaced. (For the purposes of initial committee formation, four members will be limited to a one-year term.)
5. The Corporate Safety Department will function as secretary for the Safety Committee. Duties will include keeping minutes of the meetings, maintaining records/files, disseminating correspondence, tracking accident/citation statistics, etc. All correspondence from committee members should be sent to the responsible Corporate Safety Department representative.
6. The Safety Committee will function as a cohesive group with all issues decided by majority vote. Once an issue is voted upon, the majority decision is final, the issue will be considered resolved, and the recommendation will be forwarded to management for final action.
7. Each operating site within the Aggregate Division will receive a safety inspection twice annually by a team from the Safety Committee. Every other month, one plant will be designated for a "blitz inspection" and will be inspected by the entire Safety Committee.
8. Upon completion of the safety inspection, the inspector will submit a list of safety/housekeeping issues to the appropriate plant superintendent at a close-out conference following the inspection and a copy to the Corporate Safety Department.
9. The plant superintendent will have a period of two weeks to complete the necessary repairs/modifications. (Issues deemed an imminent safety hazard will be addressed immediately.) Upon completion of the repairs, the plant superintendent will sign off on the list indicating the date each item was completed and send a copy to the Safety Committee.
10. Any controversy concerning the inspections or validity of the safety issues cited by an inspector can be appealed to the entire Safety Committee at their next scheduled meeting.
11. The Safety Committee will meet six times per year to review inspection reports, accident reports, and MSHA citations; determine future plant inspection assignments; discuss training needs; and review disciplinary action for safety violations.
12. All members will be provided with a copy of 30 CFR (MSHA regulations) and be trained in its use and interpretation. Additional training on conducting safety inspections will be provided to new members by the Corporate Safety Department.

13. Emphasis on targeted safety initiatives will change periodically based on the regulatory atmosphere and MSHA Program Policy changes (e.g., "The Thirteen Rules to Live By").

The bimonthly safety inspections of the various operations by the Safety Committee provided an opportunity for several new sets of eyes looking for safety issues, housekeeping problems, and potential MSHA violations. A minor safety problem can be overlooked if a person passes it on a day-to-day basis. The mind eventually accepts that the issue is normal and not out of place. That is what makes observations by individuals from another location invaluable. On several occasions an MSHA inspection would occur within days of a Safety Committee inspection with no MSHA citations issued, even though the committee had found several potential safety problems. This type of interplant inspections also gives the inspecting personnel an opportunity to see how other operations are managed, maintained, and kept clean. It becomes a shared learning experience for all concerned. I personally used the committee meetings to evaluate the leadership skills of the hourly committee members for possible future promotion to supervisor and several were eventually promoted to supervisor. A manager should be constantly on the lookout for talented individuals within the organization.

In Chapter 18, we talk extensively about the value of repair before failure in the sections on plant and mobile equipment maintenance. The same idea applies to safety and accidents; identify and repair workplace hazards before they cause accidents. Hazard recognition is not always obvious to every employee, especially new employees; it is a subject that needs to be extensively covered in safety training. A proactive approach to recognizing hazards is a key element to establishing a culture of safety. Much like maintenance checklists discussed in Chapter 18, a series of safety checklists is an easy way to keep employees focused on hazard awareness. MSHA requires that a safety inspection be completed by an experienced, competent person prior to each shift. This inspection needs to be documented and retained in a file for review by MSHA officials upon request. Figure 17.3 is an example of a four-page safety inspection form utilized by our Safety Committee during plant inspections.

Safety rules and regulations must be defined and enforced by management. Letting safety infractions, even minor ones, go without consequences will exacerbate safety violations and undermine the concept of safety culture. If one employee is permitted to be on the property without the proper PPE, other employees will follow this example, especially if the offending employee is in management. No one likes wearing safety glasses if they have no need of refractive correction, nor is it enjoyable wearing a reflective vest in the hot, humid summer air. I once had an employee that worked in the mobile maintenance shop who lost the vision in one eye when he was hammering on a metal object without safety glasses or face shield and a shard of steel embedded in his cornea. Not long after returning to work, I encountered him working on a piece of equipment with no safety glasses. I said: "Didn't you learn anything about safety glasses from losing sight in your left eye?" His reply was: "I hate those damned safety glasses!" He received an immediate 3-day suspension for a second offense violating the same safety rule. This finally got his attention and may have ultimately saved his other eye. This was a small step in establishing a culture of safety at that specific operation. Discipline need not be vindictive or punitive; at times it can be educational.

A comprehensive drug and alcohol program needs to be established and enforced at all times on mine property. Drugs and alcohol have no place around blasting, heavy equipment, moving mechanical parts, or power tools and are also regulated by MSHA. The danger of an

Maintenance Shop

Inspection Item	OK	Not OK	Comments
General Housekeeping			
Chemical / Solvent Storage			
Overhead Cranes / Hoists			
Oil Storage			
Fuel Storage			
Bench Grinders			
Oxy-Acetylene Storage			
Welding Techniques			
Haz-Com Labeling of Containers			
Fire Extinguishers			
Power Tools			
Hand Tools			
Lighting			
Electrical			
Face Shields, goggles, welding hoods			
Material Safety Data Sheets			

Are all employees utilizing their Personal Protective Equipment (Safety glasses, hard hats, steel-toed shoes, hearing protection)? ☐ Yes ☐ No

Is fall protection available at the plant? ☐ Yes ☐ No

Is a lock-out / tag-out system available at the plant? ☐ Yes ☐ No

Are records of annual grounding & continuity tests available and up-to-date? ☐ Yes ☐ No

Are records of compressor tank testing available and up-to-date? ☐ Yes ☐ No

Are all subcontractors (drillers, blasters, stripping crews, service techs, etc.) in compliance with MSHA regulations? ☐ Yes ☐ No

Are life jackets available and utilized around water? ☐ Yes ☐ No

Are all required First Aid supplies available on site? Yes No

Are pre-shift plant inspections completed and documented daily? Yes No

Other Comments:

____

____

____

____

____

____

____

Mobile Equipment

Equipment	OK	Not OK
Model		
Unit #		
Brakes		
Steering		
Back Up Alarm		
Lights		
Horn		
Seat Belt		
Glass		
Wipers		
Pre-shift Inspection		

Comments:______________________

Equipment	OK	Not OK
Model		
Unit #		
Brakes		
Steering		
Back Up Alarm		
Lights		
Horn		
Seat Belt		
Glass		
Wipers		
Pre-shift Inspection		

Comments:______________________

Equipment	OK	Not OK
Model		
Unit #		
Brakes		
Steering		
Back Up Alarm		
Lights		
Horn		
Seat Belt		
Glass		
Wipers		
Pre-shift Inspection		

Comments:______________________

Equipment	OK	Not OK
Model		
Unit #		
Brakes		
Steering		
Back Up Alarm		
Lights		
Horn		
Seat Belt		
Glass		
Wipers		
Pre-shift Inspection		

Comments:______________________

Equipment	OK	Not OK
Model		
Unit #		
Brakes		
Steering		
Back Up Alarm		
Lights		
Horn		
Seat Belt		
Glass		
Wipers		
Pre-shift Inspection		

Comments:______________________

Equipment	OK	Not OK
Model		
Unit #		
Brakes		
Steering		
Back Up Alarm		
Lights		
Horn		
Seat Belt		
Glass		
Wipers		
Pre-shift Inspection		

Comments:______________________

Equipment	OK	Not OK
Model		
Unit #		
Brakes		
Steering		
Back Up Alarm		
Lights		
Horn		
Seat Belt		
Glass		
Wipers		
Pre-shift Inspection		

Comments:______________________

Equipment	OK	Not OK
Model		
Unit #		
Brakes		
Steering		
Back Up Alarm		
Lights		
Horn		
Seat Belt		
Glass		
Wipers		
Pre-shift Inspection		

Comments:______________________

Equipment	OK	Not OK
Model		
Unit #		
Brakes		
Steering		
Back Up Alarm		
Lights		
Horn		
Seat Belt		
Glass		
Wipers		
Pre-shift Inspection		

Comments:______________________

Safety Inspection Form

Date:______________ Plant ______________________ Inspected by:______________________

General Operation

Inspection Item	OK	Not OK	Comments
Overall Appearance			
Entrance			
Office			
Scales			
Roadways			
Road signs			
Traffic Flow			

Pit/Quarry Area

Inspection Item	OK	Not OK	Comments
Pit Access			
Highwalls / Pit Slopes			
Haul Roads			
Drainage			
Warning Signs			
Berms			
Traffic Flow			
Dust Control			
Lake banks			
Dredge			
Service Boat			
Sump			

Primary Crushing Plant

Inspection Item	OK	Not OK	Comments
Ramp			
Feed hopper			
Feeder			
Primary Crusher			
Conveyors			
Motor Control Center			
Electrical			
Walkways			
Stairs/ladders			
Fire Extinguishers			

Dry Plant

Inspection Item	OK	Not OK	Comments
Scalping Screen			
Secondary Crusher			
Tertiary Crusher			
Sizing Screen			
Conveyors			
Transfer points			
Dust controls			
Motor Control Center			
Electrical			
Walkways			
Stairs / Ladders			
Fire Extinguishers			

Wash Plant

Inspection Item	OK	Not OK	Comments
Wash Screens			
Conveyors			
Transfer points			
Sand Screw			
Coarse Mat'l. Washer			
Log Washer			
Sand Classification Tank			
Pump Tank			
Slurry Pump			
Sediment Ponds			
Plumbing / Pipes			
Motor Control Center			
Electrical			
Walkways			
Stairs / Ladders			
Fire Extinguishers			

FIGURE 17.3 Example of a safety inspection form

employee under the influence of drugs or alcohol while at work needs no explanation. No matter your personal feelings on drug or alcohol use, it has no place in the work zone. Most companies have a zero-tolerance policy for drugs and alcohol while on company property. Drug testing should be done on an annual basis, a random basis, and whenever an accident is reported. An employee may not abuse drugs and alcohol during working hours but could still have a substance abuse problem that can be identified by drug testing. A program to assist employees with substance abuse issues enabling them to get counseling and rehabilitation is a helpful employee benefit. I was coming home late from work (7:00 PM) after a very long day and was rear-ended at a stop sign by a teenage driver while driving a company vehicle. After a police investigation, I called the company human resources manager to inform him of the accident and he said "You need to get to a clinic and get a drug test." I said "Are you kidding me? I was the one rear-ended; it was the kid's fault, not mine!" He wouldn't budge on the company policy so I had to spend an additional 2 hours dealing with drug testing issues that evening.

Tracking and reporting accidents is an important part of any safety program, but this can be taken another step forward by tracking near-miss incidents that could likely have caused an accident. These incident reports can be utilized to identify accident trends and indicate where additional training or equipment modifications might be warranted. A standardized reporting format was developed so that all the pertinent information was available for analysis by the safety department on a single form. Figure 17.4 is an example of an accident/incident report form. The one we utilized had a grid on the reverse side indicating compass directions so a representative diagram of the accident could also be included. Accidents and incidents should be routinely investigated by the safety manager and include interviews with the personnel involved, any witnesses, responsible supervisors, equipment maintenance personnel, and equipment manufacturers if necessary. Many accidents are very complex with multiple contributing factors and in those cases a root-cause analysis should be completed to determine the ultimate cause of the accident. Once an accident investigation is completed and the root cause or causes have been determined, the investigation results should be shared with the employees. Sharing the accident investigation results can aid in the prevention of similar accidents in the future by other employees, perhaps saving a life or loss of a finger.

As a standard for reporting injury rates for comparison from location to location and industry to industry, the term *incidence rate* was developed. The formula for incidence rate is

$$\text{incidence rate (IR)} = \frac{\text{number of reportable injuries} \times 200{,}000 \text{ work hours}}{\text{number of work hours during the period in question}}$$

For example, two reportable injuries occurred during 115,000 total work hours would calculate to

$$\text{IR} = \frac{2 \times 200{,}000}{115{,}000} = 3.48$$

Using this standard, injury rates can easily be compared by safety professionals, insurance companies, MSHA, and various different industries, as well as between individual plant operations. Incident (injury) rates for the various mining sectors in the United States between 2016 and 2019 can be seen in Figure 17.5.

MSHA routinely publishes what it calls fatalgrams on every fatality that occurs within the mining industry and follows up with complete published investigation reports. These fatalgrams and reports of investigations are an excellent source of educational materials for safety meetings held with employees. MSHA's investigations inevitably conclude a failure of management to provide sufficient training, or sufficient PPE, or inadequate maintenance procedures. Accident prevention is definitely a prime management responsibility but not all accidents are

Accident/Incident Report Form

Plant:		Date:		Time:	AM PM
Person(s) Involved		Occupation		Yrs. Experience	
Equipment Involved:					
Witnesses:					
Weather Conditions:		Lighting Conditions:			
Description of Incident:					
Location of Incident:					
Description of Property Damage:					
Injuries:					
MSHA Reportable:		Estimated Property Damage:	$		
How to Prevent a Similar Incident:					
Reported by:		Date:			

FIGURE 17.4 Example of an accident report form

the fault of management. The majority of accident investigations I have been part of were the direct result of the employee(s) involved either being distracted or not considering the potential hazards of what they were doing at the time. This is something I have been known to refer to as a rectal cranial inversion. This is also the reason a culture of safety needs to be developed and instilled into the employees through training, hazard identification, management attitudes, and more training so that safety is always first on everyone's mind. It is easy to understand new employees having an accident because they are new to the job and inexperienced, but many older, experienced employees including managers and supervisors are also involved in accidents. It seems that the more experienced employees can become complacent and not believe an accident can possibly happen to them.

Job safety analyses (JSAs) are a good training tool for employees. A JSA is a detailed safety analysis for an individual job or task that is prepared by experienced individuals that details the safety risks, hazard recognition, hazard avoidance methods, and employee responsibilities. These are usually very detailed and require considerable time and effort to generate. Examples of JSAs for common jobs in the industry are usually available through state association safety committees or NSSGA.

Over my 42-year career in the industry, I had the unfortunate experience to be involved in two different fatality investigations. One was an employee of a plant under my management

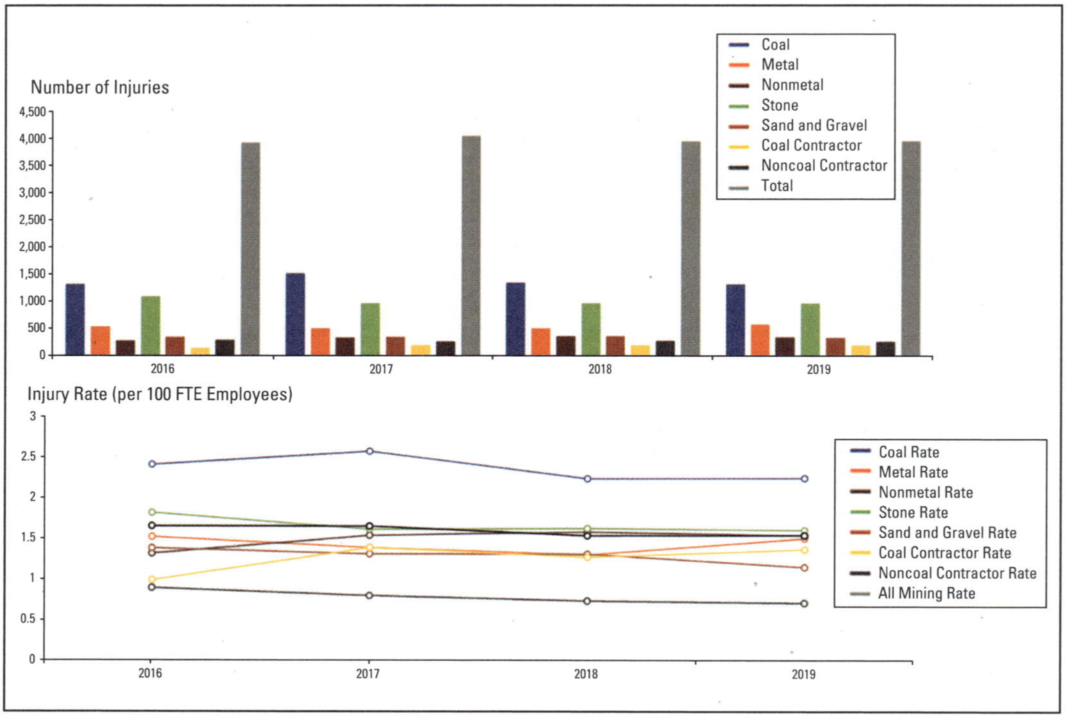

Courtesy of CDC and NIOSH

FIGURE 17.5 National Institute for Occupational Safety and Health statistics on lost time injuries in mining, 2016–2019

and another an employee of a drilling contractor at one of my operations. I never again want the experience of talking to an employee's widow after a fatal accident. In the first case, the employee violated several safety rules, which resulted in the accident. An elevated conveyor belt coming out of the quarry feeding the surge pile on the surface had ripped and fallen to the ground. A group of maintenance employees had recovered the belt and were pulling it back onto the structure utilizing a crane on the surface to pull up the belt. The crane was attached to the conveyor belt with a clamping device consisting of two pieces of steel plate with the belt sandwiched in between and three bolts through the conveyor belt clamping the plates. The employee involved had installed the clamping device himself while the belt was on the ground and was following the progress of the belt from the conveyor walkway. For some unknown reason, when the two ends of the belt came together, the employee left the protection of the walkway and climbed on top of the conveyor belt. His foreman, who was standing on the ground surface above him, immediately yelled at him to "get the hell off the belt." Shortly after the employee climbed on top of the conveyor belt, the clamping device pulled from the end of the belt and the belt crashed to the ground along with the employee, who fell approximately 42 feet (13 meters) to the hard limestone floor below. Upon later investigation, it was determined that the belt clamp had been improperly installed only inches from the edge of the belt and one of the three bolts had not penetrated the belt at all. Standard policy for clamp installation was 10–12 inches (25–30 centimeters) from the end of the conveyor belt with all three bolts drilled through the belt carcass.

This was a tragic accident that could have been prevented by (1) properly installing the belt clamping device, (2) the employee not leaving the safety of the conveyor walkway, (3) the employee not climbing onto the top of the belt, and (4) the employee utilizing the proper fall protection when leaving the walkway. The employee was experienced and had completed this same task on several previous occasions. Prior to beginning the project, the supervisor had discussed the obvious hazards with the employees and assigned individual tasks, and the employees acknowledged that they knew how to proceed. Why was the clamp improperly installed by an employee who had done so correctly on several previous occasions? Why did the employee leave the walkway? Why did he climb on top of the conveyor belt? Why did the employee not utilize the proper fall protection equipment that he had been trained to use and was available on the bench level below? No one will ever know the answers to these questions because he was no longer available to interview. The accident happened on a Saturday morning and I'll never forget the next Monday when another, younger, employee who was good friend of the individual involved came to my office and cried like a baby when I informed him of what had happened.

The second fatality was of a contract drill operator drilling blast holes on second shift. I received the phone call at approximately 10:00 PM while I was attending a meeting about 2 hours away from the quarry. The drill operator had over 10 years of experience operating in various quarries throughout the state and had drilled this same location on dozens of previous occasions. By the time I arrived on-site, the driller had been taken to the morgue but the truck-mounted drilling rig was sitting vertically, resting on the back of the truck, one bench (55 feet [17 meters]) lower than where it had been drilling. The drill was operating on an active mining face, the edge of which had been bermed prior to allowing the drill to enter the area. The berm consisted of crushed rock of various sizes and was about 30 inches (76 centimeters) tall. Several employees were operating other equipment in the quarry at the time but no one saw anything unusual until the drill went backward off the highwall. The quarry employees immediately went to render aid but were too late to be of any assistance.

As usual in any fatality, MSHA sent two special investigators to the site the following day and closed down the operation until the investigation could be completed. As the investigation proceeded, it was determined that the truck had driven backward off the highwall and went through the established berm. This set off alarm bells in my mind, as berms are designed to prevent just this sort of accident from occurring and it would take considerable force to drive through a berm of that height even in the dark. All of the employees that were present at the time were interviewed and numerous measurements and diagrams of the site completed. The MSHA inspectors were as perplexed as I was concerning the circumstances of the accident. We even went so far as constructing a similar berm on level ground and attempting the drive over it with the same model drill. This could only be accomplished driving through the berm at a fairly high rate of speed using a running start.

The owner of the drilling firm discovered some papers in the cab of the truck that indicated that the victim had been calculating his household finances and evidently had some significant financial problems. It also came to light that the individual had a developmentally handicapped son and was having marital struggles at the time as well. After 4 days of intensive investigation, we came to the conclusion that all of the evidence suggested that the individual had most likely committed suicide. Because there was no way of conclusively proving the suicide theory and the fact that the widow would receive no life insurance settlement if it were ruled to be a suicide, the cause was of death was listed as accidental. Unusually, the company received no citations or fines from MSHA as a result of this investigation.

No industrial function is worth a human life. That is why the safety program at any aggregates plant needs to be top priority among management. Much progress has been made in safe working conditions over the last 100 years, but we should never be satisfied or become complacent until the fatality rate is zero.

CHAPTER 18

Operations Management

Like all businesses, the aggregates industry needs experienced, capable, motivated managers to inspire its workforce and direct its operations. To survive in today's marketplace, aggregate operations need to be cost competitive and profitable, as well as customer, quality, and employee focused. Operations management needs to focus on safety, efficiency, cost control, maintenance, asset management, and employee motivation. I learned my management style from experience and observing the managers I reported to over the years. I attempted to emulate the good that I observed and ignore the bad traits, or at least what I considered to be negative traits. Ego is something that needs to be left buried in the depths of your subconscious. No matter your years of experience or your degree of education, you can always learn from others if you keep your eyes, ears, and mind open.

I'm going to focus this chapter on management from the viewpoint of the operations manager (OM), assuming management of multiple plants, each having an individual plant manager reporting to the OM. All the information will be able to be extrapolated to the individual plant level and plant manager. To be an effective manager, you have to set goals for your employees and your operations. I started one of my jobs as the Manager of Business Development and Plant Systems, which was a bit of a nebulous title that even I didn't really understand, but the owner of the company liked it. I was working for an asphalt producer/highway contractor that had a few aggregate plants that were managed by an executive vice president (VP) and were fairly low priority in the scheme of the entire business. Each operation had a plant supervisor with varying degrees of education and experience. They were given very little guidance, no financial goals to work toward, and no production goals to attain. The supervisors were criticized when their employees' overtime exceeded a certain level, an accident occurred, or they ran out of a certain product needed by the asphalt plant on site. Needless to say, employee morale was poor and productivity severely lacking.

The existing OM at the time was qualified and motivated but was a long-term employee of the company and had never been exposed to operations budgets, production budgets, productivity enhancement, and so on, that the larger producers routinely employed. He had been successful in navigating the corporate culture emanating from his managing VP. I was somewhat appalled when I asked to see the plant budgets and the OM said that "the boss [VP] didn't believe in budgets; budget was not in his vocabulary." I thought to myself: *This is a company that makes money in spite of itself.* I must add, at this point, that it was and still is a very successful company and all companies have their own corporate culture that continues evolving with time. One of the first things I did while working with the OM was to establish production "goals" for each plant, being careful not call them production budgets. We came up with reasonable goals (not terribly aggressive on the first go-round) for each product on a monthly basis for each operation.

We decided to roll out the "goals" at the largest production facility first and arranged to meet with the plant manager the next morning. Somehow the VP got wind of our plan, was quite perturbed, and called the OM in the morning and demanded that we not start the meeting until he personally arrived. Needless to say, the meeting began on a very tense note. We rolled out our production plan to the plant manager, who was happy to be given a goal to work toward. Before the meeting was over, the VP decided that maybe this was not such a bad idea. Thus began the evolution of the aggregate division's revised corporate culture.

BUDGETING

Let's begin with budgets as a managerial tool. Frankly, producing annual budgets is a pain in the backside! But they are a very necessary management tool in planning a successful mining operation. To begin the budgeting process, input from the marketing/sales department is necessary. What are the anticipated sales on a monthly basis for every product for the coming year? Budgets can be based on the calendar year or the company's fiscal year, if they are not the same. Sales budgets are, in most cases, based on past experience plus knowledge of any large construction jobs proposed for the market area that have been contracted or can reasonably be expected to be won. In most cases there is some guesswork involved; the tonnages are never exact, but they must be reasonable estimates. These should always be reviewed by marketing/sales management for accuracy prior to submitting them to operations. An example of a typical sales budget can be seen in Figure 18.1. When the sales budgets are complete, the plant managers put together a production budget, which is an estimate of how many total tons and individual product tons need to be produced to meet the projected sales demand. Existing inventory levels and next year's ending inventories need to be factored into the production budget.

The existing inventory plays a role in the upcoming year's budget by determining how much additional tonnage of each product needs to be produced. And an ending inventory for the next year's budget needs to be identified as well. Numerous questions will arise during this stage in the budgeting process. This is the time to decide if a second shift is required to produce the needed sales or if more personnel are needed for the next production season. Is the plant capable of producing the budgeted sales mix without producing excessive inventories of other products? How do we deal with a sales mix that requires one and a half shifts of operation? What effect will overtime have on the costs for next season? Will we need to remove one shift due to a lack of sales? When these production budgets are finally approved by the OM and integrated with the other plant budgets, the cost budgeting process can begin.

Cost budgets can be very involved with minute detail as to what portion of the costs were incurred doing various tasks in each part of the plant, for example, quarry, primary plant, crushing plant, washing plant, stockpiling, load-out, or they can be much simpler with less detail. One company I worked for had a budget that consisted of a 26-page-long Microsoft Excel spreadsheet, which was extremely daunting to a new manager (and many of the older ones). Corporate management needs to decide which costs are important to track and have a chart of accounts set up to be utilized to put the correct costs in the correct budget category. Each invoice received, each time sheet generated, needs to be assigned to a cost center tied to the budget and ultimately the plant statements. Some companies like to track direct versus

Proposed Sales Budget 2015																		
	Dry			Washed			Rip Rap			AgLime			Resale			Total		
Months	Sales Dollars	Sales Units	ASP	Sales Dollars	Sales Units	ASP	Sales Dollars	Sales Units	ASP	Sales Dollars	Sales Units	ASP	Sales Dollars	Sales Units	ASP	Sales Dollars	Sales Units	ASP
January	$57,098	9,300	$6.14	$115,958	2,166	$53.54	$14,375	925	$15.54	$7,680	800	$9.60	$0	0	#DIV/0!	$195,110	13,191	$14.79
YTD	$57,098	9,300	$6.14	$115,958	2,166	$53.54	$14,375	925	$15.54	$7,680	800	$9.60	$0	0	#DIV/0!	$195,110	13,191	$14.79
February	$137,853	22,325	$6.17	$171,635	22,635	$7.58	$18,450	1,322	$13.96	$2,880	300	$9.60	$0	0	#DIV/0!	$330,818	46,582	$7.10
YTD	$194,950	31,625	$6.16	$287,593	24,801	$11.60	$32,825	2,247	$14.61	$10,560	1,100	$9.60	$0	0	#DIV/0!	$525,928	59,773	$8.80
March	$295,250	51,300	$5.76	$293,858	37,835	$7.77	$23,725	1,005	$23.61	$14,400	2,100	$6.86	$900	50	$18.00	$628,133	92,290	$6.81
YTD	$490,200	82,925	$5.91	$581,450	62,636	$9.28	$56,550	3,252	$17.39	$24,960	3,200	$7.80	$900	50	$18.00	$1,154,060	152,063	$7.59
April	$316,093	50,650	$6.24	$428,818	54,980	$7.80	$33,950	2,725	$12.46	$19,200	1,800	$10.67	$1,200	75	$16.00	$799,260	110,230	$7.25
YTD	$806,293	133,575	$6.04	$1,010,268	117,616	$8.59	$90,500	5,977	$15.14	$44,160	5,000	$8.83	$2,100	125	$16.80	$1,953,320	262,293	$7.45
May	$433,950	71,600	$6.06	$652,516	84,260	$7.74	$43,450	3,525	$12.33	$4,800	500	$9.60	$1,200	75	$16.00	$1,135,916	159,960	$7.10
YTD	$1,240,243	205,175	$6.04	$1,662,784	201,876	$8.24	$133,950	9,502	$14.10	$48,960	5,500	$8.90	$3,300	200	$16.50	$3,089,236	422,253	$7.32
June	$483,526	79,960	$6.05	$802,455	101,456	$7.91	$49,513	2,556	$19.37	$480	50	$9.60	$1,200	75	$16.00	$1,337,174	184,097	$7.26
YTD	$1,723,769	285,135	$6.05	$2,465,239	303,332	$8.13	$183,463	12,058	$15.22	$49,440	5,550	$8.91	$4,500	275	$16.36	$4,426,410	606,350	$7.30
July	$542,150	90,100	$6.02	$865,458	113,000	$7.66	$49,180	4,005	$12.28	$240	25	$9.60	$900	50	$18.00	$1,457,928	207,180	$7.04
YTD	$2,265,919	375,235	$6.04	$3,330,696	416,332	$8.00	$232,643	16,063	$14.48	$49,680	5,575	$8.91	$5,400	325	$16.62	$5,884,337	813,530	$7.23
August	$528,026	91,550	$5.77	$955,408	121,935	$7.84	$42,850	3,525	$12.16	$240	25	$9.60	$780	40	$19.50	$1,527,304	217,075	$7.04
YTD	$2,793,945	466,785	$5.99	$4,286,104	538,267	$7.96	$275,493	19,588	$14.06	$49,920	5,600	$8.91	$6,180	365	$16.93	$7,411,641	1,030,605	$7.19
September	$537,249	90,375	$5.94	$871,078	110,135	$7.91	$39,605	3,000	$13.20	$19,200	2,000	$9.60	$780	40	$19.50	$1,467,911	205,550	$7.14
YTD	$3,331,194	557,160	$5.98	$5,157,181	648,402	$7.95	$315,098	22,588	$13.95	$69,120	7,600	$9.09	$6,960	405	$17.19	$8,879,552	1,236,155	$7.18
October	$448,911	73,056	$6.14	$601,175	76,335	$7.88	$34,875	2,875	$12.13	$105,600	11,000	$9.60	$600	25	$24.00	$1,191,161	163,291	$7.29
YTD	$3,780,105	630,216	$6.00	$5,758,356	724,737	$7.95	$349,973	25,463	$13.74	$174,720	18,600	$9.39	$7,560	430	$17.58	$10,070,714	1,399,446	$7.20
November	$294,138	48,800	$6.03	$485,603	63,527	$7.64	$25,075	2,075	$12.08	$84,480	8,800	$9.60	$600	25	$24.00	$889,895	123,227	$7.22
YTD	$4,074,242	679,016	$6.00	$6,243,959	788,264	$7.92	$375,048	27,538	$13.62	$259,200	27,400	$9.46	$8,160	455	$17.93	$10,960,609	1,522,673	$7.20
December	$208,839	33,995	$6.14	$204,358	26,135	$7.82	$7,300	600	$12.17	$9,600	1,000	$9.60	$0	0	#DIV/0!	$430,096	61,730	$6.97
YTD	$4,283,081	713,011	$6.01	$6,448,316	814,399	$7.92	$382,348	28,138	$13.59	$268,800	28,400	$9.46	$8,160	455	$17.93	$11,390,705	1,584,403	$7.19

FIGURE 18.1 Example of a typical sales budget spreadsheet

indirect costs in their budgeting processes. Direct costs would include actual invoices for materials, labor for production and maintenance, wear parts, tools, power costs, drilling and blasting coats, depletion, and so on. Indirect costs would consist of management salaries, employee fringe benefits, equipment depreciation, taxes, insurance, and such; costs not directly related to the cost of production. Other companies prefer to track maintenance and production costs separately, or straight time labor versus overtime labor. This is again a function of what the corporate management wants to emphasize, manage, and track.

This may a good time to define some relevant financial accounting terms. Investopedia .com defines depreciation as follows (Tuovila 2022):

> Depreciation is an accounting method of allocating the cost of a tangible or physical asset over its useful life or life expectancy. Depreciation represents how much of an asset's value has been used up. Depreciating assets helps companies earn revenue from an asset while expensing a portion of its cost each year the asset is in use. If not taken into account, it can greatly affect profits. Businesses can depreciate long-term assets for both tax and accounting purposes. For example, companies can take a tax deduction for the cost of the asset, meaning it reduces taxable income. However, the Internal Revenue Service (IRS) states that when depreciating assets, companies must spread the cost out over time.

InvestingAnswers.com gives the definition of depletion allowance as "a tax deduction allowed in order to compensate for the depletion or 'using up' of natural resource deposits such as oil, natural gas, iron, timber, (aggregates) etc. The allowance is a form of cost recovery for capital investment [cost of land containing the mineral resource] which, unlike income, is not taxable" (Tracy 2020). In essence, depletion is a form of depreciation of the mineral reserve value.

Budgets should be formatted to directly relate to the plant statements that most aggregate companies provide to their plant management on a monthly basis. Some producers may not want their plant personnel to have all of the budgetary information (with revenue included) but the plant personnel need to know the cost budgets and where their plant stands in relation to that budget. At a minimum, the budgets should include the following categories:

- Variable costs
  - Stripping
  - Drilling
  - Blasting
  - Hourly labor
  - Maintenance materials (parts, supplies)
  - Power (electricity)
  - Fuel
  - Equipment rental
  - Depletion/royalties
- Fixed costs
  - Supervision labor
  - Taxes
  - Office labor

- Office supplies
- Depreciation
- Miscellaneous expenses

These can all be expanded for more detailed cost budgeting by subdividing the variable costs into the various plant functions like quarry (pit), primary crushing, crushing and screening, washing/classifying, and load-out. Detailed budgets are good for tracking your costs but make sure that the details are relevant and have a purpose; not just detail for the sake of details. Plant statements should be published monthly and reflect the budgeted costs and actual costs, and provide a recap of where the plant stands with regards to what was budgeted. This can include revenue or revenue can be withheld, if preferred. At one company where I worked for a number of years, the accounting department put out a monthly superintendent's report that included all of the cost detail but no revenue information and thus no actual profit and loss (P&L) data. The entire plant statement that included the revenue, with profit, and loss was only seen by upper management.

Although tracking the monthly costs is important, the actual cost per ton produced is the most significant number, not the total cost. Every cost line item on a budget and P&L statement needs to be followed by the cost per ton for that line item. In all plant statements always divide the cost by the tons produced that month; if an annual statement, divide the cost by the annual production. Aggregates are sold by the ton and priced to the customer by the ton; therefore, the cost per ton of production, as the unit cost, is especially significant. Lowering the unit cost per ton is a function of increased efficiency and can be accomplished by good management techniques, upgrading equipment, preventive maintenance, and so on. Depending on the level of detail of the budget or statement, this could break down to the cost per ton of each individual product or just the cost per ton of the total production.

The unit cost of production of aggregates products decreases with additional production. Certain costs are fixed and will not change, but the majority of the costs vary with the volume produced. Fixed costs include management salaries, depreciation, amortization, interest, employee benefits, and so on. Variable costs are those directly related to producing the product, like power, fuel, labor, equipment maintenance, explosives, and lubricants, and are directly related to the number of tons produced. Variable costs are inversely proportional to production volume; the higher the production volume the lower the unit cost. Producing lower than expected production volumes because of lagging sales will increase the unit cost because the fixed costs have to be spread over fewer production units. If the production is not decreased when sales lag, then excessive inventories will be produced, which adds to the production costs.

Figure 18.2 shows an example of a very simplified monthly plant statement containing fictional cost figures, halfway through the production year. In the year-to-date variance column, you can see that the plant costs are 7% over budget (1.07) and the revenue is 2% under budget (0.98). The combination of these two variables results in a profit that is 17% less than budgeted (0.83). Minor fluctuations from budget can have significant effects on plant profitability. Changes in budgetary costs and sales must be periodically updated (forecast) during the budget year based on actual conditions. If the sales department was expecting to get a large job that failed to come to fruition, that could affect the entire annual budget, the required production, the inventories, and so on. Conversely, a major equipment failure in the processing plant can have a similar devastating effect on the budget.

Month/Year:	Jun-01										
	(YTD = Year to Date)	Budget Cost	Budget Cost per Ton	Actual Cost	Actual Cost per Ton	% Variance Actual Versus Budget	YTD Budget	YTD Budget per Ton	YTD Cost	YTD Cost per Ton	YTD % Variance YTD Actual Versus Budget
Production Tons		**125,000**		**137,200**		**1.10**	**485,000**		**503,726**		**1.04**
Variable Costs											
	Stripping	$18,755.00	$0.150	$22,488.25	$0.164	1.20	$80,955.00	$0.17	$84,388.92	$0.17	1.04
	Drilling	$21,250.00	$0.170	$24,696.00	$0.180	1.16	$82,450.00	$0.17	$83,618.52	$0.17	1.01
	Blasting	$46,250.00	$0.370	$51,760.00	$0.377	1.12	$179,450.00	$0.37	$193,755.74	$0.38	1.08
	Hourly Labor	$106,250.00	$0.850	$111,369.20	$0.812	1.05	$426,800.00	$0.88	$459,901.84	$0.91	1.08
	Maintenance Materials	$77,500.00	$0.620	$59,778.33	$0.436	0.77	$435,350.00	$0.90	$378,801.95	$0.75	0.87
	Power (Electricity)	$28,750.00	$0.230	$37,516.22	$0.273	1.30	$14,755.00	$0.03	$145,073.09	$0.29	9.83
	Fuel	$38,750.00	$0.310	$48,295.30	$0.352	1.25	$185,300.00	$0.38	$208,622.33	$0.41	1.13
	Equipment Rental	$5,700.00	$0.046	$9,475.35	$0.069	1.66	$34,200.00	$0.07	$29,846.22	$0.06	0.87
	Depletion/Royalties	$65,000.00	$0.520	$71,344.00	$0.520	1.10	$252,200.00	$0.52	$261,937.52	$0.52	1.04
Total Variable Cost		**$408,205.00**	**$3.266**	**$436,722.65**	**$3.183**	**1.07**	**$1,691,460.00**	**$3.49**	**$1,845,946.13**	**$3.66**	1.09
	Supervision Labor	$15,833.33	$0.127	$16,725.42	$0.122	1.06	$94,999.98	$0.20	$101,716.35	$0.20	1.07
	Taxes	$6,250.00	$0.050	$5,100.75	$0.037	0.82	$37,500.00	$0.08	$31,612.57	$0.06	0.84
	Office Labor	$5,250.00	$0.042	$5,575.00	$0.041	1.06	$31,500.00	$0.06	$33,627.52	$0.07	1.07
	Office Supplies	$2,300.00	$0.018	$3,475.00	$0.025	1.51	$13,800.00	$0.03	$11,416.77	$0.02	0.83
	Miscellaneous Expense	$8,300.00	$0.066	$2,367.89	$0.017	0.29	$48,000.00	$0.10	$33,725.16	$0.07	0.70
Total Fixed Costs		**$37,933.33**	**$0.303**	**$33,244.06**	**$0.242**	0.88	**$225,799.98**	$0.47	**$212,098.37**	$0.42	**0.94**
Total Costs		**$446,138.33**	**$3.569**	**$469,966.71**	**$3.425**	1.05	**$1,917,259.98**	$3.95	**$2,058,044.50**	$4.09	**1.07**
Tons Sold		87,325.00		74,616.50		0.85	463,150.00		448,256.50		0.97
Revenue		$589,433.75	$6.750	$516,346.18	$6.920	0.88	$3,126,262.50	$6.75	$3,066,074.46	$6.84	0.98
Profit/(Loss)		**$143,295.42**		**$46,379.47**		0.32	**$1,209,002.52**		**$1,008,029.96**		**0.83**

FIGURE 18.2 Example of a simple monthly plant statement

Capital expenditures are another area that needs to be budgeted annually. Major improvements designed to increase productivity, replacement of worn-out plant equipment, upgrades of mobile equipment, or purchases of additional reserves that require large capital expenditures that need to be depreciated or depleted per the relevant taxing authority rules also need to be budgeted. In companies with multiple plants, there will be competition for the available capital. Projects will need to be financially justified with analyses such as cash flow, return on investment, net present value, internal rate of return (IRR), and years to payback. Projects from different plants will be evaluated competitively for the best financial returns. If corporate management can achieve a 12% IRR on Plant C's project while the project from Plant A provides an IRR of 7%, then Plant C will get priority for the available capital. A certain percentage of capital should be retained for routine replacement of worn-out equipment.

The larger pieces of machinery (wheel loaders, trucks, crushers, screens, conveyors, etc.) need to be assigned either equipment numbers or asset numbers that maintenance and parts charges can be attributed to and depreciation charged against. Cost records on individual equipment are essential to justify the additional capital investment needed for new or replacement equipment. If an excessive amount of maintenance dollars are being spent on a particular crusher, screen, or piece of mobile equipment, then a replacement or an upgrade is more easily justifiable. When an asset is fully depreciated (no longer on the books) it is easier to justify a replacement piece because this equipment has essentially been paid for and the only remaining costs are operation and maintenance. Equipment costs should be tracked by total cost and cost per operated hour. Some producers keep their equipment in a separate cost center and bill the plant an hourly rate for the equipment rental. We'll talk later in the chapter about tracking equipment downtime to help justify upgrades in equipment.

INVENTORY MANAGEMENT

Once the raw ore is extracted from the ground, crushed, screened, washed, classified, stockpiled, and approved by quality control (QC), it is considered to be inventory. Inventory is an asset that has a monetary value and something that needs to be managed because the turnover of inventory is the primary source of revenue (Figure 18.3) By definition,

$$\text{ending inventory} = (\text{beginning inventory} + \text{production}) - \text{sales}$$

Sales, production, and inventory must be tracked on a daily basis. This can be done at the plant level, or in multiplant operations at the management level with individual plants reporting the information daily. Computerized weighing and ticketing systems can be networked to automatically report sales volumes.

It is not always easy to determine exact production levels of each product produced unless you have conveyor belt scales on each individual product conveyor. Therefore, inventory levels are usually approximated based on the projected production level of those products. The fact that inventory has a value makes it necessary to periodically measure the stockpiles to confirm the volumes, tonnages, and value. Stockpiles can be physically measured, but climbing on the piles can be a safety hazard and should be avoided whenever possible. Laser scanners can be utilized from a distance to get accurate stockpile volumes or aerial photogrammetry completed by an outside aerial photography firm can be utilized. The newest (and very reasonably priced) innovation for stockpile management and inventories is the use of drone technology. Aerial photography generated by drones can be downloaded utilizing computer-aided design (CAD)

FIGURE 18.3 Inventory in a stockpile has monetary value

software and very accurate stockpile volumes, in cubic yards, can be determined. Once the pile volume is calculated, it will need to be converted into tonnage (back to the concept of tons as unit costs). This conversion is fairly simple if the unit weight (discussed in Chapter 7) of each product is predetermined. Certain products may be stockpiled by trucks, the tires of which compact the underlying material. If this is the case, an additional compaction factor may need to be added to calculate accurate tonnages. An example of a typical inventory spreadsheet can be seen in Figure 18.4.

Inventories should be audited (measured) on an annual basis or at any time significant discrepancies appear. Inventory levels should always be available to the marketing/sales force so that they can determine what jobs they can reasonably bid and how to manage orders for specific products. Many companies establish a value for each ton of inventory and make a negative adjustment to the plant statements as the inventory is sold and a positive adjustment as production is entered. With this type of accounting process, it is especially important for the plant managers to monitor and adjust their production and inventories regularly.

I once heard a story about a longtime plant superintendent that was first introduced to inventory deductions on his plant statements as a result of a change of company ownership and accounting methods. When the annual inventory audits were completed and the inventory at his plant calculated he was very close to 1 million metric tons short on his stockpiles. He had been radically overstating his production for a long period of time to look good to his superiors. This of course resulted in a major loss on the P&L statement for his plant. When he was called on the carpet by the new owners, he managed to shrug his shoulders and replied: "Ground losses, I suppose." As I recall, the superintendent in question "retired" not long after the inventory audit. When I was OM, I made a point of driving the stockpile areas once a month with my plant superintendents and an updated inventory listing to make sure we were all focused on the value of inventory and making the necessary adjustments.

ENDING INVENTORY		Plant:		Date:	12/15/10		
Product	Measured Cubic Yards	Conversion Factor	Pounds per Cubic Foot	Compaction Factor	Tons	Tons in Computer System	Inventory Variance
A Rip Rap	400.0	1.12	83.0	1.0	448.0	649	(201.0)
B Rip Rap	651.0	1.12	83.0	1.0	729.1	922	(192.9)
C Rip Rap	13,042.0	1.15	85.2	1.0	14,998.3	13,489	1,509.3
D Rip Rap	11,761.0	1.15	85.2	1.0	13,525.2	3,547	9,978.2
#2	21,569.0	1.16	85.9	1.0	25,020.0	20,883	4,137.0
#3			0.0	1.0	0.0		0.0
#4	14,469.0	1.2	88.9	1.0	17,362.8	11,577	5,785.8
#4 (roofing)			0.0	1.0	0.0		0.0
#5			0.0	1.0	0.0		0.0
#57	107,969.0	1.2	88.9	1.0	129,562.8	105,984	23,578.8
#67	8,133.0	1.24	91.8	1.0	10,084.9	5,691	4,393.9
#467	1,040.0	1.22	90.4	1.0	1,268.8	6	1,262.8
#78	15,716.0	1.23	91.1	1.0	19,330.7	16,584	2,746.7
#8	29,066.0	1.23	91.1	1.0	35,751.2	20,067	15,684.2
#9	25,821.0	1.23	91.1	1.0	31,759.8	30,975	784.8
#11			0.0	1.0	0.0		0.0
#304	59,215.0	1.44	106.7	1.1	89,533.1	77,145	12,388.1
#310		1.4	103.7	1.0	0.0		0.0
#411	14,029.0	1.4	103.7	1.0	19,640.6	25,522	(5,881.4)
#617			0.0	1.0	0.0		0.0
#53			0.0	1.0	0.0		0.0
#73			0.0	1.0	0.0		0.0
Pea Gravel			0.0	1.0	0.0		0.0
DGA			0.0	1.0	0.0		0.0
Fill Sand			0.0	1.0	0.0		0.0
Natural Sand			0.0	1.0	0.0		0.0
Concrete Sand			0.0	1.0	0.0		0.0
Mason Sand			0.0	1.0	0.0		0.0
Grits			0.0	1.0	0.0		0.0
Concrete Mix			0.0	1.0	0.0		0.0
Pipe Fill			0.0	1.0	0.0		0.0
Manufactured Sand	93,625.0	1.17	86.7	1.1	120,495.4	123,448	(2,952.6)
Screenings	1,336.0	1.19	88.1	1.0	1,589.8	2,597	(1,007.2)
AgLime	2,266.0	1.17	86.7	1.0	2,651.2	1,067	1,584.2
Total	**406,015.0**				**533,751.7**	**460,153**	73,598.7

FIGURE 18.4 Typical inventory spreadsheet

MANAGEMENT INFORMATION SYSTEMS

Some form of management information system (MIS) is necessary to plan and control the essential elements of production performance at individual plants, communicate expectations to plant managers, and evaluate individual performance at the plant level. Earlier in the section on budgeting we discussed the monthly plant statement that showed variance from budget in production, costs, and sales. Tracking these variances is only one portion of an effective MIS. Other information is necessary to keep management informed and individual plant expectations highlighted. To determine actual productivity, you must first know your plant's maximum production performance. This will entail the actual measurement of peak possible plant production.

There are various forms of measuring plant performance but conveyor belt scales are by far the most commonly used measuring device. A belt scale is a measuring device that replaces a standard conveyor idler and measures actual tonnage moved by the conveyor belt. Belt scales are very accurate as long as they are calibrated on a regular basis (at intervals recommended by the manufacturer) and set to zero at the beginning of each shift. Chart recorders can be attached to the belt scales to retain a permanent record of the daily production. Belt scales contain an integrated conveyor idler that measures belt speed while a load cell measures the weight. These two measurements together calculate the metric tons per hour moving over the belt and keep a cumulative total of the tonnage produced for each shift. Remote digital readouts in the plant operator's compartment indicate instantaneous metric tons per hour and cumulative metric tons produced. It is a good idea to also have redundant remote readouts in the plant superintendent's office so the superintendent has the ability to monitor the plant performance at all times. Multiple belt scales can be installed in a processing plant but it is essential to monitor the total plant feed at a minimum. If surge piles are utilized in the process, a belt scale feeding the surge and another measuring the surge reclaim belt can allow monitoring of the various stages of the process, for example, primary production, secondary production, and wash plant production. For inventory control purposes a belt scale on each individual product belt is ideal, but the cost/benefit of the expense of numerous scales also needs to be considered.

If the belt speed is known, the actual production can be monitored by stopping the conveyor, taking a sample off of the conveyor belt, and getting an accurate weight of the sample. To get an accurate measurement, the length of the belt sample taken must be recorded and 100% of the material for that specific length must be retained for measurement. This should include sweeping the fines from the belt with a brush to retain the entire sample weight. The problem with this method is that you only get a sample of that specific window of time, no cumulative shift totals are available, and plant production is interrupted during this process. In a situation where the conveyor belt happens to be underpowered and is unable to restart under load, someone will be required to shovel off the belt prior to restarting the plant.

Once a belt scale is in place, the 100% production capacity of the processing plant needs to be measured, which will require a process known as benchmarking. Never become complacent and take the plant manufacturer's estimate of capacity for granted. These estimates are usually very conservative to compensate for the manufacturer's comfort level. The benchmarking process can be difficult and time consuming when approached and done correctly, but the results are invaluable. This process should start at the feed to the primary crusher. As much potential feed as possible needs to be stockpiled near the crusher and fed into the plant at the highest feed rate possible. The feeder control rate needs to be set at 100% while the material is fed by truck or loader into the feeder and the plant is set up at the highest capacity processing mode. The plant needs to be operated at this 100% rate until something within the process fails, for example, a conveyor belt stops due to an overload, or a crusher becomes plugged, or a chute plugs. When this initial production interruption (plugged chute, for example) is found and repaired, the entire process is begun once again. This benchmarking process is repeated until the plant can function at the 100% feed rate without interruption. We will call this 100% capacity feed rate the primary raw feed (PRF).

The best source of information in the benchmarking process is usually the plant operators. When benchmarking a high-capacity limestone quarry, as we lined up four fully loaded 50-metric-ton haul trucks to start the process, I asked the plant operator what he thought was going to happen. He grudgingly responded that "the discharge chute beneath the scalping screen will begin to back up, eventually it will plug and everything will overflow on to the ground below making one hell of a mess for me to clean up." Halfway through the second truck load that was what inevitably happened. It was a very simple fix that entailed cutting out the restriction in the discharge chute with an acetylene torch that took less than an hour to complete. This simple fix resulted in a 12% increase in plant capacity. After the repair I asked the same employee why this restriction had never been removed, and he responded with "nobody ever asked before today." A lesson for the local plant superintendent and for us all: **talk to your employees!** They are the ones that deal with the daily issues and run the plant at their "comfort level" so they don't make extra work for themselves shoveling up the spillage.

Once the 100% PRF is documented for each operating mode, plant productivity (also referred to as plant availability) can be tracked on a shift-by-shift basis. Productivity is defined as

$$\text{productivity} = \frac{\text{actual tonnage produced}}{\text{PRF} \times \text{hours run}} \times 100$$

As an example, a plant with a PRF of 750 metric tons per hour ran for 10 hours and produced 5,850 metric tons; therefore

$$\text{productivity} = \frac{5{,}850 \text{ metric tons}}{750 \text{ metric tons per hour} \times 10 \text{ hours}} \times 100$$

$$\text{productivity} = \frac{5{,}850 \text{ metric tons}}{7{,}500 \text{ metric tons}} \times 100$$

$$\text{productivity} = 78\%$$

Productivity is an excellent measure of plant performance. Using this particular example, the plant ran at a rate 22% less than it was capable of producing on that particular shift. There are numerous reasons a plant may not operate at 100% capacity. Many are related to variable performance in the pit or quarry relating to deposit issues, mobile equipment, or weather conditions; other reasons could be maintenance problems with processing equipment or a workforce shortage. Aggregates processing is not like a well-controlled factory process producing the same widget from the same raw material in an enclosed, climate-controlled environment. A reasonable productivity (availability) expectation for an aggregate operation is in the range of 85%–90% of PRF.

Maintaining productivity records provides a basis for determining if there are problems within the production process, whether they are because of lack of maintenance, improper equipment, poor management, or something else. When poor productivity is encountered, a more detailed analysis of the problem by management is warranted. *The key to achieving optimum performance is consistency in productivity.* By graphing the daily or shift productivity versus expected (85%–90%), trends can be easily determined. Figure 18.5A indicates a process that is out of control and needs to be re-evaluated as soon as possible. Figure 18.5B shows consistency of production with regard to the 85% expectation level. Productivity for each section of the plant (primary, secondary, wash plant) can be determined and tracked separately. This is especially useful when surge systems are utilized between the plant sections.

Other parameters that should be tracked and recorded are total production, production rate (tons per hour), labor hours (both straight time and overtime), total scheduled hours, maintenance hours, actual operated hours, hours of downtime, and tons per labor-hour. These are especially useful when comparing production from multiple different plants. Each of these individual parameters can be graphed for ease of analysis.

Downtime is the time within the scheduled production hours that the plant does not operate. It is critical to track the amount of downtime and the reason for the downtime to be able to analyze problem areas within the process. For example, if a specific conveyor becomes overloaded multiple times in a week causing the plant to not operate while it is cleaned off

Plant:
Month:
Year:

Date	% Productivity
1	90%
2	88%
3	62%
4	71%
5	84%
6	
7	
8	87%
9	56%
10	71%
11	84%
12	86%
13	77%
14	
15	84%
16	75%
17	66%
18	88%
19	67%
20	71%
21	
22	82%
23	69%
24	77%
25	86%
26	88%
27	
28	82%
29	77%
30	88%
31	81%

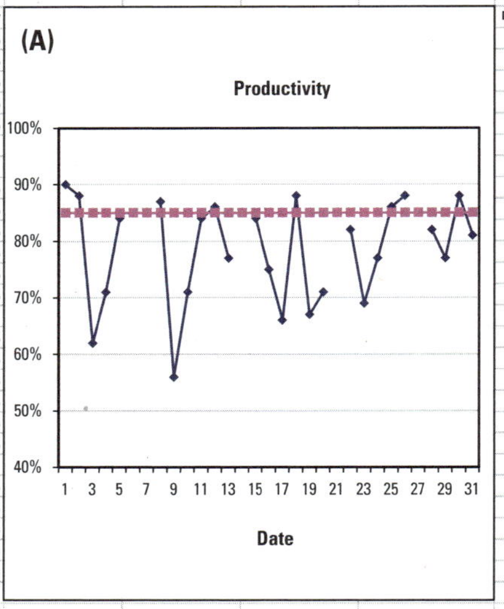

Plant:
Month:
Year:

Date	% Productivity
1	90%
2	88%
3	88%
4	92%
5	91%
6	
7	
8	87%
9	82%
10	91%
11	84%
12	86%
13	84%
14	
15	84%
16	80%
17	91%
18	88%
19	86%
20	83%
21	
22	82%
23	79%
24	87%
25	86%
26	88%
27	
28	82%
29	86%
30	88%
31	81%

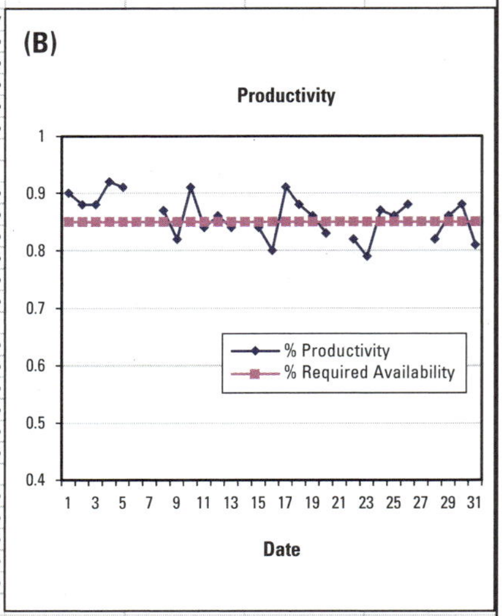

FIGURE 18.5 (A) Poorly controlled process, and (B) controlled process

and restarted, then an updated drive needs to be considered, or if the recirculating load from a crusher is overloading a screen, it's time to adjust the crusher settings. Downtime can also affect quarry or pit production and should be tracked for each piece of mobile production equipment. If a haul truck or wheel loader is only operating 60% of the time because of a transmission overheating, it's time for a transmission rebuild and/or a new truck. **Nothing costs an aggregates producer more than unscheduled downtime.** When a processing plant fails to operate during its scheduled production shift, the sum of all the operating costs does not change, but the cost has no divisor (per metric ton) because no production results. Every hour without production incurs the same cost as the production hours and significantly increases the overall unit cost of the material.

Accurate shift production reports need to be kept by the plant operators (Figure 18.6). These reports need to reflect the starting and ending shift times, the total production tons, products produced, scheduled versus operated hours, downtime, and the reasons for the downtime. Plant personnel need to be trained on the importance of and reasons for accurate reporting and record keeping. Production reports should be kept as simple as possible while providing the necessary information. The causes for downtime need to be documented and the cumulative downtime hours for each category recorded to determine the required remedies for each downtime issue. These data can be used to justify replacement of equipment causing the operating issues. Downtime costs money and, when documented, may easily quantify the savings necessary to purchase a new or replacement piece of equipment. Typical downtime reasons that should be recorded are weather (lightning); blasting; electrical problems; no plant feed (pit related); mechanical issues related to crushers, screens, conveyors, feeders, screws, pumps, and classifiers; plugged chutes; plugged crushers; screen overloads; and other. When

Plant Operator's Report

Plant:______________________________ Date:______________________

Start Time:____________ Stop Time:______________ Shift: 1 2 3 (Circle one)

Scheduled Hours:______________________ Operated Hours:______________________

Gross Production: ________________tons Truck/Bucket Count:____________ loa

Belt Scale: Starting tons :_________________ Ending tons: ______________________

Products Produced: (✓Check Box, List other products as needed)

Product	✓	Product	✓	Product	✓	Product	✓
Rip Rap		#78		Screenings			
#1		#8		Man. Sand			
#2		#9		Concrete Sand			
#3		#11		Asphalt Sand			
#4		#12		Filter Sand			
#5		#304 / #53		Fill Sand			
#57		#411 / #73		Ag Lime			
#67		#310					

Unscheduled Downtime

Time			Downtime	Comments
From	**To**	**Total**	**Reason**	**(List Equipment)**
:	:			
:	:			
:	:			
:	:			
:	:			
:	:			
:	:			
:	:			
:	:			

Downtime Reasons:

Weather	Mechanical Conveyor
Blasting	Mechanical Screw/Logwasher
Electrical	Mechanical Classifier
Mechanical Screen	Chutes Plugged
Mechanical Crusher	Crusher Plugged
Mechanical Feeder	No Plant Feed (Pit related downtime)
Mechanical Water System	Other

Operator:______________________________

FIGURE 18.6 Example of a typical plant operator's report

mechanical issues occur, the piece of equipment affected needs to be identified, for example, tertiary crusher, scalping screen, Conveyor #6.

Tracking metric tons per labor-hour is another measure of efficiency that is especially useful when comparing different aggregate operations. Although no two aggregate operations are exactly alike, because of differences in the deposit, length of haul to the plant, different plant and mobile equipment, and so on, metric tons per labor-hour can be utilized as a relative comparison of plant efficiencies. If an individual aggregates facility is producing 65 metric tons per labor-hour and another 30 metric tons per labor-hour with no obvious explanation, the reasons for the lower rate should be investigated and analyzed. The tons per labor-hour calculation removes the variations in plant capacity, type of equipment, and deposit from the efficiency calculation.

Because the terms relating to operating hours can become confusing, the following terminology is defined:

- M&R hours = Scheduled maintenance and repair hours
- Standby hours = Planned nonoperating hours (start-up/shutdown, plant inspection, safety meetings)
- Scheduled production hours = Actual hours operated + unscheduled downtime
- Unscheduled downtime = All downtime not included in scheduled M&R hours
- Total hours = Scheduled M&R + standby + scheduled production hours

We discussed a production budget being completed, prior to cost budgeting, based on existing inventories and projected sales budgets from the sales department. From this production budget, monthly production projections will be derived and the daily production needs to be recorded. Personally, I am a very visual learner and find that maps, blueprints, CAD diagrams, and graphs are very useful tools. One tool in particular that is extremely useful in tracking actual monthly production versus budgeted production is the daily production graph. I required all of my plants to keep a daily graph of cumulative monthly production and year-to-date production versus the budget. This is a simple task with an Excel spreadsheet where the daily budget forms a straight line from Day 1 to Day 31. Actual production is entered daily for the previous day and totaled to get cumulative monthly production, which is added to the graph as a different colored line. This is a very simple visual tool that instantly shows if the production is ahead or below budget by whether the cumulative line is above or below the budget line.

The first thing I asked to see when entering the plant office was the monthly production graph (Figure 18.7). This gave me an instant impression of what was being accomplished at that operation. A savvy plant manager can utilize this tool to plan maintenance projects and stay on track with budget projections and productivity levels. If a large maintenance project requiring 2 days of downtime is pending on the 25th of the month and the graphs show that the plant has already met or exceeded the monthly production budget, that would be a good time to schedule the project. If, on the other hand, the plant is behind budget, now may not be the appropriate time. Figure 18.7 is an example of a monthly production graph where the cumulative daily production (blue line) is on schedule and slightly ahead of budget (red line).

Monthly Production

Plant:
Month:
Year:
Budget Tons: 145,000

Date	Tons Produced	Cumulative Tons	Budget
1	5,857	5,857	4,677
2	5,322	11,179	9,355
3	4,100	15,279	14,032
4	5,775	21,054	18,710
5	6,103	27,157	23,387
6	5,116	32,273	28,065
7	0	32,273	32,742
8	5,122	37,395	37,419
9	5,377	42,772	42,097
10	5,375	48,147	46,774
11	2,441	50,588	51,452
12	5,210	55,798	56,129
13	5,877	61,675	60,806
14	0	61,675	65,484
15	5,915	67,590	70,161
16	6,125	73,715	74,839
17	5,858	79,573	79,516
18	6,214	85,787	84,194
19	5,817	91,604	88,871
20	5,327	96,931	93,548
21	0	96,931	98,226
22	5,588	102,519	102,903
23	6,132	108,651	107,581
24	6,104	114,755	112,258
25	5,998	120,753	116,935
26	5,775	126,528	121,613
27	5,400	131,928	126,290
28	0	131,928	130,968
29	6,066	137,994	135,645
30	5,818	143,812	140,323
31	6,177	149,989	145,000

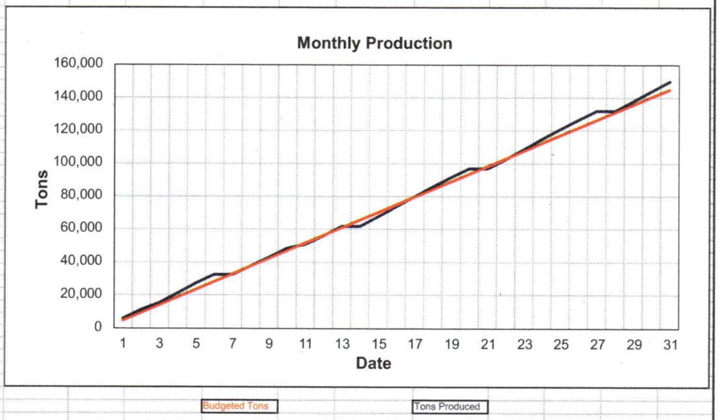

FIGURE 18.7 Monthly production graph (red line = budget, blue line = actual)

Focusing on total costs only, without considering productivity, tons per labor-hour, repairs and maintenance costs, production rates, and unit costs (cost/metric ton), gives only a partial picture of plant performance and provides no means of evaluating any potential for improvement.

PLANT MAINTENANCE

Routine maintenance must be scheduled on a regular basis when the plant is not in operation. These maintenance hours should be tracked separately and not categorized the same as operated (production) hours. Aggregate processing plants are a complex combination of mechanical components that, if not properly maintained, will break down at the worst possible time. Large plants requiring high production rates will most likely require one entire shift of daily maintenance to sustain the level of productivity necessary for a profitable operation. The plant operators, plant supervisor, and/or maintenance supervisor should maintain an ongoing, continually updated list of maintenance items that need to be addressed, prioritizing the maintenance issues that have the potential to shut down the process. High-priority items that could cause a plant shutdown should be addressed immediately on the next maintenance shift. Many plants operate on night shift (second and/or third) and complete the necessary maintenance on the day shift (first) where they have the advantage of daylight and better availability to parts suppliers, equipment rental, and such. Major maintenance projects can require multiple shifts or even multiple days and need to be planned around sales demands and inventory levels. In climates where sales are seasonal, many operations plan major maintenance projects for the winter months when demand is lowest and production personnel are available to assist in completion of the required projects.

Every piece of equipment in a processing plant will come with an operating and maintenance manual outlining the specific intervals between preventive maintenance. The manuals will specify how often to grease bearings, change oil in the gearbox, or replace drive belts. These intervals need to be strictly adhered to and integrated into the plant's maintenance schedule. By tracking the hours a plant operates, a schedule can be developed when routine maintenance needs to occur. This can be easily accomplished using a laptop where the daily and cumulative operating hours can be tracked and the maintenance intervals entered with alerts or reminders generated by the computer. Preventive maintenance, predictive maintenance, and repair before failure are the keys to cost-effective, efficient production. Never let maintenance items go unattended for the result will be a costly equipment failure that could easily have been avoided by completing routine maintenance tasks on schedule.

Individual maintenance and safety checklists should be developed for each piece of equipment based on the manufacturer's recommendations. These should be simple and easy to fill out with check marks to provide guidelines for what the employees need to be focusing on when inspecting the equipment. If an item on the list is not checked as "OK," an explanation is necessary to identify the problem. Major problems need to be brought to the attention of management as soon as possible. The plant and/or maintenance supervisor needs to review the checklists on a shift-by-shift basis and assign any problems to the list of pending maintenance projects with a priority level assigned. The checklists can be fairly generic to the type of equipment rather than individual pieces of equipment, for example, cone crushers, horizontal shaft impact crushers, screens, and conveyors. Employees should be educated/trained on the reason for and importance of regular maintenance and inspections of equipment. Periodic

Cone Crusher Maintenance Check List

Plant:____________ Crusher:____________ Date:__________

Item	OK ✓	Needs Work ✓	Comments	Priority (high/ low)
Motor				
Motor bearings				
Drive belts				
Drive sheaves				
Oil level				
Oil pressure				
Oil flow				
Oil temperature				
Oil pump				
Springs				
Accumulator				
Feed chute				
Discharge chute				
Feeder				
Feeder motor				
Feeder drive belts				
Walkway				
Handrails				
Crusher action				
Abnormal noise				
Abnormal bounce				
Crusher setting				
Feed distribution plate				
Choked feed				
Main frame				
Dust emissions				
Dust suppression spray				
Electrical feed				
Electrical wiring				

Was the oil level checked? (circle one) Yes No
Was oil changed on this shift? (circle one) Yes No
If yes, was an oil sample taken for analysis? (circle one) Yes No
Crusher Setting: ______ inches

Checked by: ____________________________

FIGURE 18.8 Typical maintenance/safety checklist for a cone crusher

re-education and encouragement of employees will be necessary to avoid complacency in the inspection process. The checklists need to be signed by the employee responsible for the process and the employees need to be held accountable for accurate inspection and reporting. An example of a maintenance checklist is shown in Figure 18.8. Deficiencies on a checklist should be followed up on immediately by plant management. Nothing is more demoralizing to employees than seeing their efforts go unacknowledged.

Many traditional managers focused on cost savings and ignored the repair-before-failure philosophy. I once had a superintendent that reported to me who was a very affable, intelligent, well-educated engineer. I visited his operation one day and found the plant not running while his employees were splicing a piece of conveyor belt into a crucial conveyor. While I watched the process, I counted nine splices in the conveyor belt where it had been patched at least eight other times. When I confronted him and asked why he didn't just buy a new belt, he said that he kept a supply of used belting for just such an occasion to save money. I then escorted him to his office and we added up the downtime in his plant resulting from conveyor belt splicing over the past year. When we tallied the downtime spent repairing conveyor belts, we could have replaced every belt in the plant and had money to spare. This was a recurring issue with the same superintendent: he would operate the plant with little maintenance until something failed and then repair it (during scheduled operating time), reasoning that he was saving money by not spending the time to do the proper preventive maintenance. I finally came to the conclusion that when things were running smoothly, he was bored and what he really enjoyed was fixing things. I attempted to educate him over the next several months on the cost advantages of preventive maintenance and repair before failure, but no matter what I tried he could never get with the program. Needless to say, I finally had to replace him. Once again, I want to reiterate that *nothing is more costly than unscheduled downtime.*

Preventive maintenance (greasing, oil changes, filters, regular inspection) is only one phase of an effective maintenance program. Predictive maintenance is a crucial step in eliminating unscheduled downtime and costly repairs. A simple tool that can be purchased for about $100 at most industrial supply houses is an infrared thermometer, also known as a heat gun. These simple tools can be used regularly to test the temperature of bearings, gear boxes, electrical equipment, and such, as an indication of impending problems. All of these mechanical and electrical items should operate at close to ambient temperature. Signs of overheating are an indication of a problem. Overheating of mechanical parts indicates excess friction, which can be indicative of a lack of lubrication, excessive internal wear, or misalignment of parts. Overheating of electrical parts usually indicates loose connections, which can cause electrical failure, result in fires, or potentially cause serious injury.

Equipment lubricated with oil (bearings, gears, etc.) needs to be changed on a regular basis as recommend by the original equipment manufacturer (OEM). Extending lubrication intervals to save time or money on lubricants is never a good idea. Samples of the used oil can be analyzed by a laboratory and reveal wear particles of various metals. Some wear metals are expected, but large percentages indicate abnormal wear and impending failure. A qualified lubricant laboratory can provide advice on wear metal levels, what metals are present in what quantities, and what parts are likely to be wearing excessively. High levels of iron particles usually indicate shaft wear, while excessive chromium is indicative of bearing wear. Many machines are equipped with magnetic drain plugs, which attract wear metals. Visual examination of the waste oil can also indicate whether excessive wear metals are present; for instance, chromium and copper leave a residual sheen on the oil. Routinely sampling and monitoring wear metals in the waste oil can predict eminent failure of the mechanism. A combination of high levels of heat and excessive wear metals is a definite recipe for impending failure.

Another very effective method of predictive maintenance is vibration analysis. This requires some very sophisticated, expensive equipment and experienced personnel to accomplish. There are companies that will come to your operation and provide this service for an hourly fee. Regular annual vibration analysis on the key plant components that have the ability to shut an operation down (crushers, screens, high-horsepower motors, etc.) is an excellent indicator of plant health. Regular monitoring of the frequency and amplitude of vibrations can identify trends that can predict equipment failures.

I was first introduced to the value of vibration analysis in the mid-1990s, when I was superintendent of a large midwestern limestone operation. It was late September of a very busy sales year when my primary plant operator came to me and said the bearings on the primary crusher were beginning to make a rattling noise, that is, "bearing chatter." It was critical for maintaining sufficient inventory going into the winter season that the plant remain in operation until the Thanksgiving break, at which time we planned to shut down for the winter. I started by having the operator keep a daily log of the time it took for the crusher rotor to stop spinning after the motor was turned off. If the trend was for the rotor to stop more quickly than normal, that was an indication of additional friction and potential bearing problems. I then hired a vibrations consultant who completed a vibration analysis of the rotor and motor bearings on the crusher. His analysis indicated substantial rotor bearing noise, especially on the drive-side bearing. I did not want to shut down the entire plant for the 3 days it would require to replace the bearings. I had the consultant come out to the plant weekly to monitor the bearings and we watched for trends in frequency changes. The bearing noise (frequency, amplitude) worsened on a weekly basis but by monitoring the situation, both ramp-down time and vibrations, we were able to make it to Thanksgiving and replace the badly worn bearings during the winter shutdown. From that day forward I routinely used this diagnostic tool to predict potential equipment failure on the larger, key plant components.

MOBILE EQUIPMENT MAINTENANCE

Mobile equipment maintenance is a subject worthy of an entire book and several have been written on the subject. I do not consider myself a qualified heavy equipment mechanic, but I have picked up a lot of the basics over my career. An efficient fleet of mobile equipment is a key component to the profitability of any aggregate operation and an effective maintenance program is absolutely necessary to keep the equipment operating at peak efficiency. Many of the same principles and procedures outlined in the previous section on plant maintenance apply to the mobile equipment fleet. Daily inspection, preventive maintenance, predictive maintenance, and repair before failure are all necessary components of an effective fleet management program. Tracking the costs of each piece of the mobile fleet on a per-operated-hour basis (back to unit cost again) is necessary to make rational equipment replacement or rebuild decisions. Costs should be categorized for each of the major components of all the mobile equipment:

- Engine
- Drive train (transmission, torque convertor, final drives)
- Steering system
- Braking system
- Hydraulic system
- Undercarriage (frame, structural components)
- Body
- Tires
- Ground-engaging tools (for bulldozers, excavators, etc.)
- Miscellaneous

Each of the above categories should include all parts, labor, rebuilds, lubricants, and supplies and should be tracked as the total category cost along with the cost per hour operated. Adding the depreciation and fuel to the maintenance gives a total owning and operating cost, which is essential to evaluating different brands and types of equipment. The last company I worked for, which included asphalt, aggregates, and highway contracting, had in excess of

3,000 pieces of equipment to maintain, track, and evaluate. Thank the heavens we were in the computer age; without computers it would be a daunting record-keeping task. As discussed earlier, a chart of accounts linked to equipment numbers and the component categories can be utilized to easily capture the costs for analyses.

Just like your automobile, preventive maintenance intervals will be outlined in the equipment maintenance manuals. Routine oil changes, filter changes, tire rotations, lubrication (greasing), and equipment inspections are necessary to maintain your fleet at peak performance. Unlike automobiles, the maintenance intervals for heavy equipment are generally defined by operated hours instead of mileage. All equipment of this type has hour meters, which should be recorded at the beginning and end of every shift. Entering the hours operated into a spreadsheet can simplify tracking when individual equipment is due for routine maintenance as well as periodic component rebuild. Use of high-quality lubricants of the specified ISO viscosity for each individual component along with good-quality filters will help prevent premature mechanical failure. Proper training of mobile equipment operators in the safe operation and maintenance of equipment is necessary to prevent equipment abuse and premature failure, and provide a safe working environment.

Predictive maintenance is also a necessary step in maintaining a fleet of mobile equipment. At every specified oil change interval, for each different component, a sample of the used oil should be carefully obtained and sent to a qualified lubricant laboratory for analysis. The sample bottle must contain the equipment number, the component (engine, transmission, right final drive), equipment hours, and the brand and viscosity of oil utilized. The oil analysis report will specify a list of chemicals and wear metals present with their quantity in parts per million as well as acidity levels, soot content, water content, coolant levels, fuel, abrasives, and viscosity. Most lab reports will have a warning system when any element exceeds certain levels and will generally specify each element present as either a normal, abnormal, or critical level. Some will identify the potential problem associated with the sample results as long as the individual component was identified on the sample label. Particle counts are often run on oil samples as well as the elemental spectroscopy to provide information on the size and concentration of the wear particles. The OEM can provide expected levels of wear metals in oil analyses for various hour meter readings on all major equipment components.

High acidity levels can indicate excessive oil oxidation, which can lead to corrosion of internal parts. Excessive soot can be an indication of improper fuel combustion or malfunctioning fuel injectors. Improper viscosity levels may be a result of the wrong oil being used at the last oil change or some form of dilution of the oil. Coolant within the engine oil is an indication of a cracked, leaking head. High levels of silica and abrasives in the oil results from dirt contamination caused by improper sampling practices, poor oil change techniques, or perhaps a problem with an oil breather. Small levels of fuel in an oil sample usually indicate a fuel pump malfunction, perhaps a leaking diaphragm, which will require replacing the fuel pump as soon as possible.

It is imperative that oil sample results be scrutinized as soon as they are received to avoid potential mechanical failures. I once had a shop foreman that, unbeknownst to me, accumulated oil sample results in a desk drawer but seldom, if ever, looked at them. I found this out after a catastrophic engine failure that cost the company almost $35,000. When I asked the foreman to see the oil sample records for this particular piece of equipment, he opened a desk drawer overflowing with test results, many of which were still in unopened envelopes and none filed in the pertinent equipment files. When we finally sorted out the mess of oil samples, it was evident that a problem with this particular engine had been progressing for some time. The trend was obvious in the sample results and the engine could have been repaired before failure at less than half the cost of a replacement engine. A replacement shop foreman reported for work about 3 weeks later.

At times it may be advisable to resample the oil for analysis on a shorter time interval than normal to confirm potential problem areas. Many lubricant laboratories will offer a cautionary warning and say change oil and resample in 50 or 100 operating hours. The overall trend in the oil analyses is very useful in predicting when the rebuild of a certain component is necessary. Increasing levels of wear metals from sample to sample indicate that wear is becoming excessive and something is inevitably going to fail. This is where the concept of repair before failure comes into effect. As wear levels begin to increase, expanded sampling intervals can show the increasing trends until the levels reach the critical point. Consulting with the OEM at this juncture can provide valuable insight on the trending oil analysis. This is probably the time to rebuild the component in question prior to a catastrophic failure. Many parts of a mechanical component (engine, transmission) can be reused if the component is repaired before failure, saving thousands of dollars. Most OEM dealers will have rebuilt or new components available for quick turnaround if the timing dictates that the equipment cannot be down for a long period of repair time.

Mobile equipment management must be a concentrated effort by the entire organization: management, supervisors, the maintenance mechanic, the lubrication technician, and the equipment operator. All mobile equipment needs to be inspected prior to each shift operated for any obvious mechanical issues and for safety reasons. A pre-shift checklist should accompany every operator and every piece of mobile equipment and should be meticulously filled out prior to operating the equipment. This checklist should include basic maintenance items like oil level, coolant level, and tire pressure, and also encompass a basic safety inspection on items like seat belts, tire wear, brakes, lights, and backup alarms (Figure 18.9). Employees need to be trained in the reasons for and importance of these checklists, both for the increased equipment longevity and especially for their own safety. A supervisor must review these checklists on a daily basis to assure the proper maintenance and safety items are prioritized and addressed. These lists must also be kept on file as a record of the shift safety inspection for later Mine Safety and Health Administration audits.

Additional information may also need to be tracked to evaluate the performance of the mobile equipment maintenance program. Caterpillar (CAT), one of the largest heavy mining equipment OEMs, has developed a series of key performance indicators (KPIs) that they recommend all mining operations adopt to assess the effectiveness of their fleet management program. These KPIs include:

- **Mean time between shutdowns (MTBS).** The average frequency of machine stoppages, which combines the effects of inherent machine reliability and the effectiveness of the equipment management organization in its ability to influence reliability through problem avoidance. CAT feels that this is the single most important measure of the equipment maintenance program performance.
- **Mean time to repair (MTTR).** The average duration of machine stoppages or repair turnaround time. This is a measurement of how quickly or slowly a machine is returned to service and the efficiency of the equipment management organization.

Operator's Log & Safety Checklist

Date: ____________________ Equipment # _________________________ Shift 1 2 3

Hours Operated: Ending Meter Reading: _______________

Starting Meter Reading: _______________

Total Hrs. Operated: _______________

Maintenance:	OK		Gallons Added
Engine Oil	□		__________
Engine Coolant	□		__________
Power Train Oil	□		__________
Hydraulic Oil	□		__________
Fuel	□		__________
Air Cleaner	□		
Drain Moisture from Air Reservoir?		□ Yes	□ No
Greased Equipment?		□ Yes	□ No

Safety Check:	OK	Needs Service	Date Repaired
Steering	□	□	__________
Brakes	□	□	__________
Lights	□	□	__________
Windows	□	□	__________
Wipers	□	□	__________
Seat Belts	□	□	__________
Back Up Alarm	□	□	__________
Horn	□	□	__________
Fire Extinguisher	□	□	__________
Tires	□	□	__________
Access Ladder	□	□	__________
Interior Clean?	□	□	__________

Comments: ____

____

Operator/Mechanic's Signature____

FIGURE 18.9 Typical mobile equipment operator's shift report

- **Availability index.** The ratio of MTBS (average shutdown frequency) to the sum of MTBS and MTTR (average shutdown duration), expressed as a percentage.
- **Percent scheduled downtime.** The percentage of total downtime hours in a given period that have been planned and scheduled. A proactive organization generally has a high percentage of scheduled downtime.
- **Maintenance ratio.** The ratio of maintenance and repair labor hours to machine operating hours, indicating the amount of effort required to keep equipment in service.
- **Asset utilization.** The percentage of time that a machine is operating (operating hours) divided by the total calendar time in the period. (In mines that do not operate 24/7, the devisor should be scheduled hours rather total calendar time.)

- **Top problems.** The distribution and impact on availability and costs of problems affecting a fleet ranked in terms of MTBS and MTTR.

Utilizing any or all of CAT's KPIs can provide an indication of the efficiency of the individual fleet equipment maintenance program. These can also be used to compare different fleet maintenance programs between various plant fleets.

The Association of Equipment Management Professionals (AEMP) provides educational materials, seminars, technical manuals, and three different levels of certifications for its membership through the AEMP Equipment Management University. The various certifications require varying levels of technical education and passing an examination to qualify. The Equipment Management Specialist Certification is designed for individuals early in their career and requires a minimum of 5 years of experience as a prerequisite. The Certified Equipment Support Professional Certification is geared toward professionals that are suppliers of heavy off-road equipment (equipment dealers), while the Certified Equipment Manager Certification demonstrates a mastery of fleet management skills and a proven ability in all fleet-related disciplines. These programs can be very beneficial in educating and developing personnel in your mobile equipment maintenance department.

One of the early production managers I worked for had a very simple management philosophy that consisted of three items: (1) Maintain all equipment on a routine schedule without exception; (2) keep the operation clean, well organized, and environmentally sound; and (3) if you do Items 1 and 2, you will achieve the production and efficiency necessary to be profitable. This is a hugely oversimplified statement but it also has a lot of merit and I reiterated it to my supervisory employees many times over the years.

HUMAN RESOURCES

When I started my career in the 1970s, the term *human resources* did not exist. Most companies had a personnel department that processed employment applications, managed payroll, withheld taxes, and sent out IRS forms. Times have certainly changed, for the better, I might add. Call the department human resources or personnel, it matters not, but your employees are your company's greatest asset. By employees I mean everyone from the CEO to the office clerk, the delivery truck driver to the QC technician. As employees, we are all cogs in the wheels spinning toward profitability. A company that is not profitable will not require employees for the long haul. Employees that are not motivated to perform their jobs efficiently will either become unemployed or run the company out of business, which in the end has the same result. This is why personnel recruitment is such an important function.

Choose your employees with great care because once hired they may become difficult to be rid of in the future. It's not enough to look at a resume and complete a 2-hour interview over lunch. It's easy to be fooled by a winning personality or to hire on the spot out of an immediate need for personnel in a tight job market. Always check a candidate's references thoroughly prior to hiring them. I worked for a small consulting firm as an area manager for about 3 years until I decided that consulting was not my cup of tea. The president of the firm wanted me to hire an environmental professional to work from my office. I reviewed numerous resumes and interviewed several candidates over a 3-month period while at the same time I was looking for an operations job within the aggregates industry. I found what I thought was a good candidate for the environmental consultant position at about the same time I accepted a new job.

I called the president of the consulting firm and told him that I was resigning to pursue the operations position, gave him 1 month notice, and said that I had also found a candidate for the environmental position. I didn't feel right hiring the candidate under the circumstances, so I forwarded his resume, and suggested that he fly the candidate to New York and personally conduct an interview. After his interview in the New York office the candidate was hired as my replacement (area manager) and he started his new job as I was starting mine. Other than a few phone conversations during the transition, I had very little contact with my replacement in the consulting office. About 4 months into my new job, a VP of the consulting firm, who was also a good friend, called me and asked if I had opened an account with Home Depot in the consulting company's name before I moved on to my new job. My response was "No, I never had a need for such an account." This was just the beginning of my replacement's interesting employment odyssey.

As it turned out, my replacement had indeed opened the Home Depot account, and in doing so forged my signature on the application. This started an internal consulting company investigation where they began by checking his references, which they had not done prior to his hiring. Not only were his references bogus, his work history was mostly fictional and the university from which he claimed both a B.S. and master's degree said he had only attended for 4 months and had never graduated. Closer scrutiny revealed that he actually never quit his previous job and was being paid by both companies simultaneously. As the investigation progressed it turned out that his wife owned a remodeling business and he was charging her decorating supplies (paint, wallpaper, etc.) to the Home Depot account he had fraudulently opened. When confronted and threatened with legal action, he attempted to blackmail his newest employer with exposing confidential information concerning a very good client. His ploy was successful; rather than compromise their client, the company dismissed him without pursuing further legal action. Be very careful who you employ and do your homework prior to hiring!

Always hire the best possible candidates for the job in question and have high expectations. A good employee will live up to the boss's expectations. I once asked a drilling contractor if he knew of any potential candidates for a very demanding plant superintendent position I had available. He looked at me said: "You expect too much of a superintendent. You want someone who can rebuild a cone crusher during the day, then change his clothes and give a talk to the Kiwanis club that same evening. Candidates like that don't exist!" My reply was: "You're right, that's exactly what I expect. If I can do it, there are others out there that can do it as well." I always tried to hire employees that I thought had the capability of being my replacement.

Not only should you thoroughly check references, work history, and education, it is also a very good idea to do a criminal background check. Numerous companies specialize in doing checks on people's criminal backgrounds on a per-case fee basis. Hiring someone with a criminal history is a decision that needs to be made prior to employment. Granted, everyone is entitled to past mistakes and their actions in the past may not reflect who they are today, but it is better to make an informed decision than be surprised after the fact.

Your employees need to be treated as the valuable assets they are. I refer to all employees, not only managers and supervisors. The hourly employee that shovels up spillage around the plant, the plant operator, the loader operator, the truck driver all play integral roles in the success of the company. The employees' attitude on a daily basis can make or break a company, especially a new company attempting to enter a new market or business. All employees need to be treated with respect and dignity, no matter if you personally like them. Make it a point to get to know their names, talk to them, and ask their opinions whenever possible. Remember the primary operator in the earlier section on benchmarking who knew exactly what the first hurdle to increasing production was going to be, but "nobody ever asked." Good managers go out of their way to praise individuals and/or teams of workers when they have accomplished something above and beyond. A great deal of employee satisfaction is feeling valued as an integral part of an organization. Satisfied employees are not only more productive employees, they are much easier to manage. Retention of employees is as important, or more so, than the initial hiring. Excessive employee turnover within an organization or specific plant is a sign of something amiss within the corporate culture. A person's perception is their reality; if employees perceive that they are not valued, whether that is management's intent or not, they will look elsewhere for employment with greater job satisfaction. A simple remark, word of encouragement, or pat on the back from the boss goes a long way toward employee satisfaction and retention!

Wages and benefits need to be competitive with or exceed those provided by the competition. Employees that feel underpaid also feel underappreciated. It is true that labor costs make up a considerable percentage of the unit cost of aggregates, but if paying a higher wage results in increased productivity because of improved employee attitude, the costs can be easily offset. An employee concerned about a spouse's or child's health condition and worried whether their insurance will cover the bills will be distracted and unable to concentrate on their duties. Health insurance benefits provided to employees should be as comprehensive as is reasonably affordable. The aggregates business is demanding, especially during the height of the construction season, requiring employees to work numerous overtime hours. Providing time off for employees to rest and rejuvenate is also necessary, especially during the busy season. Competitive paid vacation and holiday plans are a necessity for everyone's mental health, employees and management alike.

An employee handbook or manual should be provided to all employees outlining their duties and responsibilities and the company's expectations. This should include a comprehensive list of working rules and expectations on tardiness, absenteeism, insubordination, discrimination, deportment, alcohol and drug usage, and treatment of fellow employees. These rules are sometimes outlined within a union contract or employee agreement rather than in a handbook. These expectations should be accompanied by any punitive management measures should the regulations be ignored. What happens if an employee is 10 minutes late for their shift? What if they are late three times within a month? How many excused absences are permitted within a certain period of time? How many unexcused absences are permitted prior to disciplinary action? These issues should all be defined in writing and the rules enforced equitably by supervision. A 90-day probationary period where further employment is conditional on performance is always a good idea when hiring new employees. Any union contract or employee agreement should always contain a management rights clause similar to the following:

> The operation of the Employer's business, direction of the work force, planning and control of operations, subcontracting of work, the right to relieve employees from duty because of lack of available work or for any other legitimate reasons, determination of the minimum qualifications of experience, health, or fitness for any job

covered hereby, and the right to appraise the qualifications of any employee including the right to hire, suspend, layoff, transfer, discipline, promote, or discharge, are vested exclusively in the EMPLOYER.

Supervisors sometimes are reluctant to enforce disciplinary measures against their employees, especially if the individual employee is a good worker and well liked. Discipline needs to be fair and equitable toward each and every employee without exception or discrimination. That is the purpose of the employee handbook, so everyone knows what is expected of them. Is an employee to be terminated for being late for shift on one occasion because of a flat tire on the way to work? Of course not, but the employee needs to cautioned and it needs to be noted in their employee file and tracked to make sure a trend in tardiness is not beginning to develop. Disciplinary action must always be documented, especially any trends in multiple offenses. Discipline should always be progressive; first offense with a warning, second offense with a second warning, third offense may require a suspension from work without pay, and so on. More serious infractions may require immediate termination, for example, theft of company property, drinking on company property, or fighting. A supervisor is doing no one any favors by ignoring disciplinary action. No one likes to discuss unpleasant subjects with employees, but ignoring offenses emboldens the offender and demoralizes the other employees. Treat all employees fairly and with respect, and enforce company expectations immediately. Meting out discipline equitably, when necessary, does not make a supervisor a mean boss. Employees will respect this as long as it is done fairly, without discrimination, and they have been informed of the expectations beforehand.

Disciplinary action should always be accompanied by education and counseling. I don't mean marriage counseling or drug abuse counseling, but counseling on why the employee is being disciplined, the nature of the infraction, and what effect their actions have on the operation and/or the morale of the other employees. This should be done in a matter-of-fact manner with as little emotion as possible. Any employee who is late for work three times in two weeks should be told that their tardiness affects the efficient start of the shift and that their fellow employees make an effort to get to work on time and resent the fact that this individual employee does not make the same effort. It has been my experience that when a good employee starts to exhibit tardiness or absentee issues that something outside of work in their personal life is usually the cause (divorce, health issues, etc.). In this type of case a leave of absence may be called for rather than the loss of a good employee. Discipline should always happen immediately following the infraction and it is wise to have another manager or supervisor present to witness the process. A witness can eliminate any controversy over what was said if future allegations of wrongdoing or charges of unfair labor practices arise.

I've had unfair labor practice charges filed against me with the U.S. Equal Employment Opportunity Commission more than once during my career, but always had sufficient documentation to have the charges dismissed. I was once charged with discrimination in hiring practices when, after interviewing a candidate for an hourly position, I hired a different qualified candidate, who happened to have a much better attitude, instead. The gentleman filing the charges said that I discriminated against him because of his Native American heritage. I was able to get the charges dropped when I provided evidence that one of the supervisors on site was half Native American and that the superintendent, to whom the candidate would have reported, was married to a Native American woman. I've never been sure what prompted the discrimination charge unless the candidate was looking for what he thought was an easy monetary settlement.

During my four-decade career, I have had to fire several employees, both salaried and hourly. Terminating an individual is never easy or enjoyable, but I have never terminated anyone that did not deserve it and never without giving them more than one opportunity. Some employees become vindictive upon termination. In one instance, a terminated supervisor was found trespassing and attempting to steal tools on a Sunday about 2 weeks after losing his job. He was charged with theft and criminal trespass by the county sheriff. In another case, the human resources manager and I confronted an hourly employee after being informed by one of his fellow employees that the person in question was using a company fuel credit card to fill his own gas tank on a fairly regular basis. After scrutinizing the credit card account we found about $1,200 worth of questionable charges over an 18-month period, principally at two different gas stations. Our security service went to the service stations and looked at surveillance camera footage but found it inconclusive. When we confronted the employee, the human resources manager told him what we suspected, the locations of the two gas stations involved, and that the sheriff's department was going to look at the surveillance footage for the times and dates of the credit card charges. The employee readily agreed to resign immediately if we agreed not to press criminal charges. About 1 week later, the employee that alerted us to the theft found all four tires on his pickup truck had been slashed with a knife during his shift.

Continuing education for employees is a valuable resource for both the employees and the company. Offering college tuition reimbursement for completing courses related or unrelated to their jobs helps everyone concerned. There are many opportunities to increase the skill set of hourly workers as well. Many OEMs offer courses covering the operation, maintenance, and repair of their specific equipment. The National Stone, Sand and Gravel Association sponsors an annual equipment show and educational event called AggOne that rotates between different cities around the country. This event provides employees continuing industry-related education, showcases the latest equipment innovations, and gives employees the opportunity to network with other industry professionals. An educated, informed employee is a more valuable employee. A few days out of town on the expense account can also be a very nice perk for a valued employee.

Some companies offer formal mentoring programs that link newer employees with more senior employees to share their experience. Mentoring need not be formalized into a program but should be able to occur naturally. Take a new employee under your wing, familiarize them with operational procedures, safe working methods, maintenance techniques, company policies … share your knowledge. Mentoring should be encouraged among the older, more experienced workforce. Although not always possible, I've invariably advocated promotion from within the organization. Observe your employees' abilities closely and always be aware of those employees that exhibit leadership skills. Promotion through the ranks from probationary employee to hourly worker to lead person to supervisor to superintendent can provide a pathway for a motivated employee to advance. Promotion from within is good for morale and motivates employees to perform at their peak levels.

Employees should be kept informed about the performance of their operations. Productivity charts, budget goals, and production expectations can be shared with the workforce as long as confidential financial data (sales, revenue) are not made public. Many employees like to be competitive and will attempt to beat the goals set for them. Give your employees a target and they will aim for it! Bonus programs for both management and hourly workers have been shown to be very effective at motivating efficiencies within an aggregates operation. Bonuses, if offered, should be tied to production goals, productivity, profitability, individual goals, and above all the plant safety record.

Regular, annual performance evaluations should be scheduled with all employees. The supervisory level should evaluate the individual performance of the hourly workforce, while the plant manager (superintendent) evaluates the performance of the supervisors, the OM evaluates the plant managers, and so on up to the COO of the company. Copies of the evaluations should become a permanent part of every employee's personnel file. Performance evaluations should include expectations for improvement and objectives for the upcoming year. These evaluations establish accountability and give employees feedback on meeting the expectations of their immediate supervisors and the goals of the company. These evaluations will determine employee performance, salary increases, promotions, educational requirements, and any areas in need of improvement. They can also be a means of listening to employees' grievances, concerns, aspirations, and such. It is imperative that the supervisor be honest and fair in the evaluation process and work with the individual employee to improve performance, boost morale, and address employee concerns.

An updated organizational chart should be available to employees so they know who reports to whom in the management hierarchy. A flat management structure is usually more efficient than numerous layers of management. Managers should work through their supervisors to direct the workforce. It is not the OM's function to direct the hourly workforce; the OM's function is to direct, mentor, and coordinate the plant managers. One of the biggest challenges in management is motivating multiple people with different personalities, thought processes, and methodologies to achieve the same goals. That is why you need to get to know your employees personally. The motivational technique that inspires one personality does not necessarily motivate another. As a manager it can be very frustrating to watch the methods used by one of your direct reports to complete a project when, if it were your project, you would have gone about it a different manner. People do not think alike, their brains do not process information the same way, but as long as the desired outcome is achieved in the necessary time frame, you must allow them to reach the ultimate goal via their own methods.

Operations management is a stimulating role that will provide an ever-changing work environment for the person wanting a significant daily challenge. The role provides an opportunity to be involved in the mine planning, stripping, drilling/blasting, transportation, processing, and customer relations; all facets of the business on a daily basis. Multiplant operations management multiplies these challenges as well as the satisfaction.

CHAPTER 19

Sales and Marketing

Sales and marketing are two concepts that are interrelated and yet separate in their individual functions. Smaller companies with limited resources are more prone to combine the two concepts and have employees with combined marketing and sales responsibilities. The tendency toward consolidation within the aggregates industry has seen the formation of fewer, much larger companies with separate but interrelated marketing and sales departments. Very few aggregates products are sold directly to the retail public, although some producers bag their products and sell them through retailers like Home Depot or Lowes to consumers looking for landscape products. This is a very small segment of the market. The vast majority of aggregate sales are directed toward industrial customers; for example, construction companies, building contractors, asphalt producers, concrete producers, state and federal agencies, architect/engineering firms, developers, and landscaping companies. The market is therefore known as business-to-business marketing (B2B), which differs considerably from B2C, or business-to-consumer marketing.

MARKETING

The function of the marketing department is to define the customer base (the market) and promote the buying or selling of the products, while the sales department contacts, interacts with, and provides individual services to customers previously identified by the marketing personnel. The importance of marketing is a concept that should be stressed by upper management and be incorporated into all aspects of an organization: management, production, quality control (QC), office personnel, transportation, human resources, sales, and accounts receivable. Everyone in the organization has responsibility for marketing both the company itself and its products.

The marketing department is responsible for analysis of potential customers in the marketplace, forecasting product demand, identifying the capabilities and deficiencies of the competition, product advertising, and identifying the various segments of the market. It is also responsible for developing a marketing plan with strategic objectives, promotion for the various products, sales and revenue forecasts, tracking of the pertinent sales/revenue results compared to the plan, and developing pricing strategies. Marketing personnel need to have a good understanding, not only of marketing, but of the entire business and the roles that all employees play in marketing, and a keen understanding of the QC process, material specifications, specifying agencies' methodologies, customer's finished products (asphalt, concrete, etc.), as well as the company's production capabilities, cost of production, inventory levels, billing procedures, and accounting methods.

Many different definitions of marketing exist and the science of marketing has evolved tremendously over the last 20 years. All the definitions of marketing that I have read have several elements in common:

- Identify the various key customer groups within the market
- Identify each customer group's specific needs
- Provide a product or service that is of value to these customer groups
- Provide the customers a better value than the competition provides
- Communicate this enhanced value to the customers
- Create a competitive advantage through strengthening customer relationships
- Identify potential new products and services to enhance the customer's needs

There are many ways to provide added value to the customer base. As mentioned in Chapter 5, location is a key marketing device for aggregates, although it is not always within your control. Quality deposits are not always strategically located within the marketplace. The cost of transporting bulk commodities like aggregates can outweigh the cost of the material itself. Therefore, being located closer to the customer than your competition is a considerable advantage in the aggregates marketplace. Price is always a marketing incentive. The ability to produce the product at a lower cost than your competitors will provide a competitive price advantage in the market. This can be accomplished through utilizing the latest, most efficient technology, maintaining your equipment for peak performance, and having the capacity to produce large volumes of materials to lower the unit costs of the product. Superior service is probably the best marketing tool. The ability to supply the customer, whatever their needs, in the expected time frame, even upon short notice, will pay monumental dividends in gaining market share. This includes providing transportation services to the customer if the material was bid at a delivered price.

Every product produced must meet or exceed all of the quality standards and specifications required by the customer base at all times. A well-informed, knowledgeable, QC staff can help provide a competitive marketing advantage. Providing information on the quality and specifications of your products as well as providing samples to the customer's QC staff for mix design purposes is an excellent marketing tool. Partnering with your customers on mix designs and laboratory services is another method of providing added value to your product. Having the latest technology (both lab equipment and information technology) available in your QC laboratory will only enhance your standing in the eyes of your customers. This should include participation of the QC staff on state association specification committees, National Stone, Sand and Gravel Association (NSSGA) technical committees, and maintaining a close working relationship with specifying agencies (ASTM, AASHTO, state departments of transportation) on the development of new specifications and test methods.

Statistically, 65%–70% of industrial customers that change suppliers do so based on a perceived indifference from the supplier's employees. This is why every employee must constantly be focused on marketing. No customer representative must ever be treated with indifference, lack of respect, or lack of urgency. Demand professionalism, a sense of urgency, and courtesy from all of your employees at all times. Provide employee training on best customer relations practices, emphasizing the importance of high-quality customer service. This training should not only be geared toward marketing, sales, and scale personnel, but also wheel loader operators, QC technicians, accounts receivable personnel, telephone operators, production managers; in essence, your entire staff. Everyone in the organization must be focused on marketing and customer service if you are to gain a service advantage over your competitors.

Although uncommon in the business world of the 21st century, a live person answering the telephone, rather than an elaborate computerized answering system that sends your customers around in circles while on hold for several minutes, will provide your customers with a sense that they are important and their time is valued. The owners of the company from which I retired always insisted on maintaining a pleasant, knowledgeable, receptionist/telephone operator and had backup personnel available to assist during breaks to ensure no caller was ever left holding a silent telephone receiver. This is the type of personal service that will give your company an edge over the competition. Make your customers feel important and needed with personal, individual service. The cost of a telephone receptionist is a very minimal expense from a larger marketing and customer service perspective.

Marketing and sales personnel need to be cognizant of the optimum sales/production mix at the processing plants. Every plant/deposit has a distinct product mix that provides the maximum profitability. Deviation from the optimum product mix (selling too much or too little of certain products) can significantly reduce the profitability of that specific operation and/or result in massive inventory imbalances that can take years to resolve (Figure 19.1). Although large inventory is a necessity to service multiple customers, inventory is money and should not remain unsold on the ground for extended periods. Luckily, unlike other products, aggregates do not spoil. Marketing and sales personnel need to work closely with production management to optimize the sales/production mix to achieve the maximum level of profitability. If sales and production are consistently out of balance, the plant design needs to be re-evaluated and capital invested into processing improvements to help balance production capabilities and inventories with sales.

Advertising and product promotion in the B2B market like aggregates can be problematic. Unlike B2C markets, where advertising is essential to draw customers to your products and away from the competition's, aggregates producers normally have a limited customer base in a localized market. Aggregates are generally a fairly local marketplace, so national media advertising (radio, TV) is not a viable option. Nor are most construction magazines because they too are nationwide publications. Participation through associate membership in statewide associations like the Ohio Contractors Association, the Illinois Asphalt Pavement Association, or the Michigan Concrete Association will provide you with access to your customers via participation in association meetings and volunteer work on technical committees. Giving technical presentations at these types of meetings can help establish your company's credibility and amplify your employees' capabilities. Sponsoring the happy hour prior to the association's annual banquet or the refreshment cart at the annual golf outing are always good marketing initiatives. Get to know your customers on a personal basis whenever possible and focus intently on their needs. Most state associations publish membership directories and newsletters that usually provide advertising opportunities to this statewide audience. Employee participation in local civic groups and chambers of commerce provides a good opportunity for exposure to a variety of business and civic leaders. Charitable donations in the form of money or aggregates material to the local schools, park districts, churches, sports teams, and such is a form of community involvement, public relations, and advertising, and is also tax deductible.

In today's electronic world, a comprehensive website outlining your company's capabilities, services, products, specifications, distribution system, and plant locations is essential.

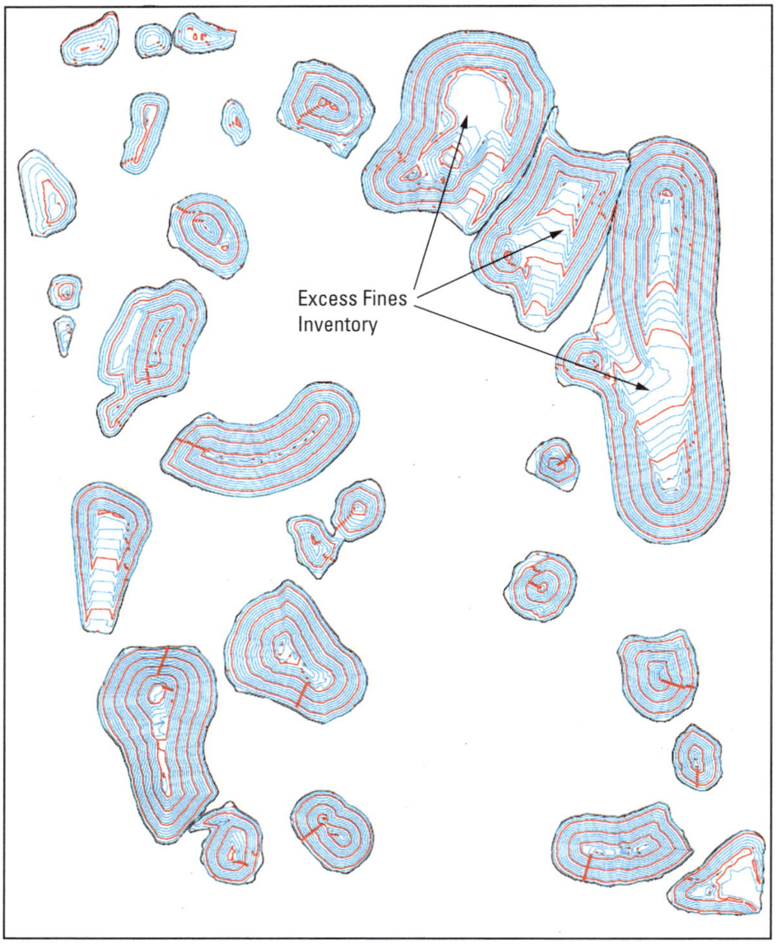

FIGURE 19.1 Inventory map with 117,000 metric tons of excess fine products

A website is especially useful to out-of-town contractors bidding on large, one-of-a-kind construction projects. A mapping application where customers can check mileage from your pit or quarry to their various jobsites is very handy to include. Products available at each location should be listed. Details concerning delivery options should also be discussed and a retail price list published on the website. It never hurts to include positive press clippings and accolades from satisfied customers. QC capabilities and services should be prominently displayed. It's probably wise to hire a professional website designer to provide expertise and flashy graphics to enable an easily utilized, pleasant, customer website experience. Make sure to update the website regularly; there is nothing more frustrating than finding out-of-date or incorrect information on a website.

Other promotional programs can include direct mail campaigns, technical newsletters, social media contacts, trade journal articles, customer luncheons, and product seminars. Direct contact with your customer's decision makers is by far the most effective product promotion and sales technique. In the business climate of the 21st century, these decision makers are very busy and their time is extremely valuable. Showing up unannounced at their place of business in the hope of getting access to them (cold calls) is generally not a wise move. Make an appointment well in advance so as not to interfere with their busy schedules. Customers will appreciate this approach of respecting their time. Brief customer surveys asking their opinions on your company's service, price, quality, delivery, sense of urgency, and reliability, and what you can do improve any on any of these items can be very helpful in developing a marketing plan. Also ask them to rate your service versus the competition's service. The key to customer surveys is brevity; remember, these decision makers' time is very valuable to them.

After developing a market plan and sales forecasts, a method of tracking the results is necessary. Tracking certain key indicators in the marketing plan is essential to identifying the effectiveness of the plan. Key items to be tracked should be:

- Actual sales versus budgeted sales
- Revenue versus budget revenue
- Average selling price versus budgeted selling price
- Number of bids won versus the total number of bids
- Which competitor won the bids your company lost
- Total number of sales inquiries
- Any and all customer complaints
- Sales and revenue of each market segment
- Percentage volume and revenue from individual market segments
- Number of sales calls by the sales department
- Results of customer satisfaction survey
- Inventories

Marketing personnel need to have a keen knowledge of their competitors' locations, production capabilities, levels of service, and pricing structure. Every possible piece of information available on your competition should be gathered, categorized, and analyzed. Check out the competition's websites regularly, watch the local newspapers for articles pertaining to your competitors, and complete periodic internet searches for news. A comparison of your operations with each competitor's location for distance to the jobsite or plant, transportation cost advantages and disadvantages, service levels, quality, dependability, and pricing will enable the marketing department to determine what must be done to win the next pending contract. It is helpful to develop a rating system of your competitors as to how they compare in all of these services.

Determining the key market groups or segments is also a function of marketing and is essential for determining pricing levels. These groups can be segmented by their potential annual sales volume ($10,000–$20,000, $20,000–$50,000, $50,000–$100,000) or by their company's main function, such as fixed plants, governmental, underground contractor, paving company, homebuilder, highway contractor, or landscaper. Or they can be segmented by a combination of both categories. Customers should always be tracked by their volume of business. Who are the top 10 customers? Which customers are in the top 20% tier? Defining the volume of a customer's business will help the sales department determine the appropriate amount of effort to put forth. Marketing should identify not only the customers, but also their purchasing strategy as well. Does the customer have a purchasing department and/or purchasing manager? Do the individual job superintendents purchase their own materials? The sales department will need to know who to contact and who determines what to purchase and when to make a purchase. A potential customer's credit and payment history is also crucial to know. I've had customers in the past that did a substantial business with us but were historically always 90+ days late in paying their bills, in which case their pricing needs to reflect their payment history.

SALES

Although marketing and sales functions definitely overlap, the sales function is one of personal customer contact and direct customer service while ascertaining and meeting the customer's needs. Price is not always the determining factor in making a sale, although all other things being equal, it has a major influence. The sales representative must have the ability to read customers and determine what is of the most importance to them. It will generally be the total package of product quality, prompt delivery, customer service, dependability, and delivered price. Certain segments of the market may have much different needs than others. An earth-moving contractor looking for compactable material to fill a large depression will be much less focused on quality issues than a ready-mix concrete customer, for example. The sales representatives need to understand the different business segments and the differing needs of each individual segment.

In today's business environment, the sales reps need to be personable, outgoing, energetic, professional, and inquisitive of the needs of their customers. They must be customer service oriented and knowledgeable of their products, their customer's products, and the interaction between these products. Sales reps should also be the eyes and ears of the marketing department passing along any intelligence gathered on the competition (e.g., prices bid on various projects, service failures). Familiarity with QC and specifications and the effect they have on their customer's business is a necessity. What effect does a change in the specific gravity of their stone product have on an asphalt mix design? How do variations in moisture affect their ready-mix customers? Material specification problems, billing issues, late deliveries, in fact any customer problem will reach the sales reps before anyone else in your organization. They are the first point of contact, liaison, and complaint department for their customers. You've never lived until you've felt the wrath of a disappointed job superintendent who has a construction crew and several million dollars' worth of equipment, expecting 15 truckloads of base products that are an hour late for the jobsite. Much to the sales rep's dismay, they will get to field that phone call, probably at 6:00 AM.

Aggregates sales reps are generally responsible for arranging and guaranteeing the transportation (trucking) to the various job sites. Many aggregates firms have their own fleet of delivery trucks and dispatcher, but during the peak portions of the construction season will require support from outside trucking companies. Coordination of transportation is usually a function of sales and can be challenging when several large jobs are happening at the same time while normal deliveries to the fixed plants must also be maintained. Personal contact with trucking company personnel and dispatchers for delivery coordination is usually the responsibility of the sales reps. Many companies also require the sales department to supervise and coordinate the weighing and ticketing of the trucks leaving the plants, although the industry trend over the last few years is to automate the scale house function (Figure 19.2).

A winning personality is not enough to be an effective sales representative. Sales personnel need to be trained in sales strategies and methodologies. The successful sales representative must be capable of making presentations, determining the customer's needs, overcoming a customer's objections, resolving complaints, and, most importantly, closing the sale. Account management techniques, time management, development of prospects, closing the sale are all learned functions that can be taught. NSSGA offers sales seminars for both new and experienced sales personnel emphasizing sales techniques, account management, customer service, time management, and sales closing techniques. Many state aggregates associations also provide periodic sales seminars.

The entire budgeting process in the industry begins with sales forecasts that turn into sales budgets on which the production department bases its annual production budgets (Figure 19.3). The sales team in cooperation with the marketing personnel generate these forecasts on an annual basis for the following year's budget process. The marketing personnel are generally responsible for tracking the actual sales in comparison to the budget on a monthly, semi-annually, and yearly basis and reforecasting sales when necessary. Sales forecasts and budgets are normally developed from a combination of past sales history (previous 5 years), market knowledge, and intelligence on upcoming jobs. Contact with customers, state departments of transportation, and county highway departments will provide insight into the next year's demand. Publications like the *Dodge Report* list upcoming large construction projects

FIGURE 19.2 Automated truck scale

Sales Budget Report
2015

Month	Dry Sales Dollars	Dry Sales Units	Dry ASP	Washed Sales Dollars	Washed Sales Units	Washed ASP	Rip Rap Sales Dollars	Rip Rap Sales Units	Rip Rap ASP	AgLime Sales Dollars	AgLime Sales Units	AgLime ASP	Resale Sales Dollars	Resale Sales Units	Resale ASP	Total Sales Dollars	Total Sales Units	Total ASP
January	$57,098	9,350	$6.11	$115,958	15,185	$7.64	$14,375	1,150	$12.50	$7,680	800	$9.60	$0	0	#DIV/0!	$195,110	26,485	$7.37
YTD	**$57,098**	**9,350**	**$6.11**	**$115,958**	**15,185**	**$7.64**	**$14,375**	**1,150**	**$12.50**	**$7,680**	**800**	**$9.60**	**$0**	**0**	**#DIV/0!**	**$195,110**	**26,485**	**$7.37**
February	$137,853	22,950	$6.01	$171,635	22,635	$7.58	$18,450	1,425	$12.95	$2,880	300	$9.60	$0	0	#DIV/0!	$330,818	47,310	$6.99
YTD	**$194,950**	**32,300**	**$6.04**	**$287,593**	**37,820**	**$7.60**	**$32,825**	**2,575**	**$12.75**	**$10,560**	**1,100**	**$9.60**	**$0**	**0**	**#DIV/0!**	**$525,928**	**73,795**	**$7.13**
March	$295,250	49,000	$6.03	$293,858	37,835	$7.77	$23,725	1,875	$12.65	$14,400	1,500	$9.60	$900	50	$18.00	$628,133	90,260	$6.96
YTD	**$490,200**	**81,300**	**$6.03**	**$581,450**	**75,655**	**$7.69**	**$56,550**	**4,450**	**$12.71**	**$24,960**	**2,600**	**$9.60**	**$900**	**50**	**$18.00**	**$1,154,060**	**164,055**	**$7.03**
April	$316,093	51,650	$6.12	$428,818	55,535	$7.72	$33,950	2,725	$12.46	$19,200	2,000	$9.60	$1,200	75	$16.00	$799,260	111,985	$7.14
YTD	**$806,293**	**132,950**	**$6.06**	**$1,010,268**	**131,190**	**$7.70**	**$90,500**	**7,175**	**$12.61**	**$44,160**	**4,600**	**$9.60**	**$2,100**	**125**	**$16.80**	**$1,953,320**	**276,040**	**$7.08**
May	$433,950	71,600	$6.06	$652,516	84,260	$7.74	$43,450	3,525	$12.33	$4,800	500	$9.60	$1,200	75	$16.00	$1,135,916	159,960	$7.10
YTD	**$1,240,243**	**204,550**	**$6.06**	**$1,662,784**	**215,450**	**$7.72**	**$133,950**	**10,700**	**$12.52**	**$48,960**	**5,100**	**$9.60**	**$3,300**	**200**	**$16.50**	**$3,089,236**	**436,000**	**$7.09**
June	$483,526	79,960	$6.05	$802,455	102,185	$7.85	$49,513	4,000	$12.38	$480	50	$9.60	$1,200	75	$16.00	$1,337,174	186,270	$7.18
YTD	**$1,723,769**	**284,510**	**$6.06**	**$2,465,239**	**317,635**	**$7.76**	**$183,463**	**14,700**	**$12.48**	**$49,440**	**5,150**	**$9.60**	**$4,500**	**275**	**$16.36**	**$4,426,410**	**622,270**	**$7.11**
July	$542,150	90,100	$6.02	$865,458	112,135	$7.72	$49,180	4,005	$12.28	$240	25	$9.60	$900	50	$18.00	$1,457,928	206,315	$7.07
YTD	**$2,265,919**	**374,610**	**$6.05**	**$3,330,696**	**429,770**	**$7.75**	**$232,643**	**18,705**	**$12.44**	**$49,680**	**5,175**	**$9.60**	**$5,400**	**325**	**$16.62**	**$5,884,337**	**828,585**	**$7.10**
August	$528,026	88,705	$5.95	$955,408	121,935	$7.84	$42,850	3,525	$12.16	$240	25	$9.60	$780	40	$19.50	$1,527,304	214,230	$7.13
YTD	**$2,793,945**	**463,315**	**$6.03**	**$4,286,104**	**551,705**	**$7.77**	**$275,493**	**22,230**	**$12.39**	**$49,920**	**5,200**	**$9.60**	**$6,180**	**365**	**$16.93**	**$7,411,641**	**1,042,815**	**$7.11**
September	$537,249	90,375	$5.94	$871,078	110,135	$7.91	$39,605	3,255	$12.17	$19,200	2,000	$9.60	$780	40	$19.50	$1,467,911	205,805	$7.13
YTD	**$3,331,194**	**553,690**	**$6.02**	**$5,157,181**	**661,840**	**$7.79**	**$315,098**	**25,485**	**$12.36**	**$69,120**	**7,200**	**$9.60**	**$6,960**	**405**	**$17.19**	**$8,879,552**	**1,248,620**	**$7.11**
October	$448,911	75,425	$5.95	$601,175	76,335	$7.88	$34,875	2,875	$12.13	$105,600	11,000	$9.60	$600	25	$24.00	$1,191,161	165,660	$7.19
YTD	**$3,780,105**	**629,115**	**$6.01**	**$5,758,356**	**738,175**	**$7.80**	**$349,973**	**28,360**	**$12.34**	**$174,720**	**18,200**	**$9.60**	**$7,560**	**430**	**$17.58**	**$10,070,714**	**1,414,280**	**$7.12**
November	$294,138	48,800	$6.03	$485,603	61,735	$7.87	$25,075	2,075	$12.08	$84,480	8,800	$9.60	$600	25	$24.00	$889,895	121,435	$7.33
YTD	**$4,074,242**	**677,915**	**$6.01**	**$6,243,959**	**799,910**	**$7.81**	**$375,048**	**30,435**	**$12.32**	**$259,200**	**27,000**	**$9.60**	**$8,160**	**455**	**$17.93**	**$10,960,609**	**1,535,715**	**$7.14**
December	$208,839	34,725	$6.01	$204,358	26,135	$7.82	$7,300	600	$12.17	$9,600	1,000	$9.60	$0	0	#DIV/0!	$430,096	62,460	$6.89
YTD	**$4,283,081**	**712,640**	**$6.01**	**$6,448,316**	**826,045**	**$7.81**	**$382,348**	**31,035**	**$12.32**	**$268,800**	**28,000**	**$9.60**	**$8,160**	**455**	**$17.93**	**$11,390,705**	**1,598,175**	**$7.13**

FIGURE 19.3 Example of a typical sales budget spreadsheet

and forecast future levels of construction on a regional basis. Forecasting and budgeting must include consideration of the existing inventory levels and therefore should also be coordinated with plant management.

PRICING STRUCTURE

The unit cost of production of aggregates products decreases with additional production. Certain costs are fixed and will not change, but the majority of the costs vary with the volume produced. Fixed costs include management salaries, depreciation, amortization, interest, and employee benefits. Variable costs are those directly related to producing the product like power, fuel, labor, equipment maintenance, explosives, parts, and lubricants and are directly related to the number of tons produced. Variable costs are inversely proportional to production volume; the higher the production volume, the lower the unit cost. Producing lower than expected production volumes because of lagging sales will increase the unit cost because the fixed costs need to be spread over fewer production units.

Selling prices are a function of the unit cost of production of the individual products plus the cost of sales plus an expected profit margin. The profit margin should be set to a predetermined rate of return based on the risk, cost of capital, and production effort required. Establishment of a bottom-line price below which the products will not be sold should be agreed upon during the budgeting process. Unlike B2C marketing, B2B markets often require negotiating prices with the customers. Price incentives for customers in the same market segments with similar sales volumes must be done equitably without discrimination under the federal antitrust regulations. Sales and marketing personnel should be educated and periodically retrained on current antitrust regulations to avoid any potential accusations of price fixing. Customer pricing must never be discussed with your competitors.

Most aggregates producers publish a price list that applies to retail customers that purchase small volumes of material for cash. These prices are the highest charged, the most profitable, and usually a fairly small percentage of the total sales volume, although some plants in rural areas can have significant cash sales. Most governmental contracts and some large projects are based on closed bidding contracts, where the contracts are advertised, sealed bids are accepted from the interested bidders and opened in a public meeting and the lowest bidder is awarded the contract. Prices in this type of contract are non-negotiable. Open, negotiated bids for large projects are also common. This process is less formal than closed bidding and discussions between producers and the job owners will take place prior to the bid date. These types of bids are not necessarily decided on the price alone, but the total perceived value of the bid (price, service, dependability, delivery, timing, etc.). The lowest price need not win these open negotiated types of bids.

Prices can be discounted based on certain factors such as early payment incentives, quantity purchases, and large, long-term projects. Payment of invoices is generally due within 30 days of receipt of the invoice. It is common practice within the industry to offer customers discounts of 2%–4% if invoices are paid within 10 days of receipt. This can be a significant incentive, especially for large customers, to pay early and this type of discount also stimulates the producer's internal cash flow. Quantity discounts for customers that regularly purchase large volumes of material like fixed plant concrete and asphalt customers are also common based on the economies of scale. Volume discounts can be offered based on purchase volumes like a reduction of $0.10/metric ton after purchasing a specific volume amount (e.g., 15,000 metric tons, 20,000 metric tons) of material. Discounts for large, long-term contracts like airports and highway reconstruction are justifiable based on the lowering of unit prices through increased production. The sales department must be careful to offer discounts equitably to individual customers based on comparable purchase volumes and segment within the market.

Producers that have a transportation disadvantage by being located further away from the plant or jobsite and are quoting a delivered price may sometimes discount the delivery portion of the total price. This is done in an effort to share the additional transport costs with the customers and offset the higher trucking fees. Regardless of whether the discount is on transportation or material charges, the producer still has a profit margin to target and maintain. Other innovative discounting initiatives can be offered as long as there is a rational method that is offered equitably to everyone in the market segment.

The aggregates business is dynamic and the market is always evolving and changing with new customers, out-of-town contractors bidding on large projects, new market segments, and evolving product specifications. To become a profitable market leader, the sales and marketing personnel need to be flexible and in tune with the changing market dynamics. This includes developing intelligence concerning the competitor's capabilities, market strategy, levels of service, quality, and pricing. Long-term customers are seeking value, not only the lowest price. Value is a result of providing quality products and quality service in a timely and professional manner. Based on superior levels of service and quality, it is possible for a market-focused, service-oriented producer to be the price setter in the marketplace, gaining consistent pricing increases on an annual basis.

CHAPTER 20

Sustainability

When I first started in the aggregates industry, no one had ever heard the term *sustainability*. My earliest employers focused exclusively on safety, resource management, employment of capital, and profitability. My first job in the industry was as a staff geologist looking for new reserves to replace depleted deposits (resource management). Environmental compliance consisted of filling out the proper applications and forms required to obtain various permits from the recently established, fledgling environmental protection agencies. Compared to many other industries, aggregates were fairly benign and under the radar, at that time, of most environmental regulators and thus enforcement. Zoning regulations were the norm in metropolitan areas, but many rural areas had minimal or no zoning codes and lacked the sophistication to enact comprehensive mining regulations. Social license meant little because the public's attitudes were laxer and industries' attitude was "we own it; we have the right."

Many definitions of sustainability exist but they all seem to agree that the basic concept is development that meets today's needs without inhibiting the ability of future generations to meet their needs as well. This is accomplished by focusing corporate efforts on not only economic prosperity but also environmental stewardship and social responsibility to enable a consistent supply of aggregate materials for now and for future generations. Most state aggregate associations have environmental committees that focus not only on regulations but also on environmental stewardship. The National Stone, Sand and Gravel Association (NSSGA) formed a sustainability task force in 2007 (NSSGA, n.d.-a).

ENVIRONMENTAL STEWARDSHIP

We've already discussed many of the environmental regulations and requirements for the aggregates industry in Chapter 6 on permitting. Environmental stewardship goes well beyond complying with federal and state regulations. Dictionary.com (n.d.) defines stewardship as "the responsible overseeing and protection of something considered worth caring for and preserving," while bizfluent.com states: "In the world of commerce, stewardship refers to taking responsibility for the business and the effects it has on the world around it" (Ahmed 2019). Environmental stewardship includes:

- Regulation compliance
- Conservation of resources
- Environmental audits
- Employee environmental awareness training
- Protection of sensitive environmental areas (wetlands, wildlife habitat, etc.)
- Mined land reclamation

- Groundwater management
- Maintenance of pollution control equipment
- Recycling

Environmental management systems (EMSs) can be a useful tool in evaluating and maintaining a culture of environmental stewardship. The International Organization for Standardization (ISO) has developed an EMS entitled ISO 14001 that is recognized as the worldwide standard and is encouraged by the U.S. Environmental Protection Agency. NSSGA also developed a template for producers to utilize to create their own internal EMS. This template contains several modules that can be adapted and modified by aggregates producers to fit their own needs. Most larger aggregate producers have environmental managers on staff to help implement EMS, assist plants in interruption of regulations, audit and inspect the plants for compliance, and assist in environmental awareness training of employees. Smaller producers may not be able to afford a full-time environmental staff, but environmental consulting firms can be utilized to fill the gap if required.

Reclamation is another form of sustainability (Figure 20.1). Aggregate sites are an interim land use and the land when adequately reclaimed can be recycled into some other useful purpose. Reclaimed quarry sites or unused reserves can be adapted for use as wildlife habitat sites and/or migration corridors for various wildlife species. Lakes produced as a result of underwater aggregates mining make excellent wildlife refuges for migratory birds, as well as excellent aquatic habitat. Many aggregate producers have partnered with the Wildlife Habitat Council in an effort to create viable habitat to attract certain specific species of plants and animals and aid in saving endangered species. I once operated a quarry consisting of 1,300 acres (326 hectares) surrounded by urban and suburban areas of a midwestern city. It contained its own unique ecosystem within the city, being home to red fox, white tailed deer, wild turkey, ground hogs, coyotes, beaver, and hundreds of species of birds and fish. It was not an officially designated wildlife habitat but it certainly could have been.

At one point in time, ospreys were considered an endangered species in the state of Ohio. The company I worked for at the time happened to have two of the 15 total breeding pairs of

Courtesy of American Aggregates

FIGURE 20.1 Limestone quarry reclaimed into parkland

osprey in the state living on two different aggregate properties. The ospreys at one site made a nest on a large, inaccessible power distribution tower. The birds were extremely punctual and returned each spring within the same two calendar days. One year the osprey pair was late and the employees became very concerned for their welfare. They showed up one day late to everyone's relief, but then the loader operator reminded me that it was to be expected because it was a leap year. The employees were fascinated and watched the birds fishing in the gravel pit lakes and raise a pair of fledgling chicks every year. The other pair first nested on the point sheave of a dragline boom and we allowed them the use of the machine for one season. The next season the employees built them a nesting platform on top of a telephone pole, erected specifically for the purpose, and relocated a portion of the previous season's nest to the platform. We ultimately entered into a cooperative program with the Ohio Division of Wildlife where they banded the new chicks fledged at this site every year (Figure 20.2). They installed a GPS tracking device on one chick and determined that the birds migrated to the Amazon River in South America where they would spend the winter. Cooperative programs like these are examples of environmental stewardship, conservation, and social responsibility.

FIGURE 20.2 Banding an osprey fledgling

Any attempt at conservation should be considered. Does that stand of trees near the western property line need to be removed or would it serve a better purpose as a buffer from the neighbors? Several midwestern states restrict the time of year to do land clearing in an effort not to interrupt the breeding season of the endangered Indiana brown bat. Do we need to pump so much water from the aquifer to the wash plant or can we clean and recycle a portion of the wash water? Are there ways of conserving fuel in the heavy equipment fleet and lower exhaust emissions at the same time? These types of decisions are all considerations over and above adhering to the environmental regulations that fall under the label of environmental stewardship.

RECYCLING

Natural resources need to be responsibly managed to include renewable resources like replanting forests immediately following timber harvests. Unlike forests, once an aggregate deposit has been mined and processed and the material sold, it cannot be replanted and regrown, but aggregates can be recycled. Construction and demolition debris (C&D) is a huge source of potential recycling of aggregates. Unlike many other parts of the world, the United States is a society of "temporary everything." A factory or shopping mall goes out of business and rather than reuse the same building it is normally torn down, removed, and the real estate developed for another purpose. For many years these materials were sent to landfills or utilized to fill depressions in the earth. C&D consists of wood products, drywall, plaster, steel, concrete, brick, tile, and asphalt, the majority of which is recyclable (Figure 20.3).

More and more wood products are being recycled from C&D as older homes and buildings are being demolished to make way for urban renewal products. Wooden structural members

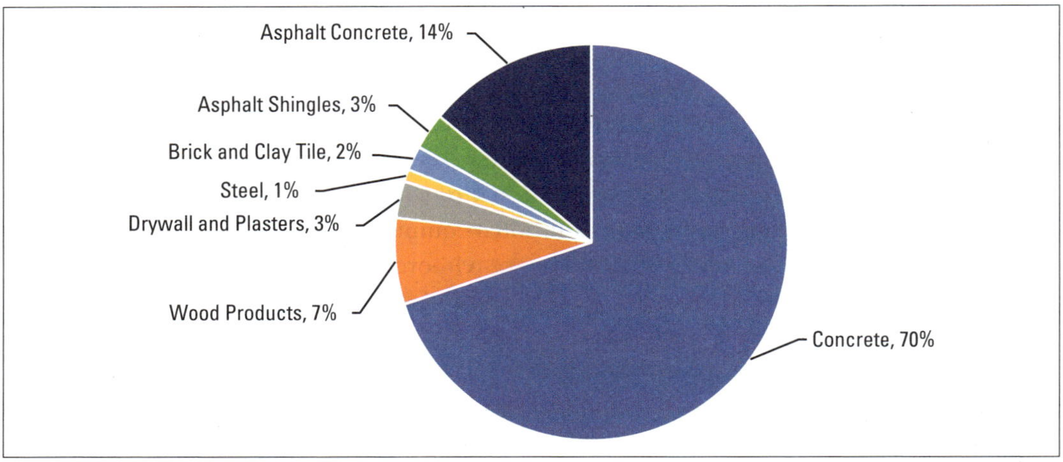

Courtesy of the USEPA

FIGURE 20.3 Construction and demolition debris materials generated in the United States in 2014

from homes built in the 19th century are being recycled into furniture and other decorative wood products. The original trees that provided this structural lumber may have been growing in the 1500s. Copper pipe and copper wiring are also lucrative recycled products. Care must be taken in the demolition of these older buildings because many contain various products made from asbestos, which require special procedures and disposal methods. Bricks can be crushed and recycled into decorative landscape stone or, if carefully removed, can be reutilized in other residential construction products or as pavers for driveways, sidewalks, and so on. Steel is easily recycled into new steel or can be utilized again if the structural members are the correct dimensions. This type of recycling is an industry unto itself.

Recycling of aggregate products consists of recycled asphalt and concrete. Asphalt pavement is one of the most recycled commodities by weight in the United States. The National Asphalt Pavement Association estimates asphalt recycling saves U.S. taxpayers more than $2.5 billion annually. Asphalt recycling has been noted as far back as 1915 but did not become economically viable until the 1970s and 1980s when the price of petroleum skyrocketed. Recycled asphalt pavement (RAP) consists of 85%–90% aggregates cemented together by various forms of bituminous material. Roadways and parking lots eventually deteriorate due to age, weathering, tire loading, or a reallocation of real estate assets. When the need to replace these pavements becomes evident, they are removed by a special milling equipment, loaded into trucks, and stockpiled for later use. Eventually the RAP is re-crushed, screened, and sized for recycling back into hot mix asphalt via specially equipped asphalt plants.

The different state departments of transportation (DOTs) specify the amount of RAP allowable in various paving products for state highways. This is usually in the range of 10%–30% for base and intermediate asphalt layers. Most states do not allow RAP to be used in the surface asphalt layers. Properly designed commercial asphalt mixes for parking lots and other non-specification uses can contain as much as 50%–60% RAP. Crushed and sized RAP can be utilized without reheating for low-traffic uses like farm lanes and driveways by putting down a layer and rolling the material. When subjected to traffic and summer heat, the layered RAP will set up to provide an adequate low-traffic pavement application. The use of recycled asphalt shingles is another application of recycling aggregates. Care must be taken in this process because many older shingles (pre-1970s) contained asbestos fillers and produce a hazardous respirable dust when shredded. Shingles need to be shredded prior to recycling into asphalt, which can be a challenging process, but they are 100% recyclable.

Recycled concrete is another sustainable reuse of aggregates. Portland cement concrete, like asphalt, is made up of 85%–90% aggregates. Concrete can be re-crushed and screened into base material for commercial usage. Additional equipment like magnets needs to be utilized in the crushing process for the removal of the reinforcing steel (rebar), but the recovered rebar is also readily recycled. Hydraulic breakers are also necessary to aid in the reduction of large slabs of concrete. Until recently, because of the cement adhering to the aggregate particles, the material could not be utilized in the same way as virgin aggregates. Recent research by the University of British Columbia School of Engineering found that "recycled concrete had comparable strength and durability (to standard concrete) after five years of being in service." The researchers suggested that "recycled concrete can be a 100% substitute for non-structural (concrete) applications" (Redling 2020). More good news for recycling and sustainability.

Recycled concrete base is generally not acceptable to state DOTs because it normally contains more fines than specification base materials, but it compacts into a usable commercial base product for utilization beneath slabs, driveways, or parking lots. We once attempted to convince the Ohio DOT of the cost benefits of recycling interstate highway concrete pavement directly on the job (Figure 20.4). We took a portable crushing plant to an interstate highway rebuild project and crushed the concrete pavement slabs as they were being removed from the old road grade. The crushed concrete was immediately graded into a base for the upcoming asphalt pavement placement. Ohio DOT liked the cost savings of this method but was concerned about excess fines in the base inhibiting drainage and also the potential for high-pH runoff into adjacent streams.

The last company I worked for prior to retirement had three portable crushing and screening plants, two in the aggregates division and one operated by the asphalt division. We decided early on that if we did not get into the recycling side of the business someone else would take that business away. These plants were kept busy year-round recycling asphalt and concrete, and

FIGURE 20.4 Recycling concrete on a highway project

FIGURE 20.5 Recycling concrete

crushing stone where needed. In a busy year we would recycle 300,000+ metric tons of asphalt and 150,000 metric tons of concrete. Many of our concrete customers utilized our crushing services to recycle the concrete washout accumulated at their ready-mix plants or rejects at the concrete precast plants. Concrete recycling is very hard on loader and truck tires because the rebar is very difficult to see on the ground (Figure 20.5). My portable plant supervisor was recycling concrete and called me one day and said: "What is going on? I've broken three blow bars in the crusher rotor in the last two days and we're getting pieces of steel that are 1 foot by 2 feet by 6 inches thick." We were operating in one of our own recycling yards and receiving broken concrete from a demolition contractor that was taking down an abandoned automobile factory built during World War II. After more inquiry, we learned that these pieces of steel had been embedded in the concrete floor to anchor the production machinery in the factory. I got together with the foreman of the demolition contractor and explained to him we couldn't continue to accept concrete with this kind of steel because it was causing severe damage to our crushing equipment. We were able to come to an accommodation where the contractor agreed to sort the steel out as they broke up the concrete.

Recycling of other materials into aggregates products will also aid in the industry's sustainability efforts. Various slag products resulting from the steel manufacturing process can be recycled into aggregate base products. Most steel slag starts as either limestone or dolomite that is crushed and utilized as a catalyst to increase the melting speed and lower the melting temperature of the steel batch. The slag consisting of melted stone and impurities from the taconite (iron ore) floats on top of the molten steel and is poured off prior to processing and cooling the molten steel. The resulting slag is a calcium aluminum silicate. Blast furnace and basic oxygen furnace slag can be crushed and screened into commercial and some state DOT products. Lightweight aggregate products can be produced from molten slag that has been sprayed with cold water to speed the cooling process. Cooling the slag quickly with water adds internal gas bubbles to the slag, producing a lighter-weight material. I once managed a slag plant where the slag was crushed to minus 8 mesh size using cage mills, with any residual steel removed with magnets, and sold as an additive to the glass manufacturing industry.

Recycling other materials at the aggregates processing plants can assist in our quest for sustainability. Waste oil should be stored and recycled through approved petroleum recycling sources or it can be burned to produce heat in specially designed waste oil furnaces. For years, worn-out heavy equipment tires were buried or left in piles to accumulate. Most tire suppliers will haul away and recycle the old tire cores when they are no longer useful. Today there are numerous companies that recycle tires into various rubber products and also recover the steel from the radial belts. Crumb rubber has been successfully recycled into asphalt pavement, as have certain types of plastic cups. Scrap steel is a commodity that is easily recycled and many scrap companies are willing to pay to take it off your hands. Trash recycling services should be utilized wherever possible to recycle aluminum cans, bottles, plastics, and so on generated by your employees.

SOCIAL RESPONSIBILITY

In Chapter 3, we discussed the need for social license. By social license, I'm referring to approval of the neighboring community to operate an aggregates business. This is first accomplished by going through the local land use and zoning approval process when starting a new operation. Part of this process is always a series of public hearings where the producer is asked to explain its reasoning for locating in the specific community, the process of mining, the traffic implications, the environmental implications, methods utilized to mitigate impact on the surrounding community, and so on. Getting this local zoning approval is only the first step in social licensing.

The responsibility to the neighbors does not end with local zoning approval. Neighbor relations is a continual process throughout the life of an aggregates operation. There will be times when a truck driver leaving your operation will have an accident or break a traffic regulation, a temperature inversion might cause an unusually high air overpressure and rattle some neighbor's windows, a local teenager will be caught trespassing, or a dry year will lower the water in a neighbor's well. These are just a few of many examples of neighbor relations issues that will be encountered and need to be dealt with. I required all my operations to fill out a standard form and forward it to me immediately whenever a neighbor complained for any reason. Depending on the severity of the complaint, I made sure that either I or one of my superintendents followed up personally. Neighbor complaints must never be ignored and should be responded to as soon as possible with courtesy and honesty. Listen to their issues with an open mind and a willingness to resolve the problem, whoever the complainant and whatever the issue. I have found that the more you honestly communicate with people, the more credible you become in their eyes and the less hostility they will exhibit.

Some producers encourage community participation by holding open houses for the neighboring community with tours, barbeques, and such on an annual basis; others hold community marathons, or Fourth of July fireworks displays with the quarry as a backdrop. Certain producers have gone as far as organizing a community advisory committee that meets several times annually with management to discuss issues of concern to the community. Many quarries offer a service of telephoning the immediate neighbors prior to initiating a blast so as not to surprise them when the overpressure and vibrations are experienced. These are all good examples of social responsibility and being a good neighbor.

Educating the public about the industry is also a part of social responsibility. Opening the property to tour groups, whether school groups or civic groups, is a service that can be used as an educational tool. Tours for elementary or secondary school students provide excellent opportunities to educate children on the importance of the industry and someday these children will be the adult audience at the zoning meeting needed to expand the reserves. One of today's students may eventually be chair of the zoning board or a county commissioner and they will remember the tour 15 years before. Volunteer to give talks at the local schools. I've always enjoyed talking to elementary school groups about geology and aggregates. Becoming active in local civic organizations like the chamber of commerce, Kiwanis club, or the Elks will enable you to meet, socialize with, and establish credibility with the local business people. As long as pricing and quality are comparable, it is always a good idea to buy from the local merchant community. Donations of money and/or materials to the local schools, food banks, churches, park districts, and so on will bring a considerable amount of goodwill and help the local community at the same time. A new park or athletic field can always use 20 metric tons of aggregates during a construction project. Hosting tours of university students or allowing graduate geology or engineering students to utilize your facilities for research projects is another facet of social responsibility. We routinely utilized college students as summer interns in our quality control (QC) laboratory, providing them with income and education while recruiting the best for future full-time employment after graduation.

I took over a newly purchased quarry in the rural Midwest and was warned by the previous plant manager that I was going to have to deal with an "extremely bitchy" neighbor across the road. He said that the best way to handle her was to ignore the old woman, which is what he did. After the first blast was set off, the phone rang and it was the woman across the road complaining about her widows shaking and the house swaying. I summoned my foreman and we immediately went to visit the complainant. We were greeted by an elderly woman (I will refer to her as Mrs. X) who explained her perception of the blast and its irritation. Before we left we heard about her deceased husband and her out-of-town children and grandchildren and were at her house for almost an hour. When we left, my foreman complained about having to listen to the old woman's life history; he was not interested! The same scenario was repeated several more times and I finally told my foreman that the next time he needed to go into town for supplies, I wanted him to call Mrs. X and ask her if she needed him to pick up anything for her while in town and to do this at least once weekly. This obviously irritated my foreman, but I explained that I thought Mrs. X was a very lonely lady and just wanted someone to talk with. He did as instructed and within a couple of weeks the complaints essentially ceased. Although he would never admit it, my foreman eventually became somewhat fond of Mrs. X.

Allowing your employees to join and participate in professional associations like the American Institute of Professional Geologists, the Association of Equipment Management Professionals (AEMP), the Forum on Geology of Industrial Minerals, ASTM International, or the International Society of Explosives Engineers allows them to broaden their horizons with additional professional contacts outside of your organization. Participation in state aggregates associations and the NSSGA provides not only political lobbying prospects but also educational and professional networking opportunities for your employees. Opportunities exist to influence state and federal regulations and material specifications by participating on committees within these organizations. By volunteering to speak and present papers at association events, you have an opportunity to influence and educate others in the industry. Support professional registration for your employees that qualify. A professional engineer (PE) or professional geologist (PG) certification adds credibility to the competence of your staff members, which is not only good for the people, it adds to the company's reputation as well.

Most aggregate operations own large tracts of real estate to have sufficient reserves to operate for a number of years. Oftentimes cultural or historical areas or artifacts may be encountered in the mining process. Most sand and gravel operations exist in the bottomlands or terraces along major rivers, which historically were the migration routes of Indigenous peoples who camped, fished, and hunted along the waterways. Encounters with village sites of the early hunter-gatherer peoples are not uncommon and should be preserved, studied, and documented for future generations. Many historic home sites have been preserved by aggregate producers over the years. Some states retain a registry of historical sites and archeological finds and require cultural evaluations prior to any excavation.

Early in my career while stripping overburden at a sand and gravel operation, a bulldozer operator unearthed a grave site. After determining that the human remains were extremely old, we contacted the University of Cincinnati Department of Anthropology and requested they evaluate the site. After seeing their enthusiasm about the artifacts, we modified our mining plan and allowed them access to the site for a period of 3 years. They began their archeological dig and spent the next three summers excavating, collecting, and categorizing the site and its artifacts (Figure 20.6). The grave of one adult and another grave containing two children between the ages of 1 and 5 were excavated along with 100 other features (firepits, trash pits, etc.). Numerous artifacts were uncovered including stone clubs, axes, arrow heads, bird points, fish net weights, and awls. It was determined that a group of nomadic hunter-gatherers from the Late Archaic Period (3,000–1,000 BCE) lived at the site during the fall and early winter season. The site was eventually placed in the National Register of Historic Places.

Social license also includes taking care of your employees in an equitable manner with fair compensation, employee benefits, and social empathy, all of which comes at a cost. As I have stated several times before, a company's employees are its greatest asset and need to be treated as such. Encouraging diversity in the workforce is another way of being socially responsible. Like most industries, especially those associated with construction, the aggregates industry has been dominated by white men, with women mostly being found in clerical, bookkeeping, or accounting positions. As societal attitudes change, more women are taking up jobs as equipment operators, plant operators, or even laborers. The lack of manual laborers in the job marketplace has been met by a dramatic increase in Spanish-speaking workers entering the industry, especially in the Southeast and Southwest, over the last two decades, requiring supervisors to become bilingual in many areas.

NEW TECHNOLOGY

New technology is another method of sustainability. Finding new ways to minimize waste and create less pollution in the mining process through the use of new technologies will also enhance our efforts to be a sustainable industry. Equipment manufacturers are producing new hybrid (diesel electric) engines for their off-highway mobile equipment. Total-electric, battery-powered underground mining equipment is finding its way to the underground mining industry, limiting emissions and exposure of miners to NO_X. Totally electric surface mining equipment will eventually follow. Under the right circumstances, it is not inconceivable that tomorrow's aggregate processing plants will be powered by wind turbines in lieu of mainline electricity.

As I write this chapter on sustainability, new technological innovations have made the news. Aramco (the Saudi Arabian Oil Company) has developed a method of curing concrete

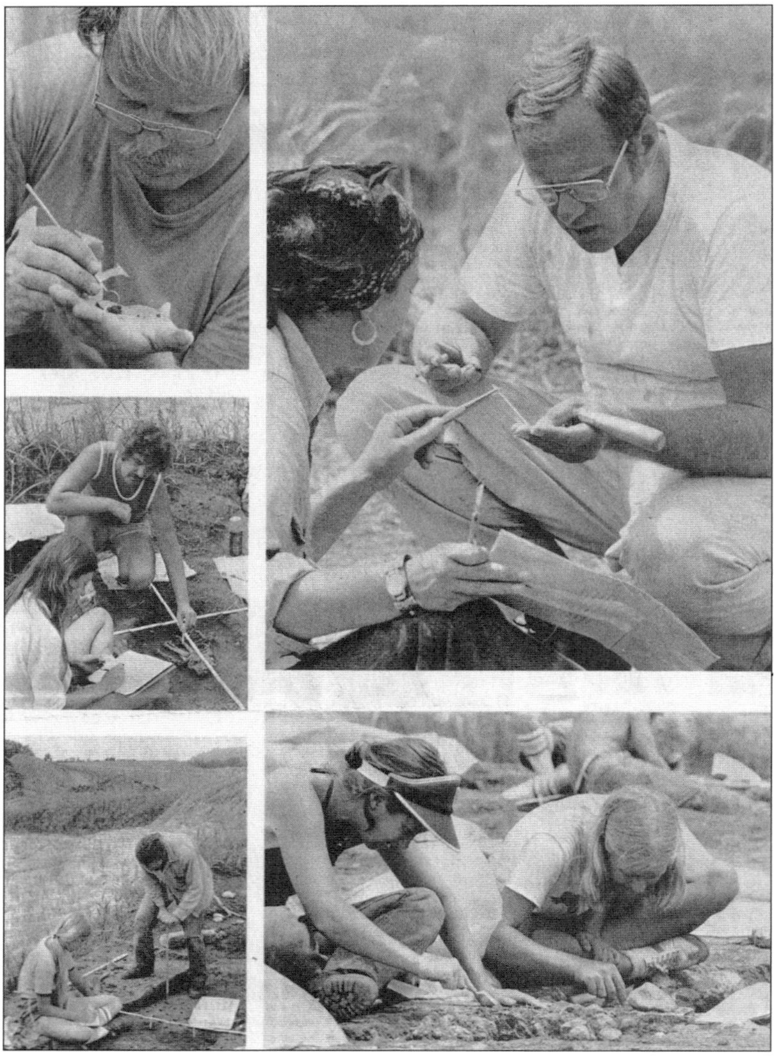

Source: *The Dravo News*, Vol. 5, No. 9, September 1979

FIGURE 20.6 Faculty and students from the University of Cincinnati Department of Anthropology excavating a Late Archaic Period site near Cleves, Ohio

by utilizing a combination of steam and carbon dioxide. This technology has been commercially used to cure precast concrete to a strength of 5,000 pounds per square inch (69 bar) (6,900 kPa) in only 3 days. The normal cure time for concrete is 28 days to reach similar strength. This curing process has also achieved the ability to store 20% CO_2 in the concrete. The fact that cement manufacture is a significant source of CO_2 emissions into the atmosphere makes this technology seem almost ironic. This new technology is more easily utilized in the manufacture of precast concrete than in massive open-air concrete pours. According to Aramco (n.d.), "if the global precast concrete industry switched to our carbon curing innovation, we could recycle up to 246 million tons of CO_2 a year—equivalent to removing emissions from 53 million cars."

Knife River Corp., a large aggregate producer in the northern and western United States, has recently partnered with Blue Planet Systems Corp. in an attempt to manufacture synthetic limestone for use as construction aggregates. Blue Planet has developed an early-stage, proprietary technology to manufacture synthetic limestone using sequestered CO_2 directly removed from the atmosphere. Knife River's specification, QC, processing, and marketing expertise is expected to enhance the technology developed by Blue Planet. Manufacturing limestone aggregates while removing CO_2 from the atmosphere is an excellent example of a sustainable new technology.

ECONOMIC PROSPERITY

Without profitability there would be no environmental stewardship programs or any social responsibility. Environmental stewardship and social outreach come at a cost to the producer. Hiring an environmental engineer or a consultant, developing an EMS, sending employees to seminars, allowing employees time to volunteer at a school all cost money and directly affect the profitability of the company. These costs are essential to establishing a sustainable industry but they cannot overshadow the fact that a company that is not profitable will not survive in the marketplace. Publicly traded companies have a fiduciary responsibility to their shareholders to manage the company in a socially and environmentally responsible manner, but they also have a responsibility to increase the value of the company through growth and profitability. Privately held companies have no shareholders to placate but still have the same environmental responsibility and require the same social license to operate. Management must strike a balance between sustainability and fiduciary responsibility by budgeting the cost of social and environmental programs while allocating funds responsibly.

APPENDIX

Useful Formulas

BLAST DESIGN FORMULAS

Borehole Diameter

$$D = \frac{H}{10}$$

where D = borehole diameter in inches; and H = bench height in feet.

Burden

$$B = 25 \text{ to } 35 \times D_e$$

where B = burden in feet; and D_e = diameter of explosive column in inches.

Depth of Stemming

$$T = 0.7 \text{ to } 1.3 \times B$$

where T = stemming in feet; and B = burden in feet.

Spacing Between Holes

$$S = 1 \text{ to } 1.8 \times B$$

where S = spacing in feet; and B = burden in feet.

Subdrilling

$$J = 0.2 \text{ to } 0.5 \times B$$

where J = subdrilling in feet; and B = burden in feet.

ELECTRICAL FORMULAS

$$1 \text{ kW} = 1{,}000 \text{ W}$$
$$\text{kWH} = \text{kW} \times \text{hours}$$
$$\text{HP (horsepower)} = 0.746 \text{ kW}$$

Ohm's Law

$$V = A \times O$$
$$V = \frac{W}{A}$$
$$A = \frac{V}{O}$$
$$A = \frac{W}{V}$$
$$O = \frac{V}{A}$$
$$W = V \times A$$

where

V = voltage (volts)
A = current (amps)
O = resistance (ohms)
W = power (watts)

EXPLORATION AND RESERVE ESTIMATION

$$\frac{\text{length of core sample}}{\text{length of drill run}} \times 100 = \text{\% core recovery}$$

$$\frac{\text{sum of (length of sound pieces > 4 inches each)}}{\text{total length of core run} \times 100} = \text{rock quality designation \%}$$

$$\text{number of acres} \times \text{thickness of sand and gravel} \times 2{,}500 \text{ metric tons/acre-foot} = \text{metric tons of sand and gravel}$$

$$\text{number of acres} \times \text{thickness of stone} \times 3{,}500 \text{ metric tons/acre-foot} = \text{metric tons of stone}$$

$$\text{number of acres} \times \text{thickness of unconsolidated overburden} \times 1{,}613 \text{ cubic yards/acre-foot} = \text{volume of overburden}$$

OPERATIONS MANAGEMENT

$$\text{ending inventory} = (\text{beginning inventory} + \text{production}) - \text{sales}$$

$$\text{productivity} = \frac{\text{actual tonnage produced}}{\text{primary raw feed} \times \text{hours run}} \times 100$$

$$\text{maintenance and repair hours} = \text{scheduled maintenance and repair hours}$$

standby hours = planned nonoperating hours (start-up/shutdown, plant inspection, safety meetings)

scheduled production hours = actual hours operated + unscheduled downtime

unscheduled downtime = all downtime not included in scheduled M&R hours

total hours = scheduled maintenance and repair + standby + scheduled production hours

PROCESSING FORMULAS

Classification

$$GPM_{slurry} = TPH_{slurry} \times \frac{4}{SG_{slurry}}$$

$$GPM_{water} = TPH_{water} \times 4$$

$$GPM_{solids} = TPH_{solids} \times \frac{4}{SG_{solids}}$$

$$SG_{slurry} = (TPH_{slurry} \times \frac{4}{GPM_{slurry}}$$

$$GPM_{solids} = 1.5 \times TPH_{solids} \text{ (when SG = 2.67)}$$

where

GPM = gallons per minute
TPH = tons per hour
SG = bulk specific gravity

Conveyor Speed

$$R = \frac{A \times B}{C}$$

$$P = \frac{R}{D}$$

$$S = P \times E \times 0.2618$$

where

A = motor rpm
B = motor sheave diameter
C = reducer sheave diameter
D = speed reducer ratio
E = diameter of head pulley
R = reducer input speed
P = drive pulley rpm
S = conveyor belt speed

Recirculating Load

$$RL = 100 \div \left(\frac{SE}{OS}\right) - 1$$

where SE = screen efficiency, %; and OS = percent of oversized material in crusher discharge.

Reduction Ratio

$$RR = D_{80} \text{ crusher feed} : D_{80} \text{ crusher discharge}$$

where D_{80} = sieve opening size (millimeters) through which 80% of the aggregate passes.

Screen Speed

$$\text{screen speed} = R \times \text{motor speed}$$

$$R(\text{ratio}) = \frac{D_{\text{motor sheave}}}{D_{\text{screen sheave}}}$$

Screening Area

$$\text{screening area (square feet)} = \frac{U}{A \times B \times C \times D \times E \times F \times G \times H \times I}$$

$$G \text{ force} = \frac{\text{rpm}^2 \times \text{throw}}{70{,}418}$$

where

U = undersize factor
A = capacity factor
B = oversize factor
C = half size factor
D = deck location factor
E = wet screening factor
F = material weight factor
G = screen open area factor
H = shape factor
I = efficiency factor

Screening Efficiency

$$E_{\text{screen deck}} = \frac{\text{metric tons per hour of undersize actually passing the screen}}{\text{metric tons per hour of underside in the feed}} \times 100$$

Stockpile Volumes

$$V = r^2 \times H \times 0.039$$

$$V = \frac{H^3 \times 0.039}{\tan \theta}$$

$$V = H(S^2 - H^2) \times 0.039$$

where

V = volume of stockpile (cubic yards)
H = height of pile
R = radius of base of the pile
θ = angle of repose
S = slope length

For radial stockpiles, the volume in cubic yards for 1° of arc of the pile with a 37° angle of repose can be calculated as follows:

$$V = \frac{H^2 \times r}{1,160} = \text{volume of a } 1^\circ \text{ arc section}$$

volume of the entire pile = (V_2 × actual degrees of arc) + volume of conical ends of the pile

PUMPING AND DREDGING FORMULAS

$$\text{head (feet)} = \text{PSI} \times 2.31$$

$$\text{PSI} = \text{head (feet)} \times 0.433$$

$$\text{velocity} = \frac{\text{GPM} \times 0.4085}{D^2}$$

$$Q = \frac{\text{velocity} \times D^2}{0.4085}$$

$$\text{BHP} = \frac{Q \times \text{TDH} \times \text{SG}}{3,960 \times \text{Eff} \times 0.95}$$

$$\text{Eff} = \frac{Q \times \text{head} \times \text{SG}}{3,960 \times \text{BHP}}$$

$$\text{metric tons/hour} = 0.25 \times \text{GPM} \times \text{head} \times \text{SG} \times \text{percent solids}$$

$$\text{GPM} = \frac{\text{TDH} \times 400}{\text{percent solids (whole number)} \times \text{SG}}$$

$$\text{TDH} = \text{FHD} + \text{static head} + \text{suction head}$$

$$\text{FHD} = \frac{0.283\left(\frac{100}{C}\right)^{1.85} \times Q^{1.85}}{D^{4.8655}}$$

$$\text{friction loss per 100 feet of pipe} = \frac{\text{line length} \times 1.1}{100 \times \text{FHD}}$$

$$\text{static head} = \text{vertical distance (center of suction to center of discharge pipe)}$$

$$\text{suction loss} = \text{typically 17 feet}$$

where

Q = flow (gallons per minute)
GPM = gallons per minute
BHP = brake horsepower
TDH= total dynamic head
FHD = friction head

D = pipe diameter
SG = specific gravity of slurry
Eff = percent efficiency (pump curve)
0.95 = drive losses
C = 120 for steel pipe, deck-mounted pump; 140 for plastic pipe; and 100 for steel pump, underwater pump with 30%–40% solids
PSI = pressure (pounds per square inch)

Speed Affinity Laws When Revolutions per Minute Change

$$Q = Q_1 \times \frac{N}{N_1}$$

$$H = H_1 \times \left(\frac{N}{N_1}\right)^2$$

$$P = P_1 \times \left(\frac{N}{N_1}\right)^3$$

$$\text{pipe diameter} = \sqrt{\frac{Q \times 0.407}{\text{velocity}}}$$

$$\text{pump impeller speed} = \sqrt{\text{TDH}} \times \frac{1{,}650}{\text{impeller diameter}}$$

where
Q_1, H_1, N_1, P_1 = known values
Q = new flow
H = new head
N = new speed (revolutions per minute [rpm])
P = new brake horsepower
TDH = total dynamic head

Typical Slurry Specific Gravities

water = 1.0
deck pump at 20% solids = 1.142
jet-assisted deck pump at 23% solids = 1.169
underwater pump (>17 feet per second) at 38% solids = 1.31

QUALITY CONTROL FORMULAS

$$\text{apparent specific gravity} = \frac{A}{A} - C$$

$$\text{bulk (dry) specific gravity} = \frac{A}{B} - C$$

$$\text{bulk SSD specific gravity} = \frac{B}{B} - C$$

$$\text{absorption (water) (\%)} = \frac{B - A}{A} \times 100$$

where

A = weight of an oven-dried sample of aggregates in air
B = weight of a saturated surface dry (SSD) sample of aggregates in air
C = weight of a saturated sample of aggregates in water

$$\%\ \text{voids} = \frac{(A \times W) - B}{A \times W} \times 100$$

where

A = bulk specific gravity
B = unit weight
W = 62.4 pounds per cubic foot (unit weight of water)

$$C_u = D_{60} / D_{10}$$

where

C_u = coefficient of uniformity
D_{60} = sieve opening size (millimeters) through which 60% of the aggregate passes
D_{10} = sieve opening size (millimeters) through which 10% of the aggregate passes

68% of the data will lie within $M \pm 1s$

95% of the data will lie within $M \pm 2s$

99.7% of the data within $M \pm 3s$

where s = standard deviation, and M = mean.

SAFETY

$$\text{incidence rate (IR)} = \frac{\text{number of reportable injuries} \times 200{,}000 \text{ work hours}}{\text{number of work hours during the period in question}}$$

UNDERGROUND MINING FORMULAS

Extraction Ratio

$$\text{extraction ratio} = 1 - \frac{A_p}{A_e}$$

where A_p = area of the pillar; and A_e = area of extraction.

Pillar Stress

$$S_p = S_o \times \frac{(r + w) \times (r + l)}{(w + l)}$$

where

S_p = postmining stress
S_o = premining stress (1.1 pounds per square inch per foot of overburden overlying the mine roof)
r = room width
w = pillar width
l = pillar length

Glossary

AASHTO: An acronym for American Association of State Highway Transportation Officials.

adit: A horizontal shaft into a hillside or highwall to access an underground mine.

aggregates: Any combination of crushed stone, sand, or gravel in their natural or processed state.

alkali: A strongly basic substance such as sodium hydroxide or sodium carbonate that has the ability to neutralize acids to form salts.

alluvial fan: A fan-shaped wedge of sediment that typically accumulates where a stream emerges from a steep canyon onto a flat-lying area.

alluvium: Unconsolidated stream-deposited sediments, including sands, silts, clays, or gravels, and any combination thereof.

ANFO: An acronym for ammonium nitrate and fuel oil, which is a blasting agent utilized in surface and underground mining.

angle of repose: The maximum angle that a soil, sediment, or other loose material can accumulate and be stable from downslope movement.

aquifer: A porous and permeable subsurface rock or sediment that stores and transmits useful quantities of water.

aquitard: A subsurface layer of soil, sediment, or rock that will not yield useful quantities of water and generally inhibits downward flow of groundwater.

ASCE: American Society of Civil Engineers.

ash (volcanic): A combination of rock, mineral, and volcanic glass fragments generally smaller than 2 millimeters in size that are blown from the vent of an erupting volcano.

back: The roof in an underground mine.

barrier pillar: An underground mine pillar with a length more than three times the width used to separate mine panels or add additional support in critical areas.

basalt: A very hard, fine-grained, dark-colored, extrusive igneous rock principally composed of the minerals pyroxene and plagioclase feldspar, also known as traprock.

belly: An excess of a particular size range in an aggregate gradation.

bench: An individual mining level in either surface or underground mines.

blinding: The slow buildup of wet or plastic material adhering to screening media that will eventually plug the screen openings.

bluffline: The point where the floor of the upper heading meets the vertical bench in an underground mine.

boulder: A sedimentary particle larger than 10 inches (256 millimeters) in size.

breast: The mining face in an underground mine.

burden: The distance from a row of blastholes to the open mining face.

burn cut: An underground blasting technique where several large-diameter drill holes are left unloaded as relief for the blastholes.

BZA: An acronym for board of zoning appeals.

calcareous: Containing calcium carbonate.

caliche: A soil or sediment layer cemented together by calcium carbonate commonly found in arid regions.

carbonate rock: A rock composed primarily of carbonate minerals (minerals containing the CO_3 anionic structure) such as limestone ($CaCO_3$) or dolomite ($CaMg\,[CO_3]_2$).

case harden: Cementation of the outermost layers of a deposit owing to natural weathering processes.

chert: A sedimentary rock composed of microcrystalline or cryptocrystalline silica (SiO_2), generally considered deleterious in aggregates.

classification: The separation of various size particles by the use of their settling velocity in water.

clastic rock: Accumulations of transported weathering debris (sediments) that have been lithified to form sedimentary rock (e.g., sandstone, siltstone, shale, conglomerate).

clay: A clastic particle of any composition that has a grain size smaller than 1/256 millimeters. Clay is deleterious in aggregates.

coastal plain: An area of low relief along the continental margin that is underlain by thick sediments that dip gently toward the ocean. The coastal plain can be mined for aggregates using deep sea marine dredging techniques.

colluvium: Loose soil and rock fragments deposited by a combination of runoff and mass wasting that accumulates at the base of a slope or outcrop.

cone of depression: A lowering of the water table around a pumping well or sump that produces a cone-shaped depression in the water table. As more water is pumped, the water level in the well falls and the cone increases in size.

conglomerate: A clastic rock type that contains rounded pebble-size particles (greater than 2 millimeters in diameter) interspersed with smaller particles (sand and silt) and bound together by a chemical cement forming rock.

contact metamorphism: Process in which the host rock is altered (metamorphosed) by heat and reactive fluids, which occurs adjacent to a dike, sill, or other body of magma.

cross-bedding: A horizontal clastic rock unit composed of inclined layers. Sand and gravel deposited on the down-current side of a bar will likely produce inclined layers.

cuttings: Small rock chips generated from drilling holes in rock.

decline: A downward sloping inclined tunnel used to access an underground mine (also known as slope).

deformation: A modification of the original shape of a rock mass in response to intense heat, stress, strain, and pressure resulting in faulting, folding, and/or inclination of the original beds.

deltaic deposits: Sediments deposited where a flowing stream enters a standing body of water such as a lake or ocean. Sediments fall out of the stream when the moving water suddenly loses velocity.

demurrage: A rental fee charged for failure to return a rented vehicle (railcar or barge) within an allotted time frame.

dewatering: The concentration of solid materials upon removal of water from a slurry; also, the removal of groundwater from an aquifer.

diorite: An intrusive igneous rock that contains a mixture of coarse-grained feldspar, pyroxene, hornblende, and sometimes quartz.

dip: The angle that a rock unit (bed), fault, joint, or rock structure makes in relation to the horizontal plane or the Earth's surface.

displacement: The relative movement of the two blocks on opposite sides of a fault, whether horizontally or vertically or a combination, usually expressed in feet.

dolomite: A mineral composed of calcium magnesium carbonate ($CaMg(CO_3)_2$); also, a sedimentary rock consisting primarily of the mineral dolomite. Most dolomite rocks are considered to have been originally limestone that was chemically altered by the magnesium in seawater to form the mineral dolomite.

DOT: An acronym for department of transportation.

drift: Underground mine opening (also known as heading).

dry banked: Above-water sand and gravel excavated dry by a wheel loader, shovel, or excavator.

eolian: Relating to the wind, for example, windblown sand deposits or sand dunes.

eskers: Meandering, subglacial stream courses that eventually filled with sand and gravel and collapse when the glacial ice melts to form sinuous ridges.

evaporation: The process by which liquid water becomes water vapor as a result of heat and/or sunshine.

evaporite: A sediment or chemical that has formed by precipitation from evaporating waters, such as gypsum, salt, and anhydrate.

evapotranspiration: Water moving naturally from a liquid to water vapor back to liquid in the normal cycle of rainfall, transpiration, and evaporation.

extraction ratio: Ratio of mined stone to the stone left behind to support the mine roof (same as recovery ratio).

FAA: An acronym for Federal Aviation Administration.

facies: The characteristics of a sedimentary rock unit that reflect the depositional environment in which it formed. Different facies can exist in the same rock formation.

fault: A fracture in the rock along which movement (displacement) occurs, generally causing an earthquake.

felsic: A light-colored igneous rock containing high percentages of feldspar, quartz, and muscovite mica (e.g., granite and rhyolite).

floodplain: Relatively level land along the banks of a stream that is inundated with water when the stream overflows its channel during a time of high flow.

floodway: The area of the floodplain that needs to be left unimpeded to allow for the free movement of floodwater and generally contains the deepest, highest-velocity flood flow.

floodway fringe: A storage area in the floodplain for generally shallow, slow-moving floodwater.

flyrock: Uncontrolled rock ejected vertically from a quarry blast.

formation: The basic rock unit of stratigraphy consisting of a continuous rock unit with a distinctive set of characteristics that make it possible to recognize and map from one outcrop or test well to another. A formation usually relates to a distinctive time range of formation. Formations may contain different facies deposited during the same time period.

friable: Very poorly cemented, able to crumble with hand pressure.

gabbro: The coarse-grained, intrusive, compositional equivalent of basalt.

geologic column: A diagram indicating the vertical sequence of rock units present beneath a specific location, with the youngest at the top of the column.

glacier: A mass of thick ice that forms on land from an accumulation and recrystallization of snow. Glaciers can grow with the accumulation of additional snow or melt and recede during times of high ambient temperature and can move in response to the force of gravity.

gneiss: A coarse-grained metamorphic rock produced as a result of regional metamorphism. The mineral grains are foliated, forming distinct color banding.

gradation: The distribution of particle sizes in an aggregate mixture defined as either the percent of the mixture retained on individual sieve sizes or the percent passing the sieves expressed as percent by weight.

granite: An intrusive, light-colored, coarse-grained, igneous rock composed primarily of the minerals quartz, orthoclase, sodium plagioclase, and muscovite mica.

hardpan: Also known as caliche, or glacial till, hardpan is a layer of soil or sediment that has been cemented together. A mining term relating to a layer of hard, deleterious material that usually separates distinct layers of sand and gravel.

headframe: A structure designed to support and direct the cables from a hoist over top of a vertical mine shaft for lifting ore or personnel.

hematite: A mineral composed of iron oxide (Fe_2O_3) that is deleterious in aggregates because it causes red staining in concrete.

highwall: Exposed mining or bench face.

hydrocyclone: A continuously operating classifier that accelerates the settling rates of particles by the use of centrifugal force.

igneous: A rock type formed by the crystallization and cooling of molten rock, either magma, lava, or volcanic ash.

intrusive: Igneous rocks that crystallize below Earth's surface, also known as magma.

isopach map: A contour map of the thickness of a deposit.

joint: A fracture in rock that, unlike a fault, shows no indication of movement (displacement). Joints are generally a result of relieving regional stresses in the rock mass.

jumbo: A chassis-mounted horizontal drilling machine designed to carry multiple drills for face drilling in underground mining.

kame: Stratified glacial sediments (sand and gravel) generally formed in contact with glacial ice that usually exhibit a hummocky topography.

karst: Subsurface drainage and associated topographic features consisting of sinkholes, enlarged joints, caves, and disappearing streams. Karst features form as a result of mildly acidic groundwater reacting with alkaline carbonate rocks (the acid-base chemical reaction).

lacustrine: Sedimentary deposits formed within a lake environment usually composed of very fine silt and clay.

lava: Extrusive molten rock that erupts onto Earth's surface and cools quickly, forming very fine-grained igneous rock or volcanic glass.

leaching: The removal of soluble portions of rocks or soil via a chemical reaction due to contact with groundwater or hydrothermal fluids.

limestone: A sedimentary rock consisting of mostly calcium carbonate ($CaCO_3$) either chemically precipitated, consisting of clastic fragments of marine organisms, or a combination of both processes.

lithology: The textural and compositional characteristics of a rock type.

load: The total amount of sediments carried by a stream, river, or glacier.

mafic: A descriptive term for igneous rock containing dark-colored minerals like amphibole, pyroxene, and olivine that are rich in magnesium and iron.

magma: Intrusive molten rock that forms beneath the surface of the Earth and cools slowly, forming larger crystals.

mineral: An inorganic solid with an ordered internal crystal structure and specific chemical composition that is naturally occurring, such as calcite, quartz, and feldspar.

mineral rights: The ownership of minerals, metals, groundwater, petroleum, and so on underlying a piece of property.

MSHA: An acronym for Mine Safety and Health Administration.

muck pile: A pile of fragmented rock resulting from blasting a quarry face or mining bench.

mucking out: Removal of the muck in an underground mine.

multibench operation: A mining operation having more than one bench or mining level. Benches can vary in height and quality.

NIOSH: An acronym for National Institute for Occupational Safety and Health.

obsidian: A rapidly cooled, dark-colored, extrusive igneous rock with the texture of glass, exhibiting conchoidal fracture.

ore: Any natural earthen material containing valuable minerals that can be profitably mined, processed, and sold.

outfall: The location of an industrial water discharge from settling pond or treatment facility.

outwash: Stratified silt, sand, and gravel deposits resulting from melting glaciers.

pegging: The plugging of screen cloth by near-sized particles, usually elongated in shape.

pH: A measure of acidity or alkalinity with a range from 0 to 14. A pH value of 7 is considered to be neutral; below 7 is acid; and above 7 is alkaline.

phyllite: A very fine-grained, foliated, metamorphic rock consisting primarily of mica.

plutonic: Intrusive igneous rocks.

point source discharge: Any discernible, confined, and discrete conveyance, including but not limited to any pipe, ditch, channel, tunnel, conduit, well, discrete fissure, and container, from which pollutants are being or may be discharged.

protolith: Original rock type prior to metamorphosis.

pumice: A very light extrusive volcanic glass similar in composition to granite.

PVC: An acronym for polyvinyl chloride.

quartz: A very hard mineral (No. 7 on the Mohs scale) with a chemical composition of SiO_2 found in sedimentary, metamorphic, and igneous rocks. Quartz is one of the most abundant minerals in the crust of the Earth.

quartzite: Metamorphosed quartz sandstone where the original sand grains are no longer discernable in the crystalline mass.

quartzose: Composed of quartz, usually in reference to sedimentary rocks.

recharge: The addition of water to an aquifer via precipitation, streamflow, and infiltration.

regional metamorphosis: Large-scale metamorphic effects usually due to the heat and pressure associated with plate tectonics, and subduction.

rhyolite: A very fine-grained extrusive volcanic rock similar in chemical composition to granite.

ribs: Pillars and exposed walls in an underground mine.

roadheader: A chassis-mounted tunnel boring machine sometimes used in mining and mine development.

rock quality designation: An empirical test of the quality of rock core indicting the degree of fracture or jointing of the rock mass being drilled.

rose diagram: A graphical representation of the direction of certain structural features like joints and faults.

runoff: Water moving across the ground without evaporating or infiltrating.

sandstone: A clastic sedimentary rock composed mainly of sand-sized particles cemented together.

scaling: Removing loose stone from the back and ribs in an underground mine.

scalping: The removal of fines from coarse material using a screen or grizzly feeder. The coarse fraction or screen overs are the scalped product.

schist: A foliated metamorphic rock containing abundant mica particles whose alignment is a result of significant pressure.

segregation: Separation of the various sizes of material within a product gradation usually as a result of gravity.

siliceous: Containing silica.

sill pillar: Layer of stone left behind between underground mining levels (also known as interburden).

siltstone: A clastic sedimentary rock composed mainly of silt-sized particles cemented together.

skip: An ore bucket designed to transport ore from an underground mine via a vertical shaft.

skip hoist: A cable hoist that raises skip buckets from a mine.

slate: Metamorphosed shale.

slip: Movement along a geological discontinuity like a joint.

slurry: Any combination of water containing solid particles in suspension.

spacing: The distance between adjoining blastholes.

spring: Groundwater interacting with and discharging onto the surface of the Earth.

stope: A cavity left behind when the ore, in this case aggregate, has been removed.

stratification: The layering of sedimentary rocks.

strike: The intersection of the plane of the surface (dipping bed, fault) with the Earth's surface. The strike is always perpendicular with the maximum dip.

suspended solids: Small particles of sediment held in suspension by the movement of water in a stream.

till: Sediments deposited in direct contact with a glacier that have not been sorted by meltwater. Till generally consists of very densely packed, hard, silty clay containing some sand and gravel.

traprock: Dark-colored, fine-grained, extrusive igneous rocks used to make aggregates; traprock usually refers to basalt.

unconformity: Two rock types of distinctively different ages in direct contact, representing a gap in the time record, usually due to a long period of erosion between formation of the different age rock types.

unconsolidated: Sediments that are not cemented together, such as glacial outwash sand and gravel.

USACE: An acronym for U.S. Army Corps of Engineers.

USBM: An acronym for U.S. Bureau of Mines (now a part of NIOSH).

USCG: An acronym for U.S. Coast Guard.

USEPA: An acronym for U.S. Environmental Protection Agency.

USGS: An acronym for U.S. Geological Survey.

V-cut: An underground blasting technique in which the blastholes are arranged in an angled pattern forming a V in the center of the drill pattern to control the burden.

vortex: A cavity or vacuum formed in the center of a whirling liquid similar to a whirlpool.

water table: The saturated zone beneath which the pores, cracks, faults, and fissures in the underlying soil or rock are entirely filled with water.

wetlands: Areas that are inundated or saturated by surface or groundwater at a frequency and duration sufficient to support, and that under normal circumstances do support, a prevalence of vegetation typically adapted for life in saturated soil conditions. Wetlands generally include swamps, marshes, bogs, and similar areas.

Z-curve: A graph of peak particle velocity versus frequency developed by the USBM to assess the potential for residential structural damage owing to ground vibrations produced from blasting.

Bibliography

48 Fed. Reg. 14153. 1983. Environmental Protection Agency: 40 CFR Part 122. National Pollutant Discharge Elimination System. Code of Federal Regulations. Washington, DC: U.S. Government Printing Office, Office of the Federal Register.

Ahmed, A. 2019. Definition of stewardship. https://bizfluent.com. Accessed September 2022.

Aramco. n.d. Managing our carbon curing. www.aramco.com. Accessed September 2022.

Arizona Rock Products Association. 2015. *Aggregate Protection Guidance*. Phoenix: Hadley & Aldrich.

ASCE (American Society of Civil Engineers). 2016. *Failure to Act, Closing the Infrastructure Investment Gap for America's Economic Future*. Reston, VA: ASCE.

ASCE (American Society of Civil Engineers). 2021. *Report Card for America's Infrastructure*. Reston, VA: ASCE. Accessed July 2022.

ASTM C88. 2018. *Standard Test Method for Soundness of Aggregates by Use of Sodium Sulfate or Magnesium Sulfate*. West Conshohocken, PA: ASTM International.

ASTM D8. 1979. *Standard Terminology Relating to Materials for Roads and Pavements*. West Conshohocken, PA: ASTM International.

Atlas Powder Company. 1987. *Explosives and Rock Blasting*, Dallas: Atlas Powder.

Ault, C.H. 1992. Geologic factors affecting the underground limestone and dolomite mines of Indiana. In *Proceedings of the 26th Forum on Geology of Industrial Minerals*. Richmond: Virginia Department of Mines, Minerals and Energy.

Ault, C.H., and Haumesser, A.F. 1990. A central Indiana model for predicting jointing characteristics in underground limestone mines. In *Proceedings of the 24th Forum on Geology of Industrial Minerals*. Bloomington: Indiana Geological & Water Survey.

Ault, C.H., Harper, D., Smith, C.R., et al. 1985a. *Faulting and Jointing in and near Surface Mines of Southwestern Indiana*. NUREG/CR-4117. Bloomington: Indiana Geological & Water Survey.

Ault, C.H., Haumesser A., and Fraser, G.S. 1985b. *Geology and Operations of the American Aggregates Corporation's Harding Street Pit, Quarry and Underground Mine Marion County, Indiana*. Bloomington: Indiana Geological & Water Survey.

Bellamy, A.G. 2016. Marine sand and gravel prospecting and seabed investigations for aggregate extraction. In *Proceedings of the 19th Extractive Industry Geology Conference*. Oxford, UK: EIG Conferences.

Bureau of Reclamation. 2001. *Engineering Geology Field Manual*, 2nd ed. Washington, DC: Bureau of Reclamation.

CDC (Centers for Disease Control and Prevention). n.d.-b. NIOSH mining. www.cdc.gov. Accessed September 2022.

CDC (Centers for Disease Control and Prevention). n.d.-a. Our mission. www.cdc.gov. Accessed September 2022.

Crawford, J., and Hustrulid, W.A. 1979. *Open Pit Mine Planning and Design*. New York: SME-AIME.

Davis, G. 1996. Using commodity price projections in mineral project valuation. *Mining Engineering Magazine* 48 (April): 67–70.

Dictionary.com. n.d. Stewardship. www.dictionary.com. Accessed September 2022.

Dougherty, H.N., and Schissler, A.P., eds. 2020. *SME Mining Reference Handbook*, 2nd ed. Englewood, CO: SME.

Duncan, D. 2021. America's 100 greatest public courses. *Golf Digest*. www.golfdigest.com. Accessed August 2022.

DuPont. 1969. *Blaster's Handbook*, 15th ed. Wilmington, DE: DuPont.

Ellis, T.R. 2002. U.S. views on valuation methodology. *The Professional Geologist* 39(6):2–9.

Fairbanks, Morse & Co. 1956. *Hydraulic Handbook*. Chicago: Fairbanks, Morse & Co.

Genovese, M. 2000. Implications of the revocation of the "Tulloch rule." Master's thesis, Virginia Polytechnic Institute and State University, Blacksburg, VA.

Global Recycling. n.d. Tire recycling riding on. https://global-recycling.info/archives/2883. Accessed August 2022.

Haumesser, A.F. 1983. Mining and pit planning from a geologist's point of view. Aggregate Production Seminar, March 21.

Haumesser, A.F. 1985. Mining and reclamation in an urban environment. Presented at the Interstate Mining Compact Commission Annual Meeting.

Haumesser, A.F. 1986. Aggregate mining and reclamation in the United States. Presented at the Ontario Ministry of Natural Resources Pits and Quarries Seminar, March.

Haumesser, A.F. 1994. Underground mining at American Aggregates' Indianapolis operations. Presented at the National Stone Association Annual Meeting.

Haumesser, A.F. 1996. Operational benchmarking. *Aggregates Manager* 1(5).

Haumesser, A.F. 1997. Benchmarking and productivity enhancement. Presented at the Pennsylvania Aggregates and Concrete Association Plant Operator's Workshop.

Haumesser, A.F. 1998. The real cost of exploration drilling. Presented at the Indiana Mineral Aggregates Association Technical Workshop.

Haumesser, A.F. 2005. Introduction. In *Environmental Compliance Manual*. Gahanna: Ohio Aggregates and Industrial Minerals Association.

Haumesser, A.F. 2010. Introduction to mechanical dredging. Presented at AGG1 Seminar, National Stone, Sand and Gravel Association.

Hazemag USA. 2003. *Impactor Model Catalog*. Uniontown, PA: Hazemag, USA.

Hoover, E.G. 1997. Valuation geology. *The Professional Geologist* 34(10):8–9.

Hustrulid, W.A. 1982. *Underground Mining Methods Handbook*. New York: SME-AIME.

Iannacchione, A., Mucho, T.P., and Prosser, L.J. 1995. Ground problems in the underground limestone industry. Presented at the Environmental Safety and Health Forum, National Stone Association, October.

Iannacchione, A. 1999. Engineering stable underground stone mines: Issues and techniques. Presented at the National Institute of Occupational Safety & Health, Underground Limestone Mine Workshop, Cincinnati, December 8.

InDOT (Indiana Department of Transportation). 2013. *Inspection & Sampling Procedures for Fine & Coarse Aggregates*. Indianapolis: InDOT.

InDOT (Indiana Department of Transportation). 2016. *Certified Aggregate Technician Manual*. Indianapolis: InDOT.

Killey, M.M. 2000. *Quaternary Glaciations in Illinois*. Geonote 3. Champaign: Illinois State Geological Survey.

KPI-JCI. 2011. *Facts & Figures*, 4th ed. Sterling, IL: Astec Industries.

Lamar, J.E. 1967. *Handbook on Limestone and Dolomite for Illinois Quarry Operators*. Bulletin 91. Urbana: Illinois Geological Survey.

Langer, W.H. 1993. *Natural Aggregates of the Conterminous United States*. Bulletin 1594. Washington, DC: U.S. Geological Survey.

Langer, W.H., and Glanzman, V.M. 1993. *Natural Aggregate, Building America's Future*. Circular 1110. Washington, DC: U.S. Geological Survey.

Langer, W.H., and Knepper, D.H., Jr. 1995. *Geological Characterization of Natural Aggregate*. Open-File Report 95-582. Denver: U.S. Geological Survey.

Lefond, S.J. 1975. *Industrial Minerals and Rocks*, 4th ed. New York: SME-AIME.

Lucius, J.C., Langer, W.H., and Ellefsen, C.J. 2007. *An Introduction to Using Surface Geophysics to Characterize Sand and Gravel Deposits*. Circular 1310. Reston, VA: U.S. Geological Survey.

Machiavelli, Niccolo. 1984. *The Prince*. New York: Bantam Classics.

Mark, C., Compton, C.S., Oyler, D.C., et al. 2002. Anchorage pull testing for fully grouted roof bolts. In *Proceedings of the 21st International Conference on Ground Control in Mining*. Morgantown: West Virginia University.

Mirsky, A., Hartke, E.J., Ault, C.H., et al. 1983. Urban and engineering geology of the Indianapolis area (Field Trip 8). In *Field Trips in Midwestern Geology*. Boulder, CO: Geological Society of America.

MSHA (Mine Safety and Health Administration). n.d. Regulations. www.msha.gov. Accessed September 2022.

National Waterways Foundation. 2017. *A Strong Inland Waterways System Delivers a Stronger Economy*. Washington, DC: National Waterways Foundation.

NIOSH (National Institute for Occupational Safety and Health). n.d. Number of active mines by sector and year, 1983–2020. wwwn.cdc.gov. Accessed August 2022.

NSSGA (National Stone, Sand and Gravel Association). n.d.-a. Environment. www.nssga.org. Accessed September 2022.

NSSGA (National Stone, Sand and Gravel Association). n.d.-b. Health and safety. www.nssga.org. Accessed September 2022.

NSSGA (National Stone, Sand and Gravel Association). 2012. *The Economic Impact of the Crushed Stone and Sand and Gravel Industries*. Alexandria, VA: NSSGA.

NSSGA (National Stone, Sand and Gravel Association). 2013. *The Aggregates Handbook*, 2nd ed. Alexandria, VA: NSSGA.

OAIMA (Ohio Aggregates and Industrial Minerals Association). 2005. *Environmental Compliance Manual*. Columbus, OH: OAIMA.

Ohio Department of Transportation. 2019. *Construction and Material Specifications*. www.transportation.ohio.gov. Accessed August 2022.

Ohio Division of Geological Survey. 2004. *Shaded Drift-Thickness Map of Ohio*. Survey Map SG-3. Columbus: Ohio Department of Natural Resources, Ohio Division of Geological Survey.

Ohio Division of Geological Survey. 2005. *Glacial Map of Ohio*. Columbus: Ohio Department of Natural Resources, Ohio Division of Geological Survey. Accessed July 2022.

Ohio Division of Geological Survey. 2006. *Bedrock Geologic Map of Ohio*. Columbus: Ohio Division of Geological Survey.

Ohio Division of Geological Survey. 2018. *2018 Report on Ohio Mineral Industries*. Columbus: Ohio Division of Geological Survey.

Parker, J. 1973a. Part 1: Practical rock mechanics for miners. *Engineering & Mining Journal* (June): 104–109.

Parker, J. 1973b. Part 2: What can be learned from surface subsidence? *Engineering & Mining Journal* (July): 70–73.

Parker, J. 1973c. Part 3: How convergence measurements can save money. *Engineering & Mining Journal* (August): 92–97.

Parker, J. 1973d. Part 4: The relationship between structure, stress, and moistures. *Engineering & Mining Journal* (October): 91–95.

Parker, J. 1973e. Part 5: How to design better mine openings. *Engineering & Mining Journal* (December): 76–80.

Parker, J. 1974a. Part 6: How to improve roof support. *Engineering & Mining Journal* (January): 90–95.

Parker, J. 1974b. Part 7: The logical way to design pillars. *Engineering & Mining Journal* (February): 67–71.

Paschall, R.H. 2000. *Appraisal of Construction Rocks*, 2nd ed. Westminster, CO: American Institute of Professional Geologists.

Pavey, R.R., Goldthwait, R.P., and Breckman, C.S. 1999. *Quarternary Geology of Ohio*, Map No. 2. Columbus, OH: Ohio Division of Geological Survey.

Pfleider, E.P. 1972. *Surface Mining*. New York: SME-AIME.

Phoenix Center for Advanced Legal & Economic Public Policy Studies. 2017. *The Economic Impact of the Natural Aggregates Industry: A National, State, and County Analysis, Industry Scoreboard*. Washington, DC: Phoenix Center for Advanced Legal & Economic Public Policy Studies.

PQ Staff. 2018. 2017 aggregate production analysis. *Pit & Quarry* (May 18). www.pitandquarry.com. Accessed September 2022.

Redling, A. 2020. Research shows recycled concrete can outperform traditional construction. *Recycling Today* (November 30) www.recyclingtoday.com. Accessed September 2022.

Rooney, L.F., and Carr, D.D. 1971. *Applied Geology of Industrial Limestone and Dolomite*. Bulletin 46. Bloomington: Indiana Geological & Water Survey.

Shrimer, F.H. 2005. Engineering geology of aggregates, British Columbia. In *Industrial Minerals with Emphasis on Western North America*. Victoria, BC, Canada: British Columbia Ministry of Energy and Mines, Geological Survey Branch. pp. 99–105.

Siskind, D.E., Stachura, V.J., Stagg, M.S., et al. 1980a. *Structural Response and Damage Produced by Airblast from Surface Mining*. Report of Investigations 8485. Minneapolis, MN: Bureau of Mines.

Siskind, D.E., Stagg, M.S., Kopp, J.W., et al. 1980b. *Structural Response and Damage Produced by Ground Vibration from Surface Mine Blasting*. Report of Investigations 8507. Minneapolis, MN: Bureau of Mines.

SME. 2017. *The SME Guide for Reporting Exploration Results, Mineral Resources, and Mineral Reserves*. Englewood, CO: SME.

Terex Corporation. 2009. *Pocket Reference*, 20th ed. Cedar Rapids, IA: Terex.

Toy, T.J., and G.R. Foster. 1998. *Guidelines for the Use of the Revised Universal Soil Loss Equation (RUSLE) Version 1.06 on Mined Lands, Construction Sites, and Reclaimed Lands.* https://citeseerx.ist.psu.edu. Accessed August 2022.

Tracy, P. 2020. Depletion allowance. https://investinganswers.com. Accessed July 2022.

Tu, J. 2007. Coal mining safety: China's Achilles' heel. *China Security* 3(2):36–53.

Tuovila, A. 2022. Depreciation. www.investopedia.com. Accessed July 2022.

University of Washington. 2018. *University of Washington Self-Efficacy Scale (UW-SES): Generic Version Users Guide.* https://uwcorr.washington.edu. Accessed August 2022.

USACE (U.S. Army Corps of Engineers). 2000. *Inland Waterway Navigation: Value to the Nation.* Washington, DC: USACE. Accessed August 2022.

USDOL (U.S. Department of Labor). 2019. National Stone, Sand and Gravel Association. www.msha.gov. Accessed September 2022.

USEPA (U.S. Environmental Protection Agency). 2017. *Visual Determination of the Opacity of Emissions from Stationary Sources.* Method 9. Washington, DC: USEPA.

USEPA (U.S. Environmental Protection Agency). 2019. *Visual Determination of Fugitive Emissions from Material Sources and Smoke Emissions from Flares.* Method 22. Washington, DC: USEPA.

USEPA (U.S. Environmental Protection Agency). 2021. What is an UST? www.epa.gov. Accessed August 2022.

USEPA (U.S. Environmental Protection Agency). 2022a. Nation secondary drinking water regulations (NSDWRs). www.epa.gov. Accessed August 2022.

USEPA (U.S. Environmental Protection Agency). 2022b. Resources and guidance documents for compliance assistance. www.epa.gov. Accessed August 2022.

USEPA (U.S. Environmental Protection Agency). 2022c. What is a wetland? www.epa.gov. Accessed August 2022.

USGS (U.S. Geological Survey). n.d.-a. Interactive geologic map. https://mrdata.usgs.gov. Accessed August 2022.

USGS (U.S. Geological Survey). n.d.-b. Mineral commodity summaries. www.usgs.gov. Accessed August 2022.

USGS (U.S. Geological Survey). n.d.-c. Mineral resources program. www.usgs.gov. Accessed July 2022.

USGS (U.S. Geological Survey). n.d.-d. What are igneous rocks? www.usgs.gov. Accessed August 2022.

USGS (U.S. Geological Survey). 2006. *FGDC Digital Cartographic Standard for Geologic Map Symbolization.* FGDC Document Number FGDC-STD-013-2006. Washington, DC: Geologic Data Subcommittee, Federal Geographic Data Committee.

USGS (U.S. Geological Survey). 2013. *Mount St. Helens, 1980 to Now—What's Going On?* Fact Sheet 2013-3014. Vancouver, WA: USGS.

Vitruvius. *Ten Books of Architecture.* Translated by M.H. Morgan, 1914. Cambridge, MA: Harvard University Press.

Wilburn, D.R., and Goonan, T.G. 1998. *Aggregates from Natural and Recycled Sources.* Circular 1176. Washington, DC: U.S. Geological Survey.

Wills, B.A., and Napier-Munn, T.J. 2006. *Wills' Mineral Processing Technology,* 7th ed. Oxford, UK: Elsevier.

World Population Review. 2022. Cincinnati, Ohio Population 2022. https://worldpopulationreview.com. Accessed August 2022.

Zoback, M.L., and Zoback, M.D. 1989. Tectonic stress field of the continental United States. In *Geophysical Framework of the Continental United States.* Edited by L.C. Pakiser and W.D. Mooney. Boulder, CO: Geological Society of America.

Index

Note: *f.* indicates figure; *t.* indicates table

F

G